Disaster Risk Reduction

Methods, Approaches and Practices

Series Editor
Rajib Shaw, Keio University, Shonan Fujisawa Campus, Fujisawa, Japan

Disaster risk reduction is a process that leads to the safety of communities and nations. After the 2005 World Conference on Disaster Reduction, held in Kobe, Japan, the Hyogo Framework for Action (HFA) was adopted as a framework for risk reduction. The academic research and higher education in disaster risk reduction has made, and continues to make, a gradual shift from pure basic research to applied, implementation-oriented research. More emphasis is being given to multi-stakeholder collaboration and multi-disciplinary research. Emerging university networks in Asia, Europe, Africa, and the Americas have urged process-oriented research in the disaster risk reduction field. With this in mind, this new series will promote the output of action research on disaster risk reduction, which will be useful for a wide range of stakeholders including academicians, professionals, practitioners, and students and researchers in related fields. The series will focus on emerging needs in the risk reduction field, starting from climate change adaptation, urban ecosystem, coastal risk reduction, education for sustainable development, community-based practices, risk communication, and human security, among other areas. Through academic review, this series will encourage young researchers and practitioners to analyze field practices and link them to theory and policies with logic, data, and evidence. In this way, the series will emphasize evidence-based risk reduction methods, approaches, and practices.

All submitted proposals are carefully reviewed by the series editor.

Publication Ethics and Publication Malpractice Statement: https://www.springernature.com/gp/policies/book-publishing-policies

Ethical, Conflict-of-Interest, and Editorial Policy Statement: https://www.springer.com/us/authors-editors/journal-author/journal-author-helpdesk/before-you-start/before-you-start/1330#c14214

José Ernesto Mancera Pineda · Andrés F. Osorio ·
Cesar Toro · Carolina Sofía Velásquez-Calderón
Editors

Climate Change Adaptation and Mitigation in the Seaflower Biosphere Reserve

From Local Thinking to Global Actions

Editors
José Ernesto Mancera Pineda
Universidad Nacional de Colombia
Bogotá, Colombia

Andrés F. Osorio
Universidad Nacional de Colombia
Medellín, Antioquia, Colombia

Cesar Toro
Ex IOCARIBE
Cartagena, Colombia

Carolina Sofía Velásquez-Calderón
Florida State University
Florida, FL, USA

ISSN 2196-4106 ISSN 2196-4114 (electronic)
Disaster Risk Reduction
ISBN 978-981-97-6662-8 ISBN 978-981-97-6663-5 (eBook)
https://doi.org/10.1007/978-981-97-6663-5

© The Editor(s) (if applicable) and The Author(s) 2025. This book is an open access publication.

Open Access This book is licensed under the terms of the Creative Commons Attribution 4.0 International License (http://creativecommons.org/licenses/by/4.0/), which permits use, sharing, adaptation, distribution and reproduction in any medium or format, as long as you give appropriate credit to the original author(s) and the source, provide a link to the Creative Commons license and indicate if changes were made.
The images or other third party material in this book are included in the book's Creative Commons license, unless indicated otherwise in a credit line to the material. If material is not included in the book's Creative Commons license and your intended use is not permitted by statutory regulation or exceeds the permitted use, you will need to obtain permission directly from the copyright holder.
The use of general descriptive names, registered names, trademarks, service marks, etc. in this publication does not imply, even in the absence of a specific statement, that such names are exempt from the relevant protective laws and regulations and therefore free for general use.
The publisher, the authors and the editors are safe to assume that the advice and information in this book are believed to be true and accurate at the date of publication. Neither the publisher nor the authors or the editors give a warranty, expressed or implied, with respect to the material contained herein or for any errors or omissions that may have been made. The publisher remains neutral with regard to jurisdictional claims in published maps and institutional affiliations.

This Springer imprint is published by the registered company Springer Nature Singapore Pte Ltd.
The registered company address is: 152 Beach Road, #21-01/04 Gateway East, Singapore 189721, Singapore

If disposing of this product, please recycle the paper.

Foreword

The long-term sustainable development and resilience to climate change of the Archipelago of San Andrés, Providencia, and Santa Catalina are of the highest priority for both local and national institutions in Colombia. Thus, from both CORALINA (the Corporation for the Sustainable Development of the Archipelago, and the local environmental authority) and the Ministry of Foreign Affairs, we welcome the publication of this book as valuable input to ongoing decision-making and policy processes.

It is gratifying to see the active collaboration of high-level researchers and practitioners from a broad range of disciplines in this book, and in particular, that many are natives or residents of the Archipelago. From their diverse fields and backgrounds, the authors included here offer insights into many physical and socioeconomic processes that are key for the Archipelago in the short, medium, and long term. The articulation between academia, communities, and public institutions reflected in this publication is increasingly important, and we invite all readers to consider both the lessons that can be learned and applied to the case of the Archipelago, and those that are relevant to other Biosphere Reserves and archipelago or small island communities around the world.

The climate crisis that we collectively face today requires both scientific evidence and political will in order to identify and implement solutions, and we strongly believe that this book can drive both.

Bogotà, Colombia

Elizabeth Taylor Jay
Vice Minister of Multilateral Affairs
of the Republic of Colombia

San Andrès Island, Colombia

Arne Britton González
Director General, Corporation for
the Sustainable Development
of the Archipelago of San Andrés,
Providencia and Santa
Catalina–CORALINA

Acknowledgments

The editors would like to thank the following individuals who acted as peer reviewers (in alphabetical order by surname) for the contribution of their time and expertise to this book:

Arturo Acero Pizarro, Ph.D., Universidad Nacional de Colombia
Michael Alexander, Ph.D., NOAA Earth System Research Laboratories
Elvira Alvarado Chacón, Ph.D., Ecomares Foundation
Christian Appendini Albrechtsen, Ph.D., Universidad Nacional Autónoma de México
Elisa Berdalet Andrés, Ph.D., Institute of Marine Sciences, Spanish National Research Council
José Antonio Bettencourt, Ph.D., NOVA University Lisbon
Sergio Cambronero, B.Sc., Universidad Nacional de Costa Rica
Diana Carolina Castaño Giraldo, M.Sc., Universidad Nacional de Colombia
Mateo Córdoba, M.A., Gran Seaflower Initiative/Franz Weber Foundation
Sandra Correa-Aristizábal, Ph.D., Universidad de Antioquia
Sharika Crawford, Ph.D., United States Naval Academy
Juan Manuel Díaz Merlano, Ph.D., Universidad Nacional de Colombia/MarViva Foundation
Jorge Diogène Fadini, Ph.D., Institute of Agrifood Research and Technology, Catalonia
Rodrigo Garza Pérez, Ph.D., Universidad Nacional Autónoma de México
Francisco de Paula Gutiérrez Bonilla, Ph.D., Universidad de Bogotá Jorge Tadeo Lozano
Purity Karuga, M.A., Africa Network for Animal Welfare
Nicolás Loaiza Díaz, M.A., Independent Researcher
Jorge Higinio Maldonado, Ph.D., Universidad de lo Andes
Germán Márquez Calle, Ph.D., Universidad Nacional de Colombia
María Fernanda Maya, M.Sc. Candidate, Blue Indigo Foundation
Bonnie McCay, Ph.D., Rutgers University
Mónica Medina, Ph.D., Pennsylvania State University
Gabriel Navas, Ph.D., Universidad de Cartagena

José Daniel Pabón Caicedo, Ph.D., Universidad Nacional de Colombia
Carlos Prada, Ph.D., University of Rhode Island
Julián Prato, M.Sc./Ph.D. Candidate, Universidad Nacional de Colombia
Vladimir Toro Valencia, Ph.D., Universidad de Antioquia
Sven Zea Sjoberg, Ph.D., Universidad Nacional de Colombia

Finally, the editors would like to thank the Corporation Center of Excellence in Marine Sciences—CEMarin (www.cemarin.org) and all of the team who contributed to the coordination and administration of this book, and its member universities: the Universidad de los Andes, the Universidad de Bogotá Jorge Tadeo Lozano, the Universidad Nacional de Colombia, the Universidad de Antioquia, the Universidad del Valle and the Justus-Liebig-Universitat, Giessen.

Contents

Introduction 1
José Ernesto Mancera Pineda, Andrés F. Osorio, Cesar Toro,
and Carolina Sofía Velásquez-Calderón

Understanding Climate Change and Its Socio-environmental Impacts in the Seaflower Biosphere Reserve

CMIP6 Ocean and Atmospheric Climate Change Projections
in the Seaflower Biosphere Reserve—Caribbean Sea—by the End
of the Twenty-First Century 11
David Francisco Bustos Usta and Rafael Ricardo Torres Parra

Reconstructing the Eta and Iota Events for San Andrés
and Providencia: A Focus on Urban and Coastal Flooding 39
Andrés F. Osorio, Rubén Montoya, Franklin F. Ayala,
and Juan D. Osorio-Cano

Rapid Remote Sensing Assessment of Impacts from Hurricane
Iota on the Coral Reef Geomorphic Zonation in Providencia 69
Hernando Hernández-Hamón, Paula A. Zapata-Ramírez,
Rafael E. Vásquez, Carlos A. Zuluaga, Juan David Santana Mejía,
and Marcela Cano

A Light Pollution Assessment in the Fringing Reefs of San Andrés
Island: Towards Reducing Stressful Conditions at Impacted Coral
Reefs 89
Andres Chilma-Arias, Sebastian Giraldo-Vaca, and Juan A. Sánchez

Ciguatera in the Seaflower Biosphere Reserve: Projecting
the Approach on HABs to Assess and Mitigate Their Impacts
on Public Health, Fisheries and Tourism 103
José Ernesto Mancera Pineda, Brigitte Gavio,
Adriana Santos-Martínez, Gustavo Arencibia Carballo, and Julián Prato

Society, Seaflower Marine Ecosystem Services, and Climate Change Adaptation

The Biosphere Reserve Concept, Seaflower, and Climate Change 127
Germán Márquez

Marine Ecosystem Services for Climate Change Adaptation and Mitigation Strategies in the Seaflower Biosphere Reserve: Coastal Protection and Fish Biodiversity Refuge at Caribbean Insular Territories 149
Julián Prato, Adriana Santos-Martínez, Amílcar Leví Cupul-Magaña, Diana Castaño, José Ernesto Mancera Pineda, Jairo Medina, Arnold Hudson, Juan C. Mejía-Rentería, Carolina Sofia Velásquez-Calderòn, Germán Márquez, Diana Morales-de-Anda, Matthias Wolff, and Peter W. Schuhmann

Climate Change Effects on Seaflower Biosphere Reserve Fishery Resources 183
Carolina Sofia Velásquez-Calderón, Adriana Santos-Martínez, Anthony Rojas-Archbold, and Julián Prato

Overcoming Iota: A Reflection on Old Providence and Santa Catalina Cultural Resilience In the Face of Disaster and Climate Change 209
Ana Isabel Márquez-Pérez

Climate Change: A Business Perspective of the Tourism Industry in the Seaflower Biosphere Reserve 233
Lorena Aldana Pedrozo and Rixcie Newball Stephens

Climate Change Education and Research

Archeology Expanded—a Multidisciplinary Approach for Natural Disaster Response 255
Víctor Andrés Pérez Bermúdez and Daniela Vargas Ariza

Taking Seaflower to the Classroom: A Proposal to Bring Sustainability Education to High Schools in an Oceanic Archipelago (Western Caribbean, Colombia) 275
Juan F. Blanco-Libreros, Sara R. López-Rodríguez, Jairo Lasso-Zapata, Beatriz Méndez, Nairo De Armas, and Margareth Mitchell-Bent

Advances and Needs in Marine Science Research in the Archipelago of San Andrés, Providencia, and Santa Catalina: A Literature Analysis 299
Camilo B. García and Johan Sebastián Villarraga

Appendix 315

About the Editors

Dr. José Ernesto Mancera Pineda is a Tenured Professor at the Universidad Nacional de Colombia, Faculty of Sciences, Department of Biology. His research interests focus on coastal zone ecology and management, with an emphasis on coastal lagoons, mangrove wetlands, and harmful algal blooms (HABs). He is currently the coordinator of the Colombian Network of Estuaries and Mangroves and President of the IOC/IOCARIBE/ANCA (Harmful Algae of the Caribbean) working group. He has been part of the IPHAB, coordinating the participation of the Caribbean region in the Global HAB Status Report. He was also a member of the Group of Experts for the preparation of the second evaluation on the state of the world's oceans by the United Nations and has written on issues related to the tropical marine environment, natural resource management, environmental hazards, and Ecological restoration. A biologist by training, he has a Ph.D. from the University of Louisiana at Lafayette, USA, and a Master's degree in Marine Biology from the Universidad Nacional de Colombia.

Dr. Andrés F. Osorio is a civil engineer, with a Master's and Ph.D. in Marine Sciences and Technologies. He is currently a tenured professor at the Universidad Nacional de Colombia where leads the OCEANICOS research group. Since January 2020, he has been the Executive Director and head of the strategic management of the Corporation Center of Excellence in Marine Sciences—CEMarin. He has been involved in international projects such as "CoastView", a project promoting vigilance of coastal systems in European countries, "Sistema de Modelado Costero" (Coastal Modeling System), a software development tool for coastal modeling, a technological strategy for climate change for Colombia's Ministry of Environment and Sustainable Development, and UNEP-Risoe, a project to incorporate environmental and development aspects into energy planning and policy worldwide. Dr. Osorio has recently been involved as an international consultant for the World Bank, the UN, and the EU, among others, in projects like: (1) Technical assistance for sustainable ocean management in Galapagos, (2) Technical assistance for sustainable ocean management in Continental Coast-Ecuador, (3) EUCDs COL01: Nature-based solutions for climate change adaptation in coastal cities and island systems in Colombia, and (4) National Ocean Policy of Panama, among others.

Dr. Cesar Toro is a physical oceanographer, with an M.Sc. in Oceanography and a Ph.D. from the University of Quebec, Canada in Physical Oceanography. Dr. Toro is a specialist in dynamical oceanography, climate change adaptation and mitigation, sustainable development, and ocean governance and policy. He also has extensive experience working for the oil and gas industry, the United Nations, and in project design and implementation, especially in developing countries and SIDS. From 2001 to 2022 Dr. Toro was head of the Subcommission for the Caribbean and Adjacent Regions IOCARIBE of the Intergovernmental Oceanographic Commission (IOC of UNESCO), based in Cartagena, Colombia. Between 2009 and 2015, he was also responsible for the UNESCO Natural Sciences Programme for the Caribbean. From IOCARIBE, he promoted the development of marine and ocean sciences and technology in the countries of the region through the strengthening of institutional capacity, coordination of intergovernmental groups, and networks of scientists and experts. He coordinates IOC of UNESCO programmes in the Latin American and Caribbean region with those of the organizations of the United Nations system, working actively with national, regional, and international agencies and entities, with the financial support of the Development Banks, the GEF, the EU, NGOs, and international development agencies. Dr. Toro has chaired a large number of expert groups and committees and has led dozens of large multidisciplinary projects. He also contributes to the UN Decade of Ocean Science for Sustainable Development 2021–2030, to the IOC's capacity development strategy and its implementation in the IOCARIBE Region-Tropical Americas region, as well as to the development of UNESCO's Science, Technology, and Innovation strategy and science policy. He has contributed as author or co-author to more than a hundred publications.

Dr. Carolina Sofía Velásquez-Calderón is an Assistant Professor at Florida State University, in the Faculty of Geography of the College of Social Sciences and Public Policy. With over a decade of expertise in Environmental and Disaster Risk Management, Dr. Velásquez has made substantial contributions in the context of small Caribbean Islands, notably San Andrés, Providencia, and Santa Catalina, Colombian Caribbean. Her career encompasses roles in renowned international organizations, research centers, and prestigious universities. She has worked at the Deutsche Gesellschaft für Internationale Zusammenarbeit (German Society for International Cooperation—GIZ), the Disaster Research Center at the University of Delaware, the US Natural Hazard Center, and the Universidad Nacional de Colombia. Dr. Velásquez's research interests revolve around post-disaster recovery, social vulnerability, water-related crises, water justice, climate change perception, and risk-informed development. She is recognized for her authorship and editorial contributions to books on risk and environmental management. Dr. Velásquez earned her undergraduate degree as a Natural Resources and Environmental Administrator. Subsequently, she completed an M.Sc. in Caribbean Studies at the Universidad Nacional de Colombia, followed by a Ph.D. in Disaster Science Management from the University of Delaware, USA.

Introduction

José Ernesto Mancera Pineda, Andrés F. Osorio, Cesar Toro, and Carolina Sofía Velásquez-Calderón

Abstract Biosphere reserves have particular, unique importance as places to study, learn, and replicate forms of disciplinary mitigation and adaptation. In the context of the worsening climate crisis, this is especially true of island and coastal biosphere reserves that disproportionately face the adverse impacts of climate change. Considering issues like biodiversity conservation, cultural diversity, and socio-culturally and environmentally sustainable economic development, biosphere reserves serve as ideal places for interdisciplinary research and to design and implement mitigation and adaptation strategies developed *from* and *for* local contexts and communities. This interdisciplinary book emphasizes the unification of the results of cutting-edge technical research with the local knowledge, struggles, and experiences of the Raizal people of the Archipelago of San Andrés, Providencia, and Santa Catalina. Combining insights from different disciplines offers insights into how best to prepare for and respond to future extreme weather events, and key inputs for decision-making by both public sector actors in the archipelago and Colombia, and any stakeholder

---

J. E. M. Pineda (✉)
Department of Biology, Faculty of Sciences, Universidad Nacional de Colombia, Bogotá, Colombia
e-mail: jemancerap@unal.edu.co

J. E. M. Pineda · A. F. Osorio
Corporation Center of Excellence in Marine Sciences, CEMarin, Bogotá, Colombia
e-mail: afosorioar@unal.edu.co

A. F. Osorio
Faculty of Mines, Department of Geosciences and Environment, Universidad Nacional de Colombia, Medellín Campus, OCEANICOS Research Group, Medellín, Colombia

C. Toro
Subcommission for the Caribbean and Adjacent Regions IOCARIBE of the Intergovernmental Oceanographic Commission (IOC of UNESCO), Cartagena, Colombia
e-mail: c.toro@toromarine.com

C. S. Velásquez-Calderón
Department of Geography, College of Social Sciences and Public Policy, Florida State University, Tallahassee, FL, USA
e-mail: csv23@fsu.edu

interested in these processes. Innovative methodologies and precise, up-to-date scientific data are crucial for effective policy-making. Hence, this book includes important results including maps, models, and ecosystem-based reconstruction methodologies focused on mangroves and coral reefs, among others.

Keywords Seaflower biosphere reserve · Archipelago of San Andrés · Providencia and Santa Catalina · Climate change · Sustainable development · Hurricanes Eta and Iota

The Seaflower Biosphere Reserve in the Colombian Archipelago of San Andrés, Providencia, and Santa Catalina (hereafter, archipelago) is one of the world's largest Biosphere Reserves since its designation by UNESCO in November 2000. Biosphere reserves have particular and unique importance as places in which to study, learn, and replicate forms of disciplinary mitigation and adaptation worldwide. According to UNESCO, they are sites for testing interdisciplinary approaches to understanding and managing changes and interactions between social and ecological systems, including conflict prevention and management of biodiversity. They are places that provide local solutions to global challenges. Biosphere reserves include terrestrial, marine, and coastal ecosystems. Each site promotes solutions reconciling the conservation of biodiversity with its sustainable use (UNESCO n.d.).

These characteristics of biosphere reserves are becoming increasingly important in the context of the worsening climate crisis, and this is especially true of island and coastal biosphere reserves that are disproportionately facing the adverse impacts of climate change. The concept of "local solutions to global challenges" was a key motivation in the development of this book, and indeed the inspiration behind its name.

There is a simultaneous need to better understand both future regional climate behavior and the most adequate adaptation and mitigation strategies that seek balanced relationships between people and nature. For our purposes, the relevance of biosphere reserves lies in them being zones where local communities and all relevant stakeholders are involved in planning and management around the three pillars of sustainable development: the environmental, the economic, and the social. In this sense, taking into account issues like the conservation of biodiversity and cultural diversity, and socio-culturally and environmentally sustainable economic development, biosphere reserves serve as ideal places to undertake interdisciplinary research and to design and implement strategies of mitigation and adaptation that are developed *from* and *for* local contexts and communities.

We should be clear that, while these characteristics of biosphere reserves are unique and important, and make much of the information and recommendations included here relevant for other regions of the world, there were several motivations specific to the archipelago in the decision to produce this book. Firstly, many of the authors included here are natives or residents of the archipelago, and many more who are based across mainland Colombia or in other parts of Latin America regularly undertake research there. Thus, for a range of professional and personal interests, the devastating events of November 2020 were one impetus behind this publication.

In November 2020, Hurricanes Eta and Iota hit the archipelago within the space of a fortnight. While the archipelago has always been affected by Atlantic hurricane seasons, the category 4[1] Hurricane Eta hit the archipelago on November 2, 2020, followed soon after by category 4 Iota on November 16, 2020—the first time a hurricane of this magnitude directly impacted this group of islands in the Southwest Colombian Caribbean—causing unprecedented damage. In the case of Providencia in particular, over 90% of the housing and infrastructure on the island was destroyed or damaged. Additionally, there was significant damage to the ecosystems on and around the islands of the archipelago, but also key lessons to be learned from the protection offered by these ecosystems to both people and infrastructure during these extreme events.

Moreover, it was not only the initial disaster itself that sparked our interest in this project. On November 18, 2020, the president of Colombia officially declared a state of emergency, with the intent of expediting the reconstruction process. Despite government assurances of rebuilding the lost or damaged infrastructure within 100 days, the entire process has come under heavy criticism from various quarters. It has been described as, at best, disorganized, and at worst, a failure. Additionally, many observers have noted that certain aspects of the response have exacerbated the crises left in the wake of Eta and Iota, as well as compounding existing structural problems in the archipelago predating these disasters. We aim not to discuss this reconstruction process, but rather to focus on providing practical and tangible tools and techniques for a sound decision-making process based on the best available science.

Additionally, in general terms, the archipelago faces many challenges, some of which are shared by mainland Colombia, and some of which are more particular to the archipelago and other small island developing states (SIDS) in the Caribbean region. As the authors in this book show, these range from environmental challenges—including climate change adaptation, biodiversity loss, and waste management—, questions of access to and use of natural resources—including fresh water—, threats from illegal activities—including illegal fishing, and the trafficking of drugs and endangered species—the resilience of existing infrastructure and public services in the face of climate change and extreme weather events, cultural preservation of the unique and constitutionally protected cultural identity of the Afro-Caribbean Raizal people of the archipelago, and the need to effectively balance economically beneficial activities like tourism with issues of sustainable resource access and use.

Thus, a combination of the disasters of November 2020, the subsequent complicated reconstruction process, and the long-term development and well-being of the archipelago and its native people and residents inspired us to ask the question of what we, as academics, could do to contribute to improving existing mechanisms of resilience, preparedness, disaster response and long-term reconstruction in the case

---

[1] It is interesting to note that, although Hurricane Iota was initially categorized as category 5, it was later downgraded to category 4. However, in both the archipelago and in general, it is still commonly referred to as category 5.

of future extreme weather events which, according to the Intergovernmental Panel on Climate Change (IPCC) will worsen in the future in both frequency and strength due to climate change.

We decided that a contribution that could have real impacts would be an interdisciplinary book with an emphasis on unifying the results of cutting-edge technical research with the local knowledge, struggles, and experiences of the Raizal people, whose culture mixes influences from various ethnic groups and is thus quite distinct to the people of mainland Colombia, and other residents of the archipelago. In a context where global phenomena like climate change are seriously impacting the lives of individuals and communities in their local settings, it is so important to communicate these local perspectives globally. We believe that addressing the very specific case of the archipelago and its recovery from the 2020 hurricane season can also generate knowledge that contributes to ongoing discussions regarding climate change mitigation and adaptation in both the archipelago, across the Caribbean, and in biosphere reserves and other SIDS generally around the world.

The different disciplines included here all offer valuable insights into how best to prepare for and respond to future extreme weather events, and are key inputs for decision-making by both public sector actors in the archipelago and Colombia, and for any stakeholder with an interest in these processes. Innovative methodologies and precise, up-to-date scientific data are crucial for effective policy-making. Hence the book includes important results including maps, models, and ecosystem-based reconstruction methodologies focused on mangroves and coral reefs among others.

Moreover, as previously mentioned, we believe it is particularly important to recognize and make visible the contributions of many islanders who work in other sectors, including fishers, school teachers, and officials from public institutions. The local knowledge included here offers vital input in terms of topics like fisheries, agriculture, tourism, and cultural practices and, conversely, we expect that the book can also help raise awareness within the communities of the archipelago regarding at-risk species and ecosystems. We believe one of the most important contributions of this book is to unify a range of knowledge in one single resource that can be a go-to reference for anyone interested in climate change mitigation and adaptation in the Seaflower Biosphere Reserve, and around the world.

Although one initial idea for the book was to produce a "white paper" with clear and precise recommendations aimed mainly at decision-makers, we later settled upon a broader approach and audience. For this reason, we have created infographics for each chapter to enhance communication of their key messages in a concise and visual way. These infographics can be found in an Appendix at the end of the book. In academia, we often write primarily for other academics and the message stays between us, but this book is an attempt to speak not only to other academics, nor to decision-makers, it is an attempt to integrate different disciplines and sectors and to make communication among them more effective.

The book is divided into three broad sections: (1) Understanding climate change and its socio-environmental impacts in the Seaflower Biosphere Reserve, (2) Society, Seaflower marine ecosystem services, and climate change adaptation, and (3) Education and research on climate change. Across these three sections, the authors examine relevant socio-environmental pathways towards collective action for adaptive capacity, and resilience. The ultimate aim is to contribute to sustainable development processes in Seaflower and other biosphere reserves worldwide, always with an emphasis on the importance of these local environments and cultures at the global level.

The first section of the book, *Understanding Climate Change and its Socio-environmental Impacts in the Seaflower Biosphere Reserve*, opens with a chapter that examines the expected behavior of seven atmospheric and oceanic variables in the Caribbean Basin and the Seaflower Biosphere Reserve during the twenty-first century, under two socioeconomic scenarios. By building an ensemble of the five models with the best oceanic resolution in the Caribbean Sea, the authors analyze expected changes in surface air temperature, sea level pressure, surface wind, precipitation, sea surface temperature, sea surface salinity, and sea level, reaching the conclusion that sea level rise in particular will modify the ecological balance in the Seaflower Biosphere Reserve and enhance flooding, thus affecting tourism and risking the disappearance of the low elevation islands.

The next chapter uses numerical modeling and field measurements to reconstruct Hurricanes Eta and Iota and their effects on the archipelago in terms of the intensity of winds and waves, and the associated coastal and urban flooding impacts. This reconstruction shows the differentiated contribution of each hazard associated with the passage of Eta and Iota through the archipelago on physical infrastructure, coastal ecosystems, and population, providing valuable input for territorial planning and decision-making regarding vulnerability and risk.

The following chapter offers a rapid remote sensing assessment of the impacts of Hurricane Iota impact on Providencia island's reef environments, using Google Earth Engine, Satellite Derived Bathymetry, and machine learning to calculate a supervised classification process that delineates six geomorphic reef units. This process is an interesting alternative for monitoring reef cover in extreme events like hurricanes and its integration can enable long-term monitoring by observing the evolution of changes over time and, therefore, provide valuable information to coastal managers and stakeholders in decision-making processes.

The next contribution assesses light pollution in the fringing reefs of San Andrés with a view to reducing stressful conditions at impacted coral reefs. Although the impacts of light pollution on marine life are significant, we still lack understanding of the brightness of natural light from bodies like the Moon. The authors implement innovative techniques and pave the way for future studies that can evaluate light pollution and its impact even more accurately. At the same time, the results of their study provide valuable artificial light management recommendations, in the context of continuous population growth on the island.

The final chapter in the first section analyzes the presence of toxic dinoflagellates along with the incidence of ciguatera in the Seaflower Biosphere Reserve. As the study shows, global warming, climate change, nutrients, and sewage discharge favor microalgal blooms, which are becoming more frequent, intense, and lasting. For these reasons, the authors evaluate the potential effects of harmful algal blooms, specifically an economic quantification of their impacts on fishing and tourism.

By proposing a conceptual model of an early warning system based on a monitoring program, the authors present a strategy to contribute to the governance and management effectiveness of different institutions in the archipelago.

Moving to the second section of the book, *Society, Seaflower Marine Ecosystem Services, and Climate Change Adaptation*, opens with a chapter by one of the most influential individuals in the designation of the archipelago as the Seaflower Biosphere Reserve. In this chapter, the author reviews the concept of biosphere reserves, the application of the concept in the case of the archipelago, and the relationship between the concept and climate change, making a clear case for deepening the implementation of the biosphere reserve concept in the archipelago in order to promote social, economic, and environmental sustainability.

The second contribution in this section focuses on the marine ecosystem services of the archipelago as strategies for climate change mitigation and adaptation, with emphasis on coral reefs and mangroves. The result of this evaluation is the presentation of different interdisciplinary management tools and recommendations of actions required by different sectors for the protection, restoration, and use of these ecosystems as Nature-based Solutions (NbS) for climate change adaptation and mitigation in the archipelago.

The next chapter analyzes the impacts of climate change on fishery resources in the Seaflower Biosphere Reserve, as the fisheries sector is one of the most affected by this phenomenon. By evaluating existing knowledge on this issue and incorporating the perceptions of fishers in the archipelago, the authors identify significant consequences for fishery resources and for the individuals who make their living through this activity, as well as policy strategies to address these risks and vulnerabilities, emphasizing the importance of aligning said strategies with fishers' priorities and enhancing the resilience of the sector. The chapter highlights the urgency of ecosystem-based and co-management policies and alternatives for the artisanal fishers of the archipelago.

The following chapter focuses on the recovery and reconstruction processes that took place in the archipelago after Hurricane Iota—an event that exacerbated existing vulnerabilities—prioritizing the manner in which the local community responded to the situation by adapting and reorganizing their ways of life. Using insights from cultural perspectives on disasters, climate change, and resilience, the author shows how cultural resilience and community processes can help islanders continue to inhabit the archipelago with well-being and autonomy, even in the context of climate change and increasingly extreme events. We note that this author, a long-term resident of the archipelago, prefers to call the islands "Old Providence and Santa Catalina" instead of their Spanish language names.

The final chapter of this second section examines the impacts of climate change on the tourism sector in the archipelago, as this is its most important economic activity with significant implications for the archipelago's resource use and territorial planning. By evaluating the knowledge and perceptions of tourism service providers regarding climate change, the authors provide relevant insights into the vulnerabilities of these actors in the face of extreme climatic events and recommendations to mitigate these vulnerabilities in the future.

The third section of the book, *Education and Research on Climate Change*, evaluates the long-term vulnerability of three archeological sites on the island of Santa Catalina in the context of climate change. As the authors explain, the archeological study of climate change offers analytical tools that allow one to record changes in the landscape, as well as the actions that have been taken to face natural disasters, in order to provide insights into the way culture interacts with climate risks and tools for designing public policy. The authors construct a vulnerability framework and historical reconstruction of the three sites studied since the sixteenth century, with important reflections on the impacts of Hurricane Iota and its impacts on both the sites and the collective memory of the archipelago.

In the penultimate contribution, the authors elaborate a proposal to bring sustainability education to high schools in the archipelago by improving the geographic understanding of the students, recognizing the need for a culturally responsive approach and the potential of geo-literacy tools and strategies as mechanisms to achieve this. The authors, including high school teachers from the archipelago, offer concrete activities and evaluations of their pilot proposal as the basis for future sustainability education directed at young people of different ages that is culturally sensitive to the particularities of the archipelago within the wider context of Colombian national education policies.

In the final chapter, the authors present a much needed literature analysis of the advances and need in marine science research in the archipelago. By constructing a database of existing literature, the authors quantify current knowledge in order to identify both the stronger and weaker areas of knowledge across different areas of the marine sciences and different types of literature, noting that dynamic aspects including responses to climate change are particularly lacking. These insights are important both for academics in the planning of future research, and for policymakers and other stakeholders given that academic research is a key input for planning processes that can help secure the current and future ecological integrity of the archipelago and the Seaflower Biosphere Reserve.

We wish to thank all of the authors and peer reviewers who have contributed their valuable time and expertise to this book. We believe that their contributions will stimulate critical and interdisciplinary thinking around biosphere reserves and the tropical island regions that are some of the most vulnerable to the impacts of climate change. By presenting new and different frameworks through which to interpret and understand localized socio-environmental impacts of climate change, we can use these as a starting point to design climate change mitigation and adaptation strategies that put the well-being of ecosystems and humans at their core.

The Editors.

Reference

UNESCO (n.d.). What are biosphere reserves? https://en.unesco.org/biosphere/about

Open Access This chapter is licensed under the terms of the Creative Commons Attribution 4.0 International License (http://creativecommons.org/licenses/by/4.0/), which permits use, sharing, adaptation, distribution and reproduction in any medium or format, as long as you give appropriate credit to the original author(s) and the source, provide a link to the Creative Commons license and indicate if changes were made.

The images or other third party material in this chapter are included in the chapter's Creative Commons license, unless indicated otherwise in a credit line to the material. If material is not included in the chapter's Creative Commons license and your intended use is not permitted by statutory regulation or exceeds the permitted use, you will need to obtain permission directly from the copyright holder.

Understanding Climate Change and Its Socio-environmental Impacts in the Seaflower Biosphere Reserve

CMIP6 Ocean and Atmospheric Climate Change Projections in the Seaflower Biosphere Reserve—Caribbean Sea—by the End of the Twenty-First Century

David Francisco Bustos Usta and Rafael Ricardo Torres Parra

Abstract Seventeen climate models from CMIP6 were examined to assess the expected behavior of seven atmospheric/ocean variables in the Caribbean Basin and the Seaflower Biosphere Reserve (SBR) during the twenty-first century, under two socioeconomic scenarios (SSP2-4.5 and SSP5-8.5). Additionally, an ensemble is made with the five models with the best oceanic resolution in the Caribbean Sea. Precipitation shows significant negative trends in most of the projected periods, while air and sea surface temperature, surface salinity and mean sterodynamic sea level (SDSL) have significant positive trends. Air temperature in SBR will probably increase by 2 °C compared to the preindustrial period after 2050 (SSP5-8.5) or 2060 (SSP2-4.5). The warming trend in the region could extend the hurricane season and/or increase hurricane frequency, affect ecosystems like coral reefs and mangroves, and intensify ocean stratification. For the same period, SDSL is expected to rise in SBR between ~24.2 and 39.9 cm. If all contributing factors are included, an increase of up to ~95 cm (SSP5-8.5) could be expected by the end of the twenty-first century. This sea level rise would modify the ecological balance and enhance flooding, affecting tourism and risking the disappearance of the low-elevation islands.

Keywords Climate change · Mean sea level rise · Archipelago of San Andrés · Providencia and Santa Catalina · CMIP6 · Caribbean Sea

---

D. F. B. Usta (✉)
Department of Oceanography, Universidad de Concepción, Concepción, Chile
e-mail: davidbustos@udec.cl

R. R. T. Parra
GEO4 Research Group on Geosciences, Department of Physics and Geosciences, Universidad del Norte, Barranquilla, Colombia
e-mail: rrtorres@uninorte.edu.co

1 Introduction

Since the last century, the Earth's climate has changed due to the increase of greenhouse gases in the atmosphere. Climate change has many negative impacts and is highly dependent on regional dynamics. Threats resulting from climate change are expected to grow, enhancing problems across the planet. Understanding the behavior of the main atmospheric and oceanic variables is of vital importance since climate is determined by different factors including the dynamics and composition of the atmosphere, the ocean, ice and snow cover, and land surface and its features as a coupled system (IPCC 2014). Coastal areas are particularly affected by climate change, since sea level rise will increase flooding, coastal erosion, storm surges, and saltwater intrusion, menacing human welfare and generating large economic losses (Tsyban et al. 1990; Nicholls and Lowe 2004; IPCC 2014). The Caribbean Sea is being affected by coastal erosion (Rangel-Buitrago et al. 2015), which is expected to increase as well as coastal flooding through the twenty-first century, as a consequence of extreme sea level increases driven by sea level rise (Torres and Tsimplis 2014).

To assess future climate behavior, Global General Circulation Models (Atmosphere–Ocean General Circulation Models or AOGCMs) have been created to simulate dynamic physical processes in the ocean, atmosphere, cryosphere, land, and Earth system interactions. These models are considered the most accurate tool to understand the response of the Earth system to different projected greenhouse gas emission rates into the atmosphere (IPCC 2014). Results from AOGCMs are developed by different institutions and research groups around the world and are evaluated in the Coupled Model Intercomparison Project (CMIP). In this chapter, the results of CMIP Phase 6 models (Meehl et al. 2014; Eyring et al. 2015) are reported.

The Intergovernmental Panel On Climate Change (IPCC) was created in 1998 by two United Nations agencies: the World Meteorological Organization (WMO) and the United Nations Environment Programme (UNEP) (Agrawala 1997). The main strategy of the IPCC is to enable political action by providing a scientific definition of the climate state and variation, including consequences and measures for adaptation and mitigation (Berg and Lidskog 2018). This is done through the preparation of Assessment Reports, which include the analysis of the latest CMIP model results. CMIP6 was created with the intention of answering the following questions: (1) How does the Earth system respond to different forcings? (2) What are the origins and consequences of systematic model biases? and (3) How to assess future climate changes given climate variability, predictability, and uncertainties in scenarios? (Meehl et al. 2014). CMIP6 is an improved version of the former CMIP5 because scientific groups were focused on developing intercomparison studies based on their own strategic goals, implying a diversity of Endorsed Model Intercomparison Projects (MIPs) to fill scientific gaps when compared to previous CMIP phases (Eyring et al. 2015).

The Caribbean Sea is in an intertropical region (Fig. 1). It is the largest marginal sea of the Atlantic Ocean with a surface extension of 2.52×10^6 km, almost twice as large as the Gulf of Mexico (Gallegos 1996). The Caribbean is connected on the northwest to the Gulf of Mexico through the Yucatan Channel and on the north and east to the Atlantic Ocean, separated by the Antilles. The Archipelago of San Andrés, Providencia, and Santa Catalina (hereafter, the archipelago) is composed of nine islands located in the Colombia Basin (Table 1). The climate is regulated by the meridional position of the Intertropical Convergence Zone (Andrade 2000) creating a windy-dry season (December–April) and rainy-warm season (August–October) (Etter et al. 1987; Angeles et al. 2010), with different regional ocean responses in the basin (Torres and Tsimplis 2012; Torres et al. 2022 in press). Eastward trade winds dominate the Caribbean, including a strong low-level jet around 15°N reaching speeds of ~12 ms^{-1} in the windy season (Andrade 2000).

The Seaflower Biosphere Reserve (hereafter SBR) is located in the Colombia Basin (Fig. 1). It has a total area of 180,000 km^2, of which 65,018 km^2 is a protected marine area. Precipitation ranges between 0.8 mm day^{-1} in the dry season to 10.6 mm day^{-1} in the rainy season. Surface air temperature varies between 26.5 °C in the dry season and 28.1 °C in the rainy season. The trade winds average speed during the year is 4.5 m s^{-1}, reducing in magnitude between September and October, and intensifying at the beginning of the year and in the month of July (Coralina-Invemar 2012). Sea surface temperature has a range between 26 °C in the dry season and 29.5 °C in the rainy season. Sea surface salinity typically presents values below 35.5 PSU,

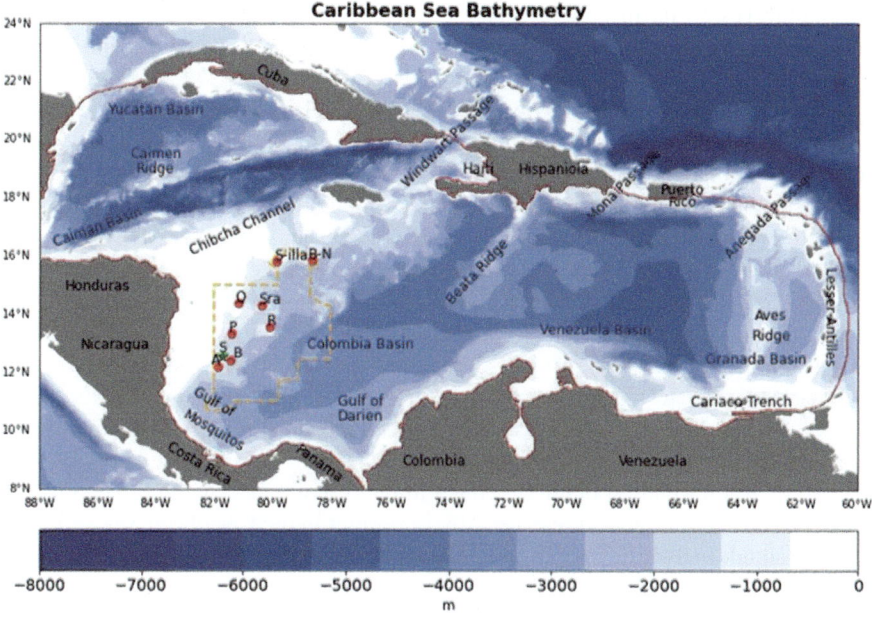

Fig. 1 Map of the Caribbean Sea. Bathymetric data from (GEBCO 2019), including the location of San Andrés (green star) and other islands (red dots) of the archipelago. Notation of islands is indicated in Table 1. The red polygon indicates Caribbean Sea limits used to obtain spatial averages of variables. The yellow polygon represents the Seaflower Biosphere Reserve (Coralina-Invemar 2012; Borrero-Pérez et al. 2019)

Table 1 Location of the islands in the archipelago (DIMAR 2019)

Name	Notation[a]	Latitude (N)	Longitude (W)	Maximum height (m)[b]	2071–2100 sterodynamic sea level rise (cm)[c]	
					SSP2-4.5	SSP5-8.5
San Andrés	S	12.55°	81.72°	100.0	24.43	39.32
Albuquerque	A	12.17°	81.84°	1.5	24.35	39.29
Bolivar	B	12.40°	81.45°	2.0	24.30	39.37
Providencia	P	13.35°	81.38°	360.0	24.61	39.40
Roncador	R	13.57°	80.08°	4.0	24.15	39.89
Serrana	Sra	14.28°	80.37°	9.8	24.37	39.71
Quitasueño	Q	14.38°	81.15°	1.5	24.68	39.02
Serranilla	S-illa	15.80°	79.83°	8.0	24.62	38.77
Bajo Nuevo	B-N	15.83°	78.67°	2.0	24.69	39.03

[a] Notation as indicated in Fig. 1
[b] Referred to mean sea level
[c] Referred to 1976–2005 sea level from M5-SBR

being affected by the freshwater contribution from the Orinoco and Magdalena rivers and rainfall (Coralina-Invemar 2012). Extreme events are dominated by hurricanes from June to November (Ortiz 2012) but are also affected by cold fronts from January to March (Ortiz et al. 2013).

Atmospheric and ocean behavior in the Caribbean Sea has been studied in the last decade, including the assessment of atmospheric temperature, pressure and wind (Montoya-Sánchez et al. 2018; Rodriguez-Vera et al. 2019; Hamed and Yunfang 2020; Bustos and Torres 2022) sea surface temperature and salinity (Ruiz et al. 2012; Beier et al. 2017) and mean sea level rise (Torres and Tsimplis 2013). Projected regional changes of these variables during the twenty-first century, using results from the CMIP5 under different radiative concentration scenarios, were recently assessed by Bustos and Torres (2021).

It is important to study projected regional and local atmospheric and ocean responses to climate change, to evaluate possible threats during the twenty-first century. Furthermore, it is necessary to understand the local risk, and its dependence on different Socioeconomic Scenarios Pathways (SSP), to develop accurate adaptation and mitigation plans to reduce impacts associated with climate change. These kinds of assessments are especially important in the Caribbean Sea and SBR, as small islands and the developing countries in the basin have constraints on adaptive capacity (Nicholls et al. 2007).

Therefore, the main objective of this chapter is to assess the projected behavior of atmospheric pressure, ambient temperature, wind, precipitation, ocean temperature, and salinity at the ocean surface, as well as mean sea level in the Caribbean Sea and SBR to the end of the twenty-first century. We used results from 17 CMIP6 models, under two Socioeconomic Pathways (SSP2-4.5 and SSP5-8.5) scenarios.

The chapter is organized as follows. Section 2 describes the datasets and methods used. Sections 3 and 4 include results and discussions of atmospheric and oceanic projections, respectively. Section 5 presents a summary and conclusions.

2 Data and Methods

We used 17 CMIP6 models (hereinafter M17) used by Chen et al. (2020), but excluding UKESM1-0-LL, MCM-UA-1-0, CAMS-CSM1-0, CNRM-CM6-1 and replacing NESM3 and MIROC-ES2L with ACCESS-CM2 and ACCESS-CM1.5. We use the "historical" run before 2014 and projections between 2014 and 2100 to assess atmospheric pressure (SLP), air temperature (Ta), precipitation (Pr), and wind at sea level, as well as sea surface temperature (SST), sea surface salinity (SSS) and mean sea level for different periods between 1850 and 2100. However, reported trends are referenced to 2005 to facilitate comparison with the CMIP5 results reported by Bustos and Torres (2021). For the projections, two scenarios are used for projections. SSP2-4.5, which is the medium-forcing scenario, with a 4.5 W m^{-2} mean radiation, a peak of emissions in 2040, and then stabilization of the rate for the rest of the century. SSP5-8.5 is the high-forcing scenario, with an 8.5 W m^{-2} mean radiation, and a peak of emissions in 2080 (O'Neill et al. 2017; Riahi et al. 2017; Gidden et al. 2019).

The SSPs are based on 5 narratives describing different levels of socioeconomic development (Riahi et al. 2017). SSP1-sustainable development, SSP2-middle of the road development, SSP3-regional rivalry, SSP4-inequality, and SSP5-fossil-fuel-driven development. Full details of the SSPs are described by O'Neill et al. (2017). Data and model descriptions were accessed from the Program for Climate Model Diagnosis and Intercomparison (PCMDI) (http://cmip-pcmdi.llnl.gov/mips/cmip6/). Model data was downloaded from the Earth System Grid Federation (https://esgf-node.llnl.gov/projects/cmip6/).

Climate models used for sea level projections do not explicitly resolve mesoscale processes in the ocean (Penduff et al. 2010; Serazin et al. 2015). Only some effects of these processes, which depend on spatial resolution, are included in the models, generating errors in the circulation that affect the regional sea level projections. Therefore, adequate sea level projections in regions like the Caribbean Sea can only be obtained with high-resolution models that are able to capture mesoscale processes (van Westen et al. 2020). Consequently, we show results from two ensembles. One using all 17 CMIP6 models, and the other using only the five models with the best oceanic resolution (0.25°–1.00°) in the Caribbean Sea (ACCESS-CM1.5, ACCESS-CM2, GFDL-ESM4, MPI-ESM1-2-HR, and MRI-ESM2-0), hereafter M5-CAR. In this chapter, we emphasize results from the latter.

The model ensemble was obtained by interpolating and averaging each model's results to a common grid (OM4 MOM6 for the five-model ensemble and T63 spectral for the 17-model ensemble) with a horizontal resolution of 1.875° × 1.25° (1.00° × 0.25°) for the atmospheric (oceanic) component to avoid errors introduced by extrapolation. Annual time series were obtained by averaging monthly data from the AOGCMs. To evaluate the Caribbean Sea and SBR regional behavior, we calculated the spatial mean and standard deviation from all model nodes in the study area (red and yellow polygons in Fig. 1, respectively). We found time series trends fitting a simple linear regression model, with an intercept using ordinary least squares. A significant error in each estimation was calculated at a 95% confidence level. Serial correlation was not used for the analysis, consequently, the 95% confidence intervals may be slightly narrower. To compute anomalies, the reference period (1976–2005) was subtracted from the end of the century (2071–2100) averaged values, with the aim of reducing the model's interannual variability that could induce bias in the results.

We compared the five-model ensemble (hereafter M5) spatial results with satellite data linearly interpolated to the model nodes (Bustos and Torres 2021). Average satellite sea surface temperature for 1993–2005 from the COBE mission (Tokyo Climate Center 2020) and 2000–2005 sea surface salinity from the Aquarius mission (Jet Propulsion Laboratory 2020) were compared. Ocean mean circulation patterns in the Caribbean were computed using monthly files from OSTM/Jason-2 absolute dynamic topography anomalies for the 1993–2005 period (NOAA 2020a) and compared to the model's mean sea level height above the geoid (SSH). Comparison showed that M5-CAR correctly reproduces the most important mesoscale features in the Caribbean (not shown).

Three main factors increase global mean sea level. First, thermal expansion (thermosteric change). Second, changes in the mass of seawater in the ocean (barystatic change). Third, global halosteric effects are far smaller than either thermosteric or barystatic changes (Griffies et al. 2016). Two variables are used in the CMIP6 AOGCMs to assess mean sea level. ZOSTOGA represents the global mean thermosteric sea level (GMTSL) that is affected by thermal expansion, representing roughly one-third to one-half of the observed global mean sea level rise in the 20th and early 21st centuries (Church et al. 2011; Gregory et al. 2013; Hanna et al. 2013). ZOS, defined as dynamic sea level (Griffies and Greatbatch 2012; Griffies et al. 2014), reflects the fluctuations due to ocean dynamics taking into account the redistribution of mass and changes in circulation (Yin 2012; Meyssignac et al. 2017). To assess total changes in local sea level, these two variables were added (Huang and Qiao 2015; Gregory et al. 2019) and reported in this chapter as sterodynamic sea level (SDSL) following van Westen et al. (2020).

AOGCMs models in CMIP6, similarly to CMIP5, do not include land-ice melting and other smaller contributions to sea level (IPCC 2014). Additionally, most CMIP6-based global climate models will have unreliable values for barystatic changes (Nowicki et al. 2016) as their dynamics are difficult to simulate (Dyurgerov and Meier 2004; Kaser et al. 2006; Henderson-Sellers and McGuffie 2012). Therefore, these changes can be estimated through the Ice Sheet Model Intercomparison Project (ISMIP6) (Nowicki et al. 2016). Finally, global halosteric effects in a CMIP simulation are associated with inaccurate estimates of ocean mass changes in these models, and represent a small fraction of the volume change that results from adding freshwater to the ocean (Wunsch et al. 2007).

Climate models often exhibit spurious trends that are unrelated to external forcing and internal climate variability, especially in oceanic variables. We calculated and removed this model drift from oceanic variables in the Caribbean using the model piControl simulation with a 500-yr length for all the models analyzed. We applied the full linear drift method recommended by Gupta et al. (2013).

3 Projections of Atmospheric Variables

3.1 Surface Air Temperature (Ta)

Spatially averaged Ta time series for the Caribbean and SBR (Fig. 2a) show large interannual variability under both SSPs projected scenarios. However, the time series are dominated by a trend after 1970. The uncertainty related to the models' internal variability increases towards the end of the century and is larger under SSP5-8.5 than SSP2-4.5 (shaded area). The Ta spatial average for the M5-CAR in 1976–2005 for the CAR (SBR) is 26.86 ± 0.19 (27.06 ± 0.19) °C (Fig. 2a). All models indicated positive and statistically significant trends for Ta in all analyzed periods (Table 2). Ta warming trends in the M5-CAR produced by the historical model run are in good agreement with regional trends determined from in-situ data (Peterson et al. 2002; Stephenson et al. 2014; Jones et al. 2016) and CMIP5 models (Bustos 2020; Bustos and Torres 2021) for the different periods.

The strongest Ta trends are for the SBR in the projected periods, with values of 2.88 ± 0.28 °C cy^{-1} (SSP2-4.5) and 3.52 ± 0.22 °C cy^{-1} (SSP5-8.5) for 2005–2050. For 2005–2100, the trends are 1.95 ± 0.11 °C cy^{-1} (SSP2-4.5) and 3.39 ± 0.08 °C cy^{-1} (SSP5-8.5). Under SSP2-4.5, the weaker trends of 2005–2100 relative to 2005–2050 are attributable to radiative emissions decline after 2040 in that scenario (O'Neill et al. 2017). However, because the trends are always positive, Ta is expected to continue to increase in the CAR and the SBR during the entire twenty-first century.

In addition to the time series, we also assessed Ta spatial behavior for the 2071–2100 averaged period from the M5-CAR. The Ta spatial average for 2071–2100 is estimated to be 28.79 ± 0.21 (29.52 ± 0.51) °C, under SSP2-4.5 (SSP5-8.5) scenarios, respectively. Differences from the reference period 1976–2005 represent a nearly homogeneous increase of 1.93 (2.66) °C under SSP2-4.5 (SSP5-8.5) scenarios (Fig. 3b, c).

We also studied Ta and SST seasonal changes, because they are important for sea level sub-regional behavior and extremes in the Caribbean (Torres and Tsimplis 2012, 2013). Regardless of SSP scenarios used, an increase in Ta is expected in all months at the end of the century (2071–2100), maintaining seasonality with maximum values between June–November (rainy season) (Fig. 4). The annual range increases from 2.05 °C in 1976–2005 to 2.09 °C (2.12 °C) in 2071–2100 for SSP2-4.5 (SSP5-8.5) scenarios respectively.

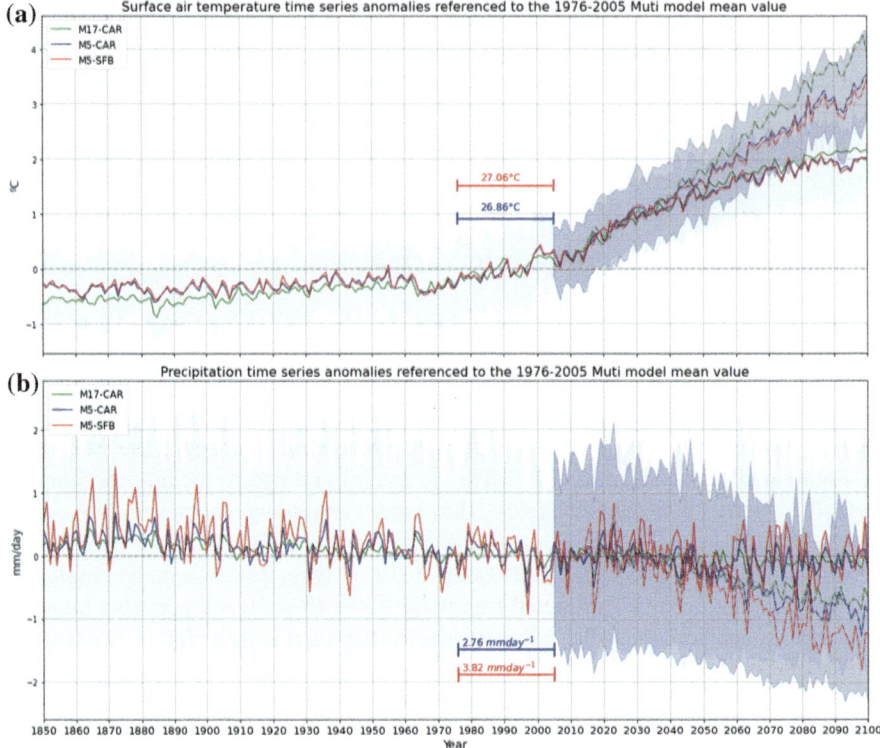

Fig. 2 Spatially averaged time series. For the Caribbean from 17 model ensemble (M17-CAR in green), 5 model ensemble (M5-CAR in blue) and SBR (M5-SBR in red). **a** Air Temperature (°C) and **b** Precipitation (mm day^{-1}) anomalies referenced to the 1976–2005 averaged period. The 1850–2014 results are from historical experiments. The 2014–2100 projection is for SSP2-4.5 (solid line) and SSP5-8.5 (dashed line) scenarios. Trends values are presented in Table 2. The 5–95% range calculated across models is included as a shaded area for M5-CAR. Clear blue for historical and SSP2-4.5, dark blue for SSP5-8.5

The preindustrial (1860–1900) Ta mean in the CAR (SBR) has a spatial mean of 26.47 ± 0.09 (26.70 ± 0.10) °C for the M5 (Fig. 2a). In line with the Paris Agreement (IPCC 2014; United Nations 2015), a 2 °C increase in Ta relative to the preindustrial period has been defined as the limit at which the planet could experience risks and impacts associated with climate change in the attempt to meet the sustainable development goals (United Nations 2015). Under SSP2-4.5, Ta in the CAR (SBR) would be near the limit established by the Paris Agreement 28.47 (28.70) °C, about 2059 (2060), whereas under SSP5-8.5 this limit would be reached after 2046 (2050) (Fig. 2a).

Taylor et al. (2018) studied air temperature and precipitation from 42 CMIP5 models of the Caribbean for the period 1861–2100 under RCP-4.5. Most models indicated that air temperature would attain an increase of 2 °C between 2033 and 2062 relative to the preindustrial period (1861–1900). Similarly, Bustos and Torres

Table 2 Adjusted trends per century for spatially averaged atmospheric and oceanic variables

Experiment		Historical		SSP2-45		SSP5-85	
Variable	Model/Per	1850–2005	1960–2005	2005–2050	2005–2100	2005–2050	2005–2100
Atmospheric variables							
Ta	M17-CAR	0.44 ± 0.04*	1.37 ± 0.23*	2.94 ± 0.22*	2.21 ± 0.10*	3.67 ± 0.20*	4.33 ± 0.09*
	M5-CAR	0.27 ± 0.05*	1.50 ± 0.27*	2.89 ± 0.26*	1.96 ± 0.12*	3.54 ± 0.20*	3.53 ± 0.08*
	M5-SBR	0.25 ± 0.05*	1.44 ± 0.31*	2.88 ± 0.28*	1.95 ± 0.11*	3.52 ± 0.22*	3.39 ± 0.08*
SLP	M17-CAR	0.08 ± 0.03*	0.05 ± 0.21	−0.04 ± 0.22	0.03 ± 0.06	0.07 ± 0.25	0.28 ± 0.07*
	M5-CAR	0.09 ± 0.07*	−0.14 ± 0.51	0.03 ± 0.43	−0.02 ± 0.14	−0.14 ± 0.40	0.45 ± 0.15*
	M5-SBR	0.07 ± 0.06*	−0.06 ± 0.47	0.01 ± 0.38	−0.02 ± 0.12	−0.10 ± 0.35	0.47 ± 0.13*
Wind	M17-CAR	0.03 ± 0.02*	0.03 ± 0.17	0.11 ± 0.10*	0.06 ± 0.05*	0.10 ± 0.12	0.24 ± 0.04*
	M5-CAR	0.01 ± 0.05	−0.14 ± 0.39	0.19 ± 0.28	0.11 ± 0.09*	−0.04 ± 0.30	0.18 ± 0.09*
	M5-SBR	0.01 ± 0.09	−0.14 ± 0.66	0.18 ± 0.44	0.10 ± 0.15	−0.17 ± 0.50	0.23 ± 0.16*
Pr	M17-CAR	−0.17 ± 0.03*	−0.25 ± 0.26	−0.09 ± 0.20*	−0.15 ± 0.07*	−0.45 ± 0.28*	−0.92 ± 0.09*
	M5-CAR	−0.19 ± 0.18*	−0.49 ± 0.48*	−0.21 ± 0.39	−0.14 ± 0.13*	−0.32 ± 0.21*	−1.16 ± 0.14*
	M5-SBR	−0.36 ± 0.13*	−0.46 ± 0.78	−0.24 ± 0.76	−0.14 ± 0.12*	−0.73 ± 0.56*	−1.79 ± 0.23*
Oceanic variables							
SST	M17-CAR	0.26 ± 0.04*	0.92 ± 0.25*	2.27 ± 0.21*	1.89 ± 0.07*	2.60 ± 0.28*	3.29 ± 0.27*
	M5-CAR	0.08 ± 0.06*	1.47 ± 0.53*	2.81 ± 0.37*	2.36 ± 0.14*	3.35 ± 0.38*	3.43 ± 0.12*
	M5-SBR	0.05 ± 0.04*	1.39 ± 0.57*	2.79 ± 0.42*	2.39 ± 0.16*	3.33 ± 0.45*	3.37 ± 0.42*
SSS	M17-CAR	0.03 ± 0.01*	0.13 ± 0.07*	0.48 ± 0.06*	0.54 ± 0.03*	0.69 ± 0.11*	1.00 ± 0.04*
	M5-CAR	0.09 ± 0.03*	0.19 ± 0.18*	0.95 ± 0.22*	0.36 ± 0.07*	1.14 ± 0.19*	1.45 ± 0.06*
	M5-SBR	0.08 ± 0.02*	0.11 ± 0.09*	0.99 ± 0.13*	0.34 ± 0.04*	1.17 ± 0.12*	1.41 ± 0.07*
SSH	M17-CAR	0.03 ± 0.01*	0.13 ± 0.07*	0.48 ± 0.06*	0.54 ± 0.03*	0.64 ± 0.11*	1.00 ± 0.04*

(continued)

Table 2 (continued)

Experiment		Historical		SSP2-45		SSP5-85	
Variable	Model/Per	1850–2005	1960–2005	2005–2050	2005–2100	2005–2050	2005–2100
	M5-CAR	−2.41 ± 0.49*	−2.11 ± 2.90	−2.62 ± 2.28*	−2.42 ± 0.79*	0.63 ± 3.13	−2.44 ± 0.98*
	M5-SBR	−2.49 ± 0.67*	0.19 ± 0.42	−1.09 ± 3.09	−1.87 ± 1.18*	0.85 ± 2.95	−0.28 ± 1.18
GMTSL	M17-CAR	2.56 ± 0.14*	5.94 ± 0.68*	20.41 ± 0.96*	22.64 ± 0.40*	20.87 ± 2.33*	32.19 ± 1.37*
	M5-CAR	3.49 ± 0.12*	6.05 ± 0.76*	25.73 ± 0.54*	30.64 ± 0.42*	28.58 ± 0.96*	42.89 ± 1.49*
SDSL	M17-CAR	1.15 ± 0.19*	4.72 ± 0.13*	24.49 ± 1.30*	24.78 ± 0.53*	22.10 ± 1.35*	22.21 ± 0.48*
	M5-CAR	1.08 ± 0.50*	3.93 ± 3.36*	23.11 ± 2.24*	28.23 ± 0.92*	29.21 ± 3.35*	40.46 ± 1.89*
	M5-SBR	1.00 ± 0.68*	6.24 ± 4.47*	24.65 ± 2.96*	28.77 ± 1.23*	32.66 ± 4.00*	42.62 ± 1.85*

For the Caribbean Sea using 17 Multi-model Ensemble (M17-CAR); 5 Multi-model Ensemble (M5-CAR); and for the Seaflower Biosphere with the 5 multi-model ensemble (M5-SBR). Two CMIP6 Socioeconomic Pathways (SSP) are assessed. Significant trends (95% confidence) are indicated with (*). Air temperature (Ta), atmospheric pressure (SLP) wind speed, and Precipitation (Pr) trends indicated in °C cy^{-1}, hPa cy^{-1}, ms^{-1} cy^{-1}, and mm day^{-1} cy^{-1}, respectively. Sea surface temperature (SST), sea surface salinity (SSS), sea surface above geoid (SSH), global averaged steric sea level (GMTSL), and local mean sterodynamic sea level (SDSL) trends indicated in °C cy^{-1}, PSU cy^{-1}, and cm cy^{-1}, respectively

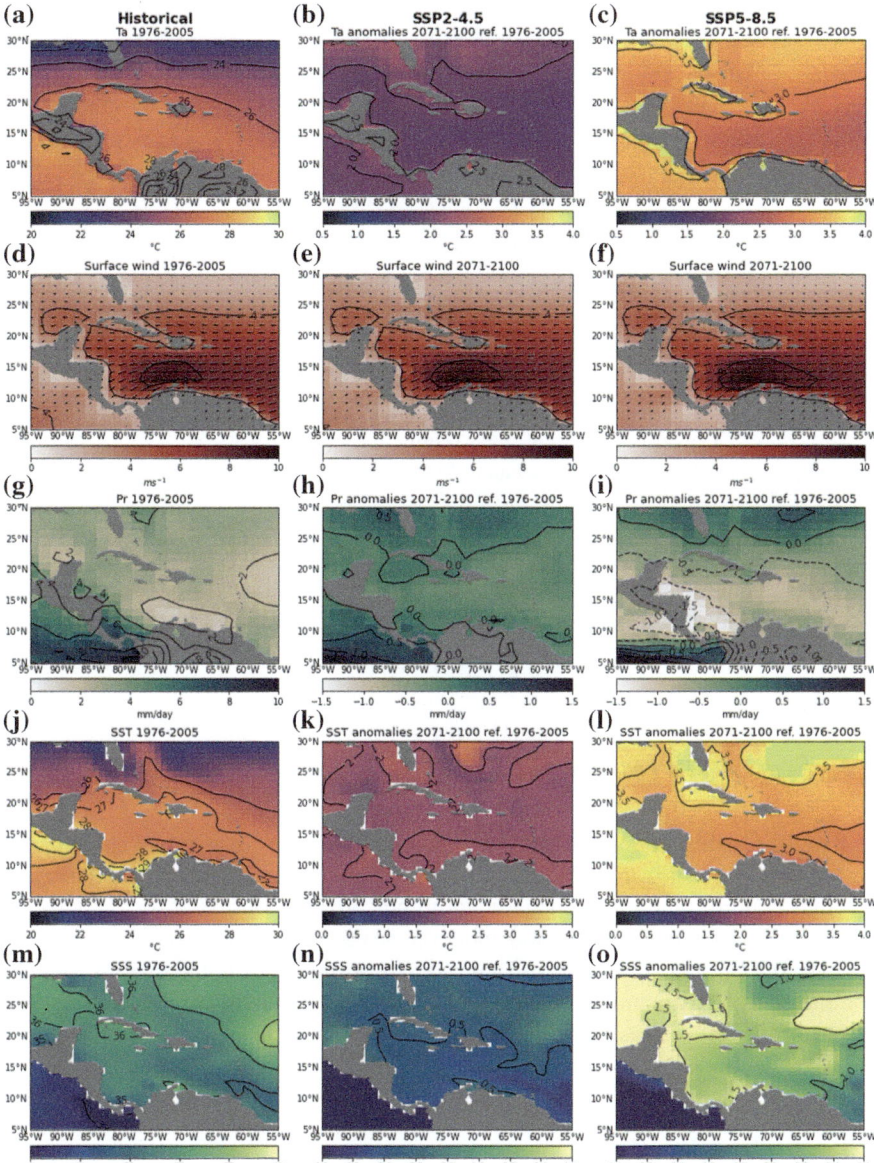

Fig. 3 Air temperature (°C, first row), wind (m s^{-1}, second row), precipitation (mm day^{-1}, third row), sea surface temperature (°C, fourth row) and sea surface salinity (PSU, fifth row) spatial behavior in the Caribbean Sea from the 5-model ensemble (M5-CAR). For the 1976–2005 averaged period (first column) after the "historical" run. Surface air temperature, precipitation, sea surface temperature and sea surface salinity anomalies for 2071–2100 relative to 1976–2005 under SSP2-4.5 are shown in the second column, except wind which shows the 2071–2050 mean behavior. Third column as second column, but under the SSP5-8.5 scenario

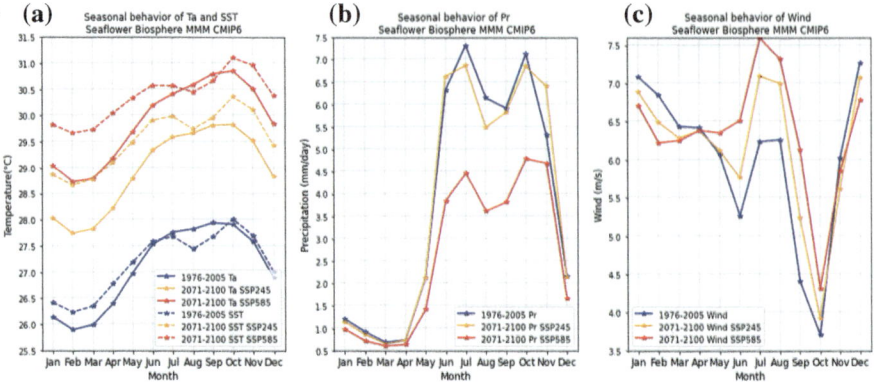

Fig. 4 Seasonal behavior for 1976–2005 and 2071–2100 averaged periods in the SBR from the 5-model ensemble (M5-SBR) under SSP2-4.5 and SSP5-8.5 projected scenarios. **a** Surface Ambient temperature (Ta) and Sea surface temperature (SST). **b** Precipitation (Pr), and **c** wind speed

(2021) found 2060 (2040) as the attainment dates for the 2 °C limit under RCP4.5 (RCP8.5) scenarios respectively. This indicates that the Ta increase according to the M5-CAR is close to the upper limit (highest trends) compared to the results from Taylor et al. (2018).

3.2 Sea Level Pressure (SLP)

SLP in the Caribbean Sea has modest trends <0.1 hPa cy^{-1} that are significant only for the 1850–2005 historical period (Table 2), indicating a steady behavior during the previous century. In addition, significant trends are only observed for CAR (SBR) for the 2005–2100 projected period under the SSP5-8.5 scenario, with values of 0.45 ± 0.15 (0.47 ± 0.13) hPa cy^{-1} (Table 2). Time series are not shown as they are dominated by climate models' interannual variability. Besides, spatial differences in time (not shown) are basin-wide coherent.

3.3 Surface Wind

Surface wind time series are dominated by large interannual variability (not shown), as was also found for sea level pressure (SLP). Significant coherent trends in wind speed are found only for 2005–2100 under SSP5-8.5, with a value of 0.23 ± 0.16 ms^{-1} cy^{-1} in M5-SBR (Table 2). The regional wind pattern shows an increase in the area of wind speed >8 ms^{-1}, without significant changes in direction for the 2071–2100 period under both SSPs scenarios, relative to 1976–2005 (Fig. 3e, f). In the 1976–2005 period, the average wind speed for the M5-CAR (M5-SBR) is 6.05 ± 0.17 (5.99 ± 0.29) ms^{-1}, with values >8 ms^{-1}, indicating the position of the Caribbean low-level jet (Fig. 3d) and resembling the typical wind behavior in the basin. The largest increase in wind speed is for 2071–2100 referenced to the 1976–2005 period, under the SSP5-8.5 in M5-SBR (spatial mean of 0.32 ms^{-1}). Similar results were found from the CMIP5 models (Bustos and Torres 2021).

We executed a seasonal analysis of long-term variation of the surface wind, since this variable dominates seasonal changes in the regional ocean circulation (Torres and Tsimplis 2012). Thus, we investigated wind time series from the dry-windy season (December–January–February) and rainy-warm season (September–October–November) using both SSPs scenarios (not shown). For the 1976–2005 period, wind speed in M5-SBR was stronger in the dry season (7.06 ± 0.53 ms^{-1}) when compared to the rainy season (4.71 ± 1.12 ms^{-1}). For the same area in the 2071–2100 period under the SSP5-8.5, the mean wind speed in the dry season is 6.56 ± 0.47 ms^{-1}, and 5.43 ± 0.90 ms^{-1} in the rainy season. Thus, M5-SBR models indicate that wind speed differences between the dry and rainy seasons will decrease without changing their seasonality (Fig. 4c). Besides, note that the maximum wind speed month in M5-SBR is projected to change from December (1975–2005) to July (2071–2100—SSP5-8.5). These results are in agreement with Bustos and Torres (2022), who also show different spatial patterns in wind intensification based on CMIP6 models, especially seen in the Caribbean low-level jet with seasonal different responses.

Costoya et al. (2019) assessed wind energy projections in the Caribbean for the twenty-first century using downscaling techniques in seven AOGCM models from CMIP5. They found that the maximum annual wind speed increase in the CAR would be ~0.4 ms^{-1} by 2100 under RCP8.5, referenced to the 2005 value. Thus, their increase in regional wind is about twice the result from the M5-CAR in the same period with the SSP5-8.5 scenario (0.18 ms^{-1} in Table 2).

Evaluating future changes in surface wind is important, because various authors have shown the dominance of wind in driving the CAR ocean circulation (Brenes and Trejos 1994; Torres and Tsimplis 2012; Montoya-Sánchez et al. 2018). We found that significant positive trends in M5-SBR for 2005–2100 (SSP5-8.5) do not result from a homogeneous wind speed increase in all months. Although wind direction and seasonal behavior seem not to have significant changes (Figs. 3d–f and 4c), seasonal variations in wind speed could significantly affect regional circulation. Therefore, future changes in the Caribbean surface wind should include a seasonal assessment.

3.4 Precipitation (Pr)

The M5-CAR and M5-SBR show significant negative trends for most of the periods and projected scenarios analyzed (Table 2). The largest trends are in the 2005–2100 period for SSP5-8.5 (-1.79 ± 0.23 mm day^{-1} cy^{-1} in M5-SBR). Regardless of the significant coherent trends, model projections of this variable have large differences (shading area in Fig. 2b); therefore, results must be treated with caution. Pr is expected to decrease in the CAR and the SBR during the twenty-first century, as trends are negative. It appears that the future radiative emission rate significantly affects Pr, as was seen for Ta, due to significantly larger trends in SSP5-8.5 when compared to SSP2-4.5 (Table 2). These results are in good agreement with the reduction of Pr in the Caribbean founded by Almazroui et al. (2021), also using CMIP6 models.

The Pr spatial average for the M5-CAR (M5-SBR) in 1976–2005 is 2.76 ± 0.22 (3.82 ± 0.34) mm day^{-1} (Fig. 3g). Similar values were found by Lee and Wang (2014) in a shorter period (1980–2005) using CMIP5 models. The Pr M5-SBR spatial average for 2071–2100 is expected to be 3.78 ± 0.34 (2.84 ± 0.51) mm day^{-1}, under SSP2-4.5 (SSP5-8.5) scenarios, respectively. Differences from the reference period 1976–2005, represent a decrease in SBR of -0.04 (-0.98) mm day^{-1} under SSP2-4.5 (SSP5-8.5) scenarios, respectively (Fig. 3h, i).

We also studied Pr seasonal changes for the SBR. All scenarios show a bimodal distribution with maximum values from June to November (rainy season). Under SSP2-4.5, the Pr annual pattern in 2071–2100 does not show a large difference (Fig. 4) when compared to the 1976–2005 behavior. The largest difference is a reduction in Pr during the rainy season for the 2071–2100 SSP5-8.5 projection. The annual range decreases from 6.64 mm day^{-1} in 1976–2005 to 6.21 (4.19) mm day^{-1} in 2071–2100 for SSP2-4.5 (SSP5-8.5) scenarios respectively. These projections under the SSP5-8.5 scenario coincide with the results of Karmalkar et al. (2013) where a reduction of Pr in the Caribbean islands was identified under the SRES A2 scenario using 15 GCMs from CMIP3 and a Regional Climate Model. These are important results, as inhabited islands in the SBR depend on the rainy season to collect fresh water for their yearly consumption. The shown reduction of precipitation in the rainy season (>35% under SSP5-8.5) should be tracked carefully, in order to define convenient mitigation plans if data starts to confirm such projections.

4 Projections of Oceanic Variables

4.1 Sea Surface Temperature (SST)

Significant and positive trends are found for SST in all experiments and projected periods for the CAR and SBR (Table 2). SST behavior was similar to Ta as expected, due to heat fluxes between the atmosphere and ocean. We found the smallest trends for 1850–2005, with a noticeable increase over 1960–2005 with values of 1.47 ± 0.53 (1.39 ± 0.57) °C cy^{-1} for the M5-CAR (M5-SBR), respectively. However, in the historical run, significant SST trends are within the bounds of the interannual variability (Fig. 5a).

Warming trends throughout the Caribbean for the second half of the twentieth century (Table 2) are consistent with those evidenced by Peterson et al. (2002) and

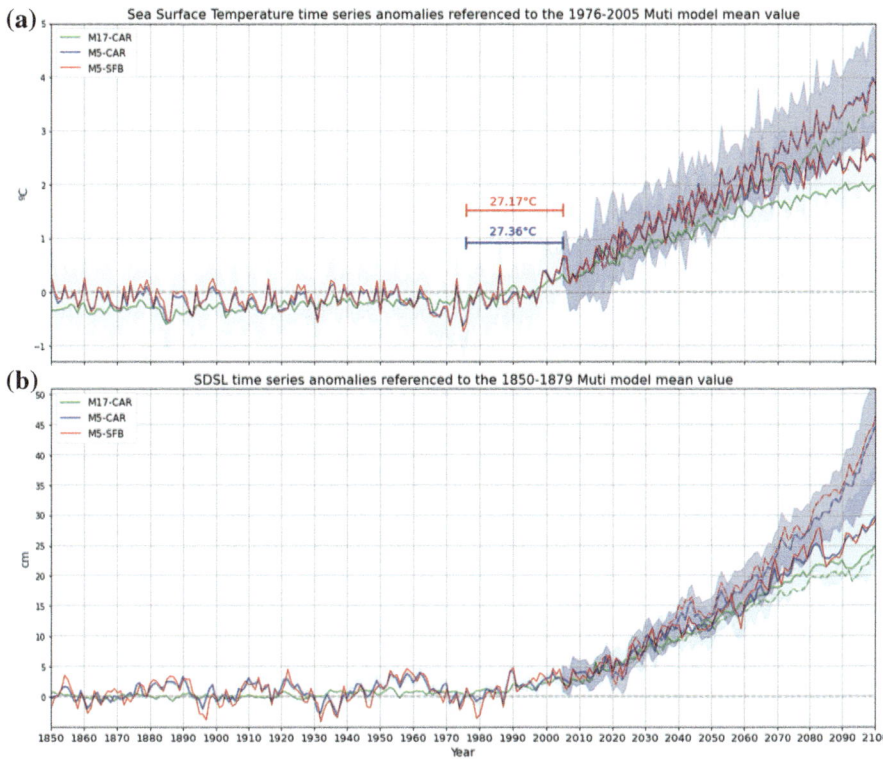

Fig. 5 Spatially averaged time series for the Caribbean Sea from the 17 model ensemble (M17-CAR in green), 5 models ensemble (M5-CAR in blue) and SBR (M5-SBR in red): **a** Sea surface temperature (°C); **b** Sterodynamic sea level (cm) anomalies referenced to the 1976–2005 period. The 1850–2014 results are from historical experiments. The 2014–2100 projection is for SSP2-4.5 (solid line) and SSP5-8.5 (dashed line) scenarios. Trend values are presented in Table 2. The 5–95% range calculated across models is included as a shaded area for M5-CAR. Clear blue for historical and SSP2-4.5, dark blue for SSP5-8.5

Villegas et al. (2021). Similarly, Antuña-Marrero et al. (2016), using information reconstructed from the International Comprehensive Ocean–Atmosphere Dataset, found a trend for 1972–2005 of 1.41 ± 0.67 °C cy^{-1}. Deser et al. (2010) found regional trends in SST with a range of 0.4–1.6 °C cy^{-1} for the period 1900–2008, using observed data and a model-based reconstruction. Trends from the historical run are smaller compared to the results of Bustos and Torres (2021) using CMIP5 models.

Significant positive trends dominated SST in all the projected periods analyzed for the Caribbean, regardless of the model or SSP scenario used (Table 2). For 2005–2050, the trend for the SBR is 2.79 ± 0.42 (3.33 ± 0.45) °C cy^{-1} under SSP2-4.5 (SSP5-8.5) respectively. For 2005−2100, the trends are between 2.39 ± 0.16 (3.37 ± 0.42) °C cy^{-1} under SSP2-4.5 (SSP5-8.5). For the projected periods, most SST trends are smaller than Ta trends. According to the IPCC (2014), SST is expected to increase globally between 0.70 and 2.40 °C by 2100 under RCP4.5. Therefore, under a similar projection scenario (SSP2-4.5) trends found for the CAR and SBR are at the maximum among these values.

Because of this positive trend, by 2071–2100 SST in the M5-CAR (M5-SBR) is expected to be 29.40 ± 0.19 (29.26 ± 0.18) °C or 30.56 ± 0.39 (30.35 ± 0.38)°C, under SSP2-4.5 or SSP5-8.5, respectively. That represents a temperature increase in the Caribbean of 2.04–3.20 °C and for SBR 2.09–3.18 °C (Fig. 3k, l), over the value for the 1976–2005 averaged period (27.36 °C in M5-CAR and 27.17 °C in M5-SBR) (Fig. 3j), depending on the radiative scenario used.

SST warming trends in the CAR produced by the SSP5-8.5 scenario run are in good agreement with regional trends determined from the 26 CMIP5 models ensemble presented by Alexander et al. (2018) with values between 2.5 and 3.5 °C cy^{-1} for the Gulf of Mexico and the Caribbean in the 1976–2099 period. Similar results were obtained by Bustos and Torres (2021).

SST is warmest between September and November and coldest in February (Fig. 4a), resembling the seasonal behavior in the Caribbean (Torres and Tsimplis 2012). Regardless of the projected scenario, all months in the 2071–2100 period are expected to be warmer than the warmest month in the 1976–2005 period. Several studies have shown diverse factors that can affect the formation of hurricanes (Goldenberg et al. 2001; Wang and Lee 2007), of which SST increase is probably the most important. Because warmer sea surface temperatures enhance tropical cyclone formation in the Atlantic (Grinsted et al. 2013), the hurricane season in the Caribbean runs from June to November (NOAA 2020b), corresponding to the warmer months. Therefore, M5-SBR projections for the end of the century show that all year round there would be sufficient heat in the ocean to permit hurricane formation, which in the future could extend the hurricane season in the basin, among other impacts.

SST increase could also affect the biosphere in the local upwelling areas. Taylor et al. (2013), using monthly observations from the CARIACO Ocean Time-Series in the southern Caribbean Sea between 1996 and 2010, identified that the SST increase would intensify ocean stratification, reducing the delivery of upwelled nutrients to surface waters, generating an ecological state of change in the planktonic system.

Bustos and Torres (2021) showed that the SST increase in the Caribbean expected from the CMIP5-ACCESS1.0 model could have severe consequences for hurricane frequency, its season length, and in coral bleaching by the end of the twenty-first century. This chapter builds on these results using a 5-model ensemble from the CMIP6. Results are consistent, although the CMIP6 shows that most of the SST trends for the projected periods are larger compared to the CMIP5-ACCESS model results.

4.2 Sea Surface Salinity (SSS)

For all experiments using the M5-CAR and M5-SBR, positive and significant SSS trends are seen in all periods (Table 2). The smallest trends are for 1850–2005 with values of 0.09 ± 0.03 (0.08 ± 0.02) PSU cy^{-1} for CAR (SBR), respectively. Nonetheless, in the historical run, significant SSS trends do not extend beyond the interannual variability (not shown). The SSS trends for 2005–2100 in the M5-CAR are 0.36 ± 0.07 (1.45 ± 0.06) PSU cy^{-1} and in SBR 0.34 ± 0.04 (1.41 ± 0.07) PSU cy^{-1} under SSP2-4.5 (SSP5-8.5) (Table 2). Therefore, SSS trends in the Caribbean Sea are sensitive to the projected radiative scenarios.

The M5-CAR mean salinity field for the 1976–2005 period is 35.51 PSU (Fig. 3m), while the expected SSS mean for the 2071–2100 period is 35.96 ± 0.10 (36.87 ± 0.22) PSU and for M5-SBR 36.19 ± 0.13 (37.07 ± 0.25) PSU under SSP2-4.5 (SSP5-8.5), respectively. The difference of ~0.91 (0.89) PSU for CAR (SBR) between the two SSP scenarios by the end of the twenty-first century indicates the sensitivity of this variable to the radiative forcing scenario used. Due to the positive trends, for 2071–2100 under SSP2-4.5, SSS increases uniformly in the basin between 0.3 and 0.6 PSU (Fig. 3n), while under SSP5-8.5, SSS anomalies are between 1.0 and 1.5 PSU (Fig. 3o), with larger values toward the Cayman Basin and the Yucatan Peninsula.

The SSS increase shown in the present study is similar to the results presented by Bustos and Torres (2021) for the CMIP5-ACCESS1.0 model. In their study, they suggested that positive SSS trends were related to SST increase, due to evaporation enhancement. In this chapter, the 5-model ensemble from CMIP6 results also shows an increase in SST which could lead to an increase in evaporation. Besides, we also include an assessment of precipitation. We show negative trends in precipitation, which would also increase SSS due to rainfall decrease and have an effect reducing local river freshwater fluxes.

4.3 Sea Level

To assess the future sea level behavior in the Caribbean Sea and SBR, we present two analyses, each one including different periods under two SSP scenarios. First, we determined SSH, GMTSL and sterodynamic sea level (SDSL) spatially averaged trends (Table 2, Fig. 5b), as well as SDSL trends spatial behavior (Fig. 6) to study regional patterns. Second, expected sterodynamic sea-level rise is presented for the SBR and the archipelago (Fig. 7, Table 1).

Trends in SSH from the historical run in M5-CAR and SBR are only significant and negative in the 1850–2005 period (Table 2). On the contrary, GMTSL trends are significant and positive for the Caribbean in all periods of the historical run, making all SDSL trends in this period significant and positive. SDSL trends in 1960–2005 are more than double the 1850–2005 trends.

All significant SSH trends for the CMIP6 projected periods in M5-CAR and SBR are negative (Table 2), contrary to CMIP5-ACCESS results (Bustos and Torres 2021). However, all GMTSL and SDSL trends in CMIP6 projected periods are positive and significant, regardless of their radiative scenario. Therefore, positive GMTSL trends prevail over negative SSH trends projected in the Caribbean. In the 2005–2100 period

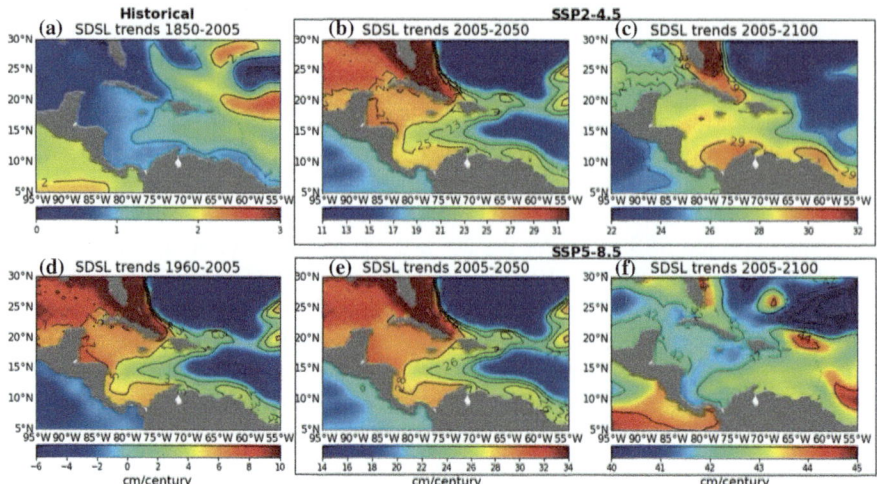

Fig. 6 Sterodynamic sea level trends (SDSL) in cm cy^{-1} for the Caribbean Sea from 5 models ensemble. Periods **a** 1850–2005 and **d** 1960–2005 from historical run. Period 2005–2050 using **b** SSP2-4.5 and **e** SSP5-8.5. Period 2005–2100 using **c** SSP2-4.5 and **f** SSP5-8.5. Isolines every 2 cm cy^{-1}. Note different trend ranges in panels

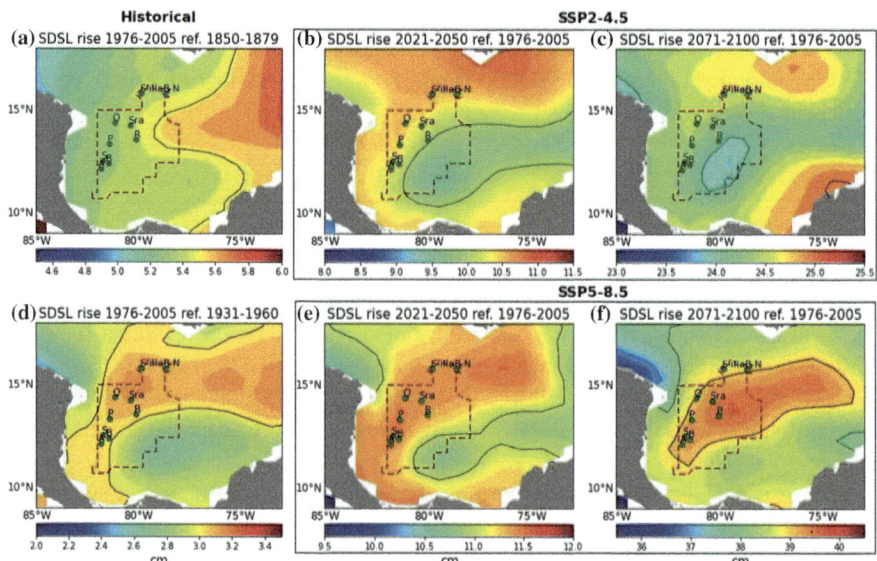

Fig. 7 Sterodynamic sea level (SDSL) rise in (cm) for the archipelago from the 5-model ensemble. Periods 1976–2005 referenced to **a** 1850–1879 and **d** 1931–1960 from the historical run. Period 2021–2050 using **b** SSP2-4.5 and **e** SSP5-8.5 projection scenarios. Period 2071–2100 using **c** SSP2-4.5 and **f** SSP5-8.5 projection scenarios. Panels **b**, **c**, **e**, and **f** referenced to 1976–2005 averaged period. Isolines every 1 cm. Note different height ranges in each panel. Brown Polygons represent the Seaflower Biosphere Reserve (Coralina-Invemar 2012; Borrero-Pérez et al. 2019)

for the Caribbean (SBR), SDSL trends under SSP5-8.5 are 40.46 ± 1.89 (42.62 ± 1.85) cm cy^{-1} respectively (Table 2). The largest contribution from GMTSL to SDSL trends in the Caribbean and SBR, compared to SSH contribution, indicates that regional sea level rise will be largely due to global ocean temperature increase rather than regional effects (Figs. 6 and 7).

SDSL trends in the Caribbean dominate the interannual variability with a clear exponential behavior under SSP5-8.5 through the end of the century (Fig. 5b). An acceleration of 1.6 ± 0.1 mm yr^{-1} cy^{-1} was reported for Cristobal sea level in the southern Caribbean, for the period 1908–2009 (Torres and Tsimplis 2013). In the same period, the M5-CAR SDSL time series shows an acceleration of 0.56 ± 0.16 mm yr^{-1} cy^{-1}. Additionally, for the 2005–2100 period an acceleration of 1.62 ± 0.17 (2.64 ± 0.21) mm yr^{-1} cy^{-1} is expected under the SSP2-4.5 (SSP5-8.5) scenario respectively. The SDSL acceleration observed in Fig. 5b, is mainly due to GMTSL acceleration with values of 1.27 ± 0.14 (2.21 ± 0.19) mm yr^{-1} cy^{-1} for the 2005–2100 period under SSP2-4.5 (SSP5-8.5) scenario, respectively. Note that this acceleration is larger in the 5-model ensemble than in the 17-model ensemble (Fig. 5b). As the main difference between the two model ensembles is the spatial resolution in the intertropical ocean, this indicates that projected GMTSL variations can depend on climate models' resolution in addition to the climate sensitivity of the models used in an ensemble.

SDSL trends are not spatially homogeneous in the study area. For 1850–2005, M5-CAR has larger trends (>1.3 cm cy^{-1}) toward the north-eastern Caribbean (Fig. 6a). On the contrary, larger trends (>5 cm cy^{-1}) for 1960–2005 appear toward the Yucatan Peninsula and in the SBR (4–5 cm cy^{-1}) (Fig. 6d). For the projected 2005–2050 period, trends range between 15–27 cm cy^{-1} under SSP2-4.5 (Fig. 6b) and 17.5–29 cm cy^{-1} under SSP5-8.5 (Fig. 6e). In both cases, trends are larger toward the Cayman Sea and smaller toward the eastern Caribbean. For 2005–2100, trends are 27–30 cm cy^{-1} under SSP2-4.5 (Fig. 6c). However, the strongest trends are for the SSP5-8.5 scenario, with a range of 41–44 cm cy^{-1} (Fig. 6f) with larger values in the Venezuela and Granada basins. Modeled SDSL trends in the Caribbean show strong spatial variability and changes with time, which coincides with observed regional sea level behavior assessed from observed data (Torres and Tsimplis 2013).

We also assessed SDSL rise in the SBR, which results from the reported trends using the M5 ensemble (Fig. 7). In a 45-year period of the projection run (2021–2050 referenced to 1976–2005), the SDSL is expected to rise in the islands of the archipelago >10.1 (>11.3) cm under SSP2-4.5 (SSP5-8.5) (Fig. 7b, e). Additionally, in a 95-year period (2071–2100 referenced to 1976–2005), the SDSL is projected to rise under the same scenarios >24.2 (>38.8) cm (Fig. 7c, f). The largest SDSL rise for this period is projected for Bajo Nuevo under SSP245 (24.69 cm) and for Roncador under SSP585 (39.89 cm), among the islands in the SBR (Table 1). However, the sea level is projected to rise in all the islands with differences <2 cm (Fig. 7). According to projections from CMIP6 of Jevrejeva et al. (2020), the GMTSL increase by 2081–2100 is expected to be between 12.8–23.6 cm (18.6–34.6 cm) relative to 1995–2014 under SSP2-4.5 (SSP5-8.5) scenarios, respectively. Reported results of SDSL rise from the SBR with the 5-model ensemble are just above the upper limit of the projections obtained in the mentioned study. Although SDSL rise includes the SSH's smaller contribution (Table 2) and the periods assessed are slightly different, results are in good agreement.

M5-CAR using CMIP6 models shows an underestimation of SDSL trends when compared to previous studies using CMIP5 results (Palanisamy et al. 2012; Gupta et al. 2013;). In the archipelago, SDSL rise from the CMIP5-ACCESS1.0 model was projected to be >14.5 (>16) cm for 2021–2050 and >32.6 (>45.5) cm for 2071–2100 under RCP4.5 (RCP8.5), respectively, referenced to the 1976–2005 period (Bustos and Torres 2021). Therefore, the 5-model ensemble from CMIP6 projects a smaller SDSL increase in the study area when compared to just one model from CMIP5, but with similar spatial patterns under each scenario. This difference is mainly due to differences in the GMTSL trend, assessed by Bustos and Torres (2021) using the ZOSGA variable, which represents the total change in global mean sea level due to thermosteric changes, water flux input and salinity influences on density. In CMIP6, the ZOSGA variable is not available, therefore in this study, we use ZOSTOGA, which only accounts for the contribution due to thermal structure changes, which partially explains the smaller GMTSL trends reported in Table 2.

Conversely, Jevrejeva et al. (2020) reported an increase and larger variance for the projected GMTSL rise for the twenty-first century, from the comparison of a 15-CMIP6 model ensemble with 20-CMIP5 models (they used ZOSTOGA in their comparison). They discuss probable causes for these differences between the CMIP, mainly associated with a new generation of climate models as well as a new set of scenarios of concentrations, emissions, and land use. Besides, they assess explanations for large uncertainties in the simulation of GMTSL. Other factors associated with the models' physics such as parametrizations, circulation, and mixing schemes can also influence the SDSL projections (Swart et al. 2019).

5 Summary and Final Remarks

Seventeen CMIP6 models were used to evaluate atmospheric and oceanographic trends over the twenty-first century in the Caribbean Sea (CAR) and the Seaflower Biosphere Reserve (SBR), under two different radiative emission scenarios SSP2-4.5 and SSP5-8.5. We selected five models with the highest oceanic resolution in the Caribbean Sea to obtain the best projections for the study area due to their better ability to resolve mesoscale processes (van Westen et al. 2020).

Surface air temperature (Ta) shows significant positive trends in M17-CAR, M5-CAR, and M5-SBR ensembles (Fig. 2a). A 2 °C increase compared to the preindustrial period, defined in the Paris Agreement as a limit for greatly increased global risks, would be expected between 2046 and 2059 (2050–2060) for the CAR (SBR), depending on the radiative scenario used. On the contrary, precipitation (Pr) shows significant negative trends in most of the periods for M5-CAR and M5-SBR, therefore a reduction of Pr in the study area is expected by the end of the century (Fig. 2b). Besides, considerable Pr reduction was identified under the SSP5-8.5 scenario in the rainy season for 2071–2100 (Fig. 4b), which can reduce freshwater availability in the SBR inhabited islands. Therefore, it is important to continue monitoring these variables, to develop opportune adaptation plans to minimize freshwater supply problems for the population of the archipelago.

Sea level pressure and surface wind show a large interannual variability with the largest trends for the 2005–2100 period under the SSP5-8.5 scenario in the CAR and SBR, but without considerable spatial changes by the end of the century (Fig. 3e, f). Besides, wind speed seasonality in the SBR is not projected to change, however, an intensification from June to October can be expected (Fig. 4c). Besides, projected changes in surface wind speed will probably have different spatial responses in the study area (Bustos and Torres 2022). As wind is a major driver of seasonal sea level changes in the region (Torres and Tsimplis 2012), such wind changes might force ocean dynamics to sub-annual variations.

Sea surface temperature (SST) is expected to increase by the end of the twenty-first century compared to the baseline period 1976–2005, between 2.03–3.20 °C in the Caribbean and 2.09–3.18 °C in the SBR under SSP2-4.5 and SSP5-8.5 scenarios respectively (Fig. 3k, f). SST trends are all positive and significant with higher values for the projected periods (Fig. 5a). Furthermore, SST seasonality will not change in the region. However, all months at the end of the century (2071–2100) will be warmer than the warmest month in the baseline period (1976–2005), regardless of the radiative scenario used (Fig. 4a). This temperature increase in the ocean could intensify coral bleaching events, extend the hurricane season, and/or increase hurricane frequency in the basin by the end of the century (Saunders and Lea 2008; Eakin et al. 2009; Bender et al. 2010; Lough et al. 2018). However, other important factors modulate the genesis and development of tropical cyclones such as wave perturbation in trade winds, upper troposphere divergence, low vertical wind shear, and air temperature profiles in the atmosphere (IPCC 2014). Therefore, only the expected SST trends shown in this study are not enough to conclude about future cyclone-related changes in the Caribbean Sea.

Similarly, sea surface salinity (SSS) trends are positive and significant in all periods, with larger values under the SSP5-8.5 scenario, coinciding with the expected reduction in precipitation and the Ta and SST increase, which will probably enhance evaporation in the region. Besides, SST and SSS positive trends will compensate for surface density changes. Trends indicate that SST warming will overcome SSS increase, therefore, surface water will become lighter, which could affect local upwelling areas due to ocean stratification intensification (Torres et al. 2022 in press), reducing the delivery of upwelled nutrients affecting planktonic systems (Taylor et al. 2013).

In the islands of the SBR, depending on the SSP scenario, SDSL is expected to rise between ~24.2 and 39.9 cm for 2071–2100 compared to the baseline period 1976–2005, due to trends up to 42.62 ± 1.85 cm yr^{-1} in the SBR (SSP5-8.5). An SDSL acceleration of 0.56 ± 0.16 mm yr^{-1} cy^{-1} is observed for the 1908–2009 period in the Caribbean from the 5-model ensemble, which coincides with previous sea level acceleration reported from a long tide-gauge time series in the region (Torres and Tsimplis 2013). Besides, an acceleration of 1.62 ± 0.17 (2.64 ± 0.21) mm yr^{-1} cy^{-1}

is found under the SSP2-4.5 (SSP5-8.5) scenario for the 2005–2100 period. This acceleration is due to GMTSL behavior, which protrudes from the model ensemble with the best oceanic resolution in the region (Fig. 5b). Therefore, the assessment of GMTSL trends might be sensitive to the model's ocean resolution in the tropics, as this is an important area for heat storage and redistribution.

We found that GMTSL dominates SDSL trends. SDSL trends from CMIP6 M5-CAR are lower in all cases when compared to the ACCESS-CMIP5 model (Bustos and Torres 2021), however, slightly different variables were used in both studies to assess GMTSL (ZOSTOGA in the former and ZOSGA in the later as discussed in Sect. 4.3). On the contrary, Jevrejeva et al. (2020) reported GMTSL (ZOSTOGA) trends increase in CMIP6 models when compared to CMIP5, and assessed probable causes for these differences. Accuracy in projections of future sea level change depends on the ability of climate models to reproduce the components of sea level rise over the twenty-first century and simulate future changes across a range of emission scenarios. Differences between CMIP5 and CMIP6 thermosteric sea level projections differ mainly due to a new generation of climate models (Eyring et al. 2015) and a new set of scenarios of concentrations, emissions, and land use (O'Neill et al. 2016). However, large uncertainties in the Caribbean and global sea level projections remain due to the climate models' limited ability to reproduce future thermosteric sea level changes.

Using a 9–model ensemble from CMIP6, Sung et al. (2021) identified for the 2081–2100 period, compared to 1986–2005, a future global sea level mean (range) increase of 0.28 m (0.17–0.38 m) and 0.65 m (0.52–0.78 m) under SSP1-2.6 and SSP5-8.5 scenarios respectively. They included the contribution from changes in ocean density (mainly from thermal expansion), land ice melting from glaciers and ice sheets, groundwater, and Global Isostatic Adjustment (GIA). From these, ocean-related processes and glacier melting are major contributors to SLR. On a global scale, the former contributes 54% (42%) and the latter 32% (52%) for SSP1-2.6 (SSP5-8.5). Therefore, the SDSL rise reported in this chapter, which only accounts for ocean thermosteric changes, will probably contribute nearly half of the total sea level rise expected for the Caribbean Sea by the end of the century. If we use this global % contribution (42%) in a coarse calculation, total sea level rise by the end of the century would be up to ~95 cm for SSP5-8.5 in the islands of the archipelago.

Such mean sea level rise will interact with the tide, sea level seasonal cycle and meteorological extremes (Torres and Tsimplis 2014), enhancing flooding and erosion, risking the complete submersion of the low elevation islands of the archipelago (height <2 m—Table 1). Besides, erosion will reduce beaches' extension, which can affect tourism, the main income for some Caribbean island countries (IPCC 2014). Sea level rise will also affect the biosphere. For example, the ecological balance around mangroves will change (Bacon 1994), affecting a large number of species around this area of high biological productivity in the SBR (Urrego et al. 2009). These ecosystems provide feeding and breeding grounds for birds, reptiles, fish, and invertebrates, including many endemic vulnerable threatened and endangered species (Prato and Newball 2016). For all these reasons, it is important to continue strengthening ocean and atmosphere monitoring in the archipelago, including projections from higher spatial resolution climate models and the use of downscaling techniques to improve the dynamics of some air-ocean mesoscale processes. In parallel to the former, there is a need to enhance nations' adaptive capacity, in order to build and improve regional and local mitigation and adaptation plans, and to reduce climate change impacts in the Seaflower Biosphere Reserve by the end of the twenty-first century.

References

Agrawala S (1997) Explaining the evolution of the IPCC structure and process. IIASA Interim Report. IR-97-032

Alexander MA, Scott JD, Friedland KD et al (2018) Projected sea surface temperatures over the 21st century: changes in the mean, variability and extremes for large marine ecosystem regions of Northern Oceans. Elem Sci Anthr 6:9–12. https://doi.org/10.1525/elementa.191

Almazroui M, Islam MN, Saeed F et al (2021) Projected changes in temperature and precipitation over the united states, central America, and the Caribbean in CMIP6 GCMs. Earth Syst Environ 5:1–24. https://doi.org/10.1007/s41748-021-00199-5

Andrade C (2000) The circulation and variability of the Colombian basin in the Caribbean sea. PhD Thesis

Angeles ME, González JE, Ramírez-Beltrán ND et al (2010) Origins of the Caribbean rainfall bimodal behavior. J Geophys Res Atmospheres 115:1–9. https://doi.org/10.1029/2009JD012990

Antuña-Marrero JC, Otterå OH, Robock A et al (2016) Modelled and observed sea surface temperature trends for the Caribbean and Antilles. Int J Climatol 36:1873–1886. https://doi.org/10.1002/joc.4466

Bacon PR (1994) Template for evaluation of impacts of sea level rise on Caribbean coastal wetlands. Ecol Eng 3:171–186. https://doi.org/10.1016/0925-8574(94)90044-2

Beier E, Bernal G, Ruiz-Ochoa M et al (2017) Freshwater exchanges and surface salinity in the Colombian basin, Caribbean Sea. PLoS ONE 12:1–19. https://doi.org/10.1371/journal.pone.0182116

Bender MA, Knutson TR, Tuleya RE et al (2010) Modeled impact of anthropogenic warming on the frequency of intense Atlantic hurricanes. Science 327:454–458. https://doi.org/10.1126/science.1180568

Berg M, Lidskog R (2018) Pathways to deliberative capacity: the role of the IPCC. Clim Change 148:11–24. https://doi.org/10.1007/s10584-018-2180-8

Borrero-Pérez GH, Benavides-Serrato M, Campos NH et al (2019) Echinoderms of the Seaflower biosphere reserve: state of knowledge and new findings. Front Mar Sci 6:188. https://doi.org/10.3389/fmars.2019.00188

Brenes A, Trejos V (1994) Changes in the general circulation and its influence on precipitation trends in Central America: Costa Rica. Ambio 23:87–90

Bustos D (2020) Atmospheric and oceanic behavior by 2100 in the Caribbean Sea based on CMIP5 climate model projections. Universidad del Norte

Bustos D, Torres R (2022) Projected wind changes in the Caribbean Sea based on CMIP6 models. Clim Dyn 60:3713–3727. https://doi.org/10.1007/s00382-022-06535-3

Bustos D, Torres R (2021) Ocean and atmosphere changes in the Caribbean Sea during the twenty-first century using CMIP5 models. Ocean Dyn 71:757–777. https://doi.org/10.1007/s10236-021-01462-z

Chen Z, Zhou T, Zhang L et al (2020) Global land monsoon precipitation changes in CMIP6 Projections. Geophys Res Lett 47:e2019GL086902. https://doi.org/10.1029/2019GL086902

Church JA, White NJ, Konikow LF et al (2011) Revisiting the earth's sea-level and energy budgets from 1961 to 2008. Geophys Res Lett 38. https://doi.org/10.1029/2011GL048794

Coralina-Invemar (2012) Atlas de la Reserva de Biósfera Seaflower. In: Archipiélago de San Andrés, Providencia y Santa Catalina. Santa Marta: Instituto de Investigaciones Marinas y Costeras "José Benito Vives De Andréis" - INVEMAR y Corporación para el Desarrollo Sostenible del Archipiélago de San Andrés

Costoya X, deCastro M, Santos F et al (2019) Projections of wind energy resources in the Caribbean for the 21st century. Energy 178:356–367. https://doi.org/10.1016/j.energy.2019.04.121

Deser C, Phillips AS, Alexander MA (2010) Twentieth century tropical sea surface temperature trends revisited. Geophys Res Lett 37. https://doi.org/10.1029/2010GL043321

DIMAR (2019) Derrotero de las Costas y áreas insulares del Caribe y Pacifico Colombiano. https://cecoldodigital.dimar.mil.co/1746/1/298_DIMAR.pdf

Dyurgerov M, Meier M (2004) Glaciers and the changing earth system: a 2004 snapshot. Institute of Arctic and Alpine Research, University of Colorado 58:1–118

Eakin CM, Lough JM, Heron SF (2009) Climate variability and change: monitoring data and evidence for increased coral bleaching stress. In: van Oppen MJH, Lough JM (eds) Coral bleaching: patterns, processes, causes and consequences. Springer, Berlin Heidelberg, Berlin, Heidelberg, pp 41–67

Etter PC, Lamb PJ, Portis DH (1987) Heat and freshwater budgets of the Caribbean sea with revised estimates for the Central American seas. J Phys Oceanogr 17:1232–1248. https://doi.org/10.1175/1520-0485(1987)017%3c1232:HAFBOT%3e2.0.CO;2

Eyring V, Bony S, Meehl G et al (2015) Overview of the coupled model intercomparison project phase 6 (CMIP6) experimental design and organization. Geosci Model Dev 8:10539–10583. https://doi.org/10.5194/gmd-9-1937-2016

Gallegos A (1996) Descriptive physical oceanography of the Caribbean Sea. In: Small islands marine science and sustainable development. American geophysical union (AGU), pp 36–55

GEBCO (2019) Gridded bathymetry data. Gridded bathymetry data. In: GEBCO 2019 Grid. https://www.gebco.net/data_and_products/gridded_bathymetry_data/gebco_2019/gebco_2019_info.html. Accessed 30 April 2021

Gidden MJ, Riahi K, Smith SJ et al (2019) Global emissions pathways under different socioeconomic scenarios for use in CMIP6: a dataset of harmonized emissions trajectories through the end of the century. Geosci Model Dev 12:1443–1475. https://doi.org/10.5194/gmd-12-1443-2019

Goldenberg SB, Landsea CW, Mestas-Nuñez AM et al (2001) The recent increase in Atlantic hurricane activity: causes and implications. Science 293:474–479. https://doi.org/10.1126/science.1060040

Gregory JM, Griffies SM, Hughes CW et al (2019) Concepts and terminology for sea level: mean, variability and change, both local and global. Surv Geophys 40:1251–1289. https://doi.org/10.1007/s10712-019-09525-z

Gregory JM, White NJ, Church JA et al (2013) Twentieth-Century global-mean sea level rise: is the whole greater than the sum of the parts? J Clim 26:4476–4499. https://doi.org/10.1175/JCLI-D-12-00319.1

Griffies SM, Danabasoglu G, Durack PJ et al (2016) OMIP contribution to CMIP6: experimental and diagnostic protocol for the physical component of the ocean model intercomparison project. Geosci Model Dev 9:3231–3296. https://doi.org/10.5194/gmd-9-3231-2016

Griffies SM, Greatbatch RJ (2012) Physical processes that impact the evolution of global mean sea level in ocean climate models. Ocean Model 51:37–72. https://doi.org/10.1016/j.ocemod.2012.04.003

Griffies SM, Yin J, Durack PJ et al (2014) An assessment of global and regional sea level for years 1993–2007 in a suite of interannual CORE-II simulations. Ocean Model 78:35–89. https://doi.org/10.1016/j.ocemod.2014.03.004

Grinsted A, Moore JC, Jevrejeva S (2013) Projected Atlantic hurricane surge threat from rising temperatures. Proc Natl Acad Sci 110:5369–5373. https://doi.org/10.1073/pnas.1209980110

Gupta AS, Jourdain NC, Brown JN et al (2013) Climate drift in the CMIP5 models. J Clim 26:8597–8615. https://doi.org/10.1175/JCLI-D-12-00521.1

Hamed I, Yunfang S (2020) Mechanism study of the 2010–2016 rapid rise of the Caribbean sea level. Glob Planet Change 191:103219. https://doi.org/10.1016/j.gloplacha.2020.103219

Hanna E, Navarro FJ, Pattyn F et al (2013) Ice-sheet mass balance and climate change. Nature 498:51–59. https://doi.org/10.1038/nature12238

Henderson-Sellers A, McGuffie K (2012) The future of the world's climate. Elsevier, Boston

Huang C, Qiao F (2015) Sea level rise projection in the South China Sea from CMIP5 models. Acta Oceanol Sin 34:31–41. https://doi.org/10.1007/s13131-015-0631-x

IPCC (2014) Climate change 2014: synthesis report. Contribution of working groups I, II and III to the fifth assessment report of the intergovernmental panel on climate change [Core Writing Team, Pachauri RK, Meyer LA (eds)]. Geneva, Switzerland

Jet Propulsion Laboratory (2020) Global sea surface salinity data sets. Obtenido de global sea surface salinity data sets. https://podaac.jpl.nasa.gov/datasetlist

Jevrejeva S, Palanisamy H, Jackson LP (2020) Global mean thermosteric sea level projections by 2100 in CMIP6 climate models. Environ Res Lett 16:014028. https://doi.org/10.1088/1748-9326/abceea

Jones PD, Harpham C, Harris I et al (2016) Long-term trends in precipitation and temperature across the Caribbean. Int J Climatol 36:3314–3333. https://doi.org/10.1002/joc.4557

Karmalkar AV, Taylor MA, Campbell J et al (2013) A review of observed and projected changes in climate for the islands in the Caribbean. Atmósfera 26:283–309

Kaser G, Cogley JG, Dyurgerov MB et al (2006) Mass balance of glaciers and ice caps: consensus estimates for 1961–2004. Geophys Res Lett 33. https://doi.org/10.1029/2006GL027511

Lee J-Y, Wang B (2014) Future change of global monsoon in the CMIP5. Clim Dyn 42:101–119. https://doi.org/10.1007/s00382-012-1564-0

Lough JM, Anderson KD, Hughes TP (2018) Increasing thermal stress for tropical coral reefs: 1871–2017. Sci Rep 8:6079. https://doi.org/10.1038/s41598-018-24530-9

Meehl GA, Moss R, Taylor KE et al (2014) Climate model Intercomparisons: Preparing for the next phase. Eos, Trans Am Geo Union 95:77–78. https://doi.org/10.1002/2014EO090001

Meyssignac B, Slangen ABA, Melet A et al (2017) Evaluating model simulations of twentieth-century sea-level rise. Part II: Regional sea-level changes. J Clim 30:8565–8593. https://doi.org/10.1175/JCLI-D-17-0112.1

Montoya-Sánchez RA, Devis-Morales A, Bernal G et al (2018) Seasonal and intraseasonal variability of active and quiescent upwelling events in the Guajira system, southern Caribbean Sea. Cont Shelf Res 171:97–112. https://doi.org/10.1016/j.csr.2018.10.006

Nicholls RJ, Lowe JA (2004) Benefits of mitigation of climate change for coastal areas. Glob Environ Change 14:229–244. https://doi.org/10.1016/j.gloenvcha.2004.04.005

Nicholls RJ, Wong PP, Burkett VR, et al (2007) Coastal systems and low-lying areas. Climate change 2007: impacts, adaptation and vulnerability. Contribution of working group II to the fourth assessment report of the intergovernmental panel on climate change. Coastal systems and low-lying areas in: climate change 2007: impacts, adaptation and vulnerability, pp 315–356

NOAA (2020a) NCEI OSTM/Jason-2 and Jason-3 Satellite Products Archive. https://www.nodc.noaa.gov/SatelliteData/jason/

NOAA (2020b) Tropical cyclone climatology. In: Tropical cyclone climatology. https://www.nhc.noaa.gov/climo/

Nowicki SMJ, Payne A, Larour E et al (2016) Ice sheet model intercomparison project (ISMIP6) contribution to CMIP6. Geosci Model Dev 9:4521–4545. https://doi.org/10.5194/gmd-9-4521-2016

O'Neill BC, Kriegler E, Ebi KL et al (2017) The roads ahead: narratives for shared socioeconomic pathways describing world futures in the 21st century. Glob Environ Change 42:169–180. https://doi.org/10.1016/j.gloenvcha.2015.01.004

O'Neill BC, Tebaldi C, van Vuuren DP et al (2016) The scenario model intercomparison project (ScenarioMIP) for CMIP6. Geosci Model Dev 9:3461–3482. https://doi.org/10.5194/gmd-9-3461-2016

Ortiz JC (2012) Exposure of the Colombian Caribbean coast, including San Andrés Island, to tropical storms and hurricanes, 1900–2010. Nat Hazards 61:815–827. https://doi.org/10.1007/s11069-011-0069-1

Ortiz JC, Otero LJ, Restrepo JC et al (2013) Cold fronts in the Colombian Caribbean Sea and their relationship to extreme wave events. Nat Hazards Earth Syst Sci 13:2797–2804. https://doi.org/10.5194/nhess-13-2797-2013

Palanisamy H, Becker M, Meyssignac B et al (2012) Regional sea level change and variability in the Caribbean Sea since 1950. J Geod Sci 2:125–133. https://doi.org/10.2478/v10156-011-0029-4

Penduff T, Melanie J, Brodeau L et al (2010) Impact of global ocean model resolution on sea-level variability with emphasis on interannual time scales. Ocean Sci 6:269–284. https://doi.org/10.5194/os-6-269-2010

Peterson TC, Taylor MA, Demeritte R et al (2002) Recent changes in climate extremes in the Caribbean region. J Geophys Res Atmospheres 107:ACL 16-1–ACL 16-9. https://doi.org/10.1029/2002JD002251

Prato J, Newball R (2016) Aproximación a la valoración económica ambiental del Departamento Archipiélago de San Andrés, Providencia y Santa Catalina. Reserva de Biosfera Seaflower., 1st edn. DIMAR, Bogotá

Rangel-Buitrago NG, Anfuso G, Williams AT (2015) Coastal erosion along the Caribbean coast of Colombia: magnitudes, causes and management. Ocean Coast Manag 114:129–144. https://doi.org/10.1016/j.ocecoaman.2015.06.024

Riahi K, van Vuuren DP, Kriegler E et al (2017) The shared socioeconomic pathways and their energy, land use, and greenhouse gas emissions implications: an overview. Glob Environ Change 42:153–168. https://doi.org/10.1016/j.gloenvcha.2016.05.009

Rodriguez-Vera G, Romero-Centeno R, Castro CL et al (2019) Coupled Interannual variability of wind and sea surface temperature in the Caribbean Sea and the Gulf of Mexico. J Clim 32:4263–4280. https://doi.org/10.1175/JCLI-D-18-0573.1

Ruiz M, Beier E, Bernal G et al (2012) Sea surface temperature variability in the Colombian Basin, Caribbean Sea. Deep Sea Res Part I: Oceanogr Res Pap 64:43–53. https://doi.org/10.1016/j.dsr.2012.01.013

Saunders M, Lea A (2008) Large contribution of sea surface warming to recent increase in Atlantic hurricane activity. Nature 451:557–560. https://doi.org/10.1038/nature06422

Serazin G, Penduff T, Grégorio S et al (2015) Intrinsic variability of sea level from global ocean simulations: spatiotemporal Scales. J Clim 28:4279–4292. https://doi.org/10.1175/JCLI-D-14-00554.1

Stephenson TS, Vincent LA, Allen T et al (2014) Changes in extreme temperature and precipitation in the Caribbean region, 1961–2010. Int J Climatol 34:2957–2971. https://doi.org/10.1002/joc.3889

Sung HM, Kim J, Lee J-H et al (2021) Future changes in the global and regional sea level rise and sea surface temperature based on CMIP6 models. Atmosphere 12(1):90. https://doi.org/10.3390/atmos12010090

Swart N, Cole J, Kharin V et al (2019) The Canadian earth system model version 5 (CanESM5.0.3). Geosci Model Dev Disc 1–68. https://doi.org/10.5194/gmd-2019-177

Taylor MA, Clarke LA, Centella A et al (2018) Future Caribbean climates in a world of rising temperatures: the 1.5 vs 2.0 Dilemma. J Clim 31:2907–2926. https://doi.org/10.1175/JCLI-D-17-0074.1

Taylor MA, Whyte FS, Stephenson TS et al (2013) Why dry? Investigating the future evolution of the Caribbean low level jet to explain projected Caribbean drying. Int J Climatol 33:784–792. https://doi.org/10.1002/joc.3461

Tokyo Climate Center (2020) Global sea surface temperature data sets. http://ds.data.jma.go.jp/tcc/tcc/products/elnino/cobesst/cobe-sst.html

Torres R, Latandret S, Salón B et al (2022) Water masses in the Caribbean Sea and sub-annual variability in the Guajira upwelling region. Ocean Dyn 73:39–57. https://doi.org/10.1007/s10236-023-01571-x

Torres RR, Tsimplis MN (2014) Sea level extremes in the Caribbean Sea. J Geophys Res Oceans 119:4714–4731. https://doi.org/10.1002/2014JC009929

Torres RR, Tsimplis MN (2012) Seasonal sea level cycle in the Caribbean Sea. J Geophys Res Oceans 117. https://doi.org/10.1029/2012JC008159

Torres RR, Tsimplis MN (2013) Sea-level trends and interannual variability in the Caribbean Sea. J Geophys Res Oceans 118:2934–2947. https://doi.org/10.1002/jgrc.20229

Tsyban A, Everett JT, Titus JG (1990) World oceans and coastal zones. Climate change: the IPCC impacts assessment. Citeseer, Canberra, Australia

United Nations (2015) Paris Agreement. United Nations, Paris

Urrego L, Polanía Vorenberg J, Buitrago M et al (2009) Mangrove zonation patterns in san Andres island (Colombian Caribbean). Bull Mar Sci 85:27–43

van Westen RM, Dijkstra HA, van der Boog CG et al (2020) Ocean model resolution dependence of Caribbean sea-level projections. Sci Rep 10:14599. https://doi.org/10.1038/s41598-020-71563-0

Villegas N, Málikov I, Farneti R (2021) Sea surface temperature in continental and insular coastal Colombian waters: observations of the recent past and near-term numerical projections. Lat Am J Aquat Res 49:307–328. https://doi.org/10.3856/vol49-issue2-fulltext-2481

Wang C, Lee S (2007) Atlantic warm pool, Caribbean low-level jet, and their potential impact on Atlantic hurricanes. Geophys Res Lett 34. https://doi.org/10.1029/2006GL028579

Wunsch C, Ponte RM, Heimbach P (2007) Decadal trends in sea level patterns: 1993–2004. J Clim 20:5889–5911. https://doi.org/10.1175/2007JCLI1840.1

Yin J (2012) Century to multi-century sea level rise projections from CMIP5 models. Geophys Res Lett 39. https://doi.org/10.1029/2012GL052947

Open Access This chapter is licensed under the terms of the Creative Commons Attribution 4.0 International License (http://creativecommons.org/licenses/by/4.0/), which permits use, sharing, adaptation, distribution and reproduction in any medium or format, as long as you give appropriate credit to the original author(s) and the source, provide a link to the Creative Commons license and indicate if changes were made.

The images or other third party material in this chapter are included in the chapter's Creative Commons license, unless indicated otherwise in a credit line to the material. If material is not included in the chapter's Creative Commons license and your intended use is not permitted by statutory regulation or exceeds the permitted use, you will need to obtain permission directly from the copyright holder.

Reconstructing the Eta and Iota Events for San Andrés and Providencia: A Focus on Urban and Coastal Flooding

Andrés F. Osorio©, Rubén Montoya©, Franklin F. Ayala, and Juan D. Osorio-Cano

Abstract Hurricanes Eta and Iota were the most intense events during the 2020 Atlantic hurricane season, and their passage caused serious infrastructure affectations and even human losses in the Archipelago of San Andrés, Providencia, and Santa Catalina due to the extreme winds, storm surge flooding, and rainfall flooding. Numerical modeling and field measurements were used to reconstruct the effects of these events on the archipelago. The simulations were conducted with WAVE-WATCHIII, SWAN, XBeach, Storm Water Management Model (SWMM), and a parametric model for hurricane winds. A differentiated contribution of each hazard on physical infrastructure, coastal ecosystems, and population is represented through: winds up to 50 m/s, significant wave heights (Hs) between 1 and 6 m in intermediate waters (around 10 m deep) associated with flood levels in the order of 2 m on the coast, and flood distances varying between 12 and 904 m. A spatial distribution of Hs and the contribution of wave run-up and storm surge in some areas of the archipelago showed the importance of mangrove and coral reef ecosystems to

---

A. F. Osorio (✉) · F. F. Ayala
Universidad Nacional de Colombia, Medellín Campus, Faculty of Mines, Department of Geosciences and Environment, OCEANICOS Research Group, Medellín, Colombia
e-mail: afosorioar@unal.edu.co

F. F. Ayala
e-mail: ffayalac@unal.edu.co

A. F. Osorio · J. D. Osorio-Cano
Corporation Center of Excellence in Marine Sciences, CEMarin, Bogotá, Colombia
e-mail: jdosori0@unal.edu.co

R. Montoya
GICI Civil Engineering Research Group, GICAMH Water Quality and Hydrological Modeling Research Group, Universidad de Medellín, Medellín, Colombia
e-mail: rmontoya@udemedellin.edu.co

J. D. Osorio-Cano
Universidad Nacional de Colombia, Caribbean Campus, Caribbean Environmental Studies Research Group, San Andrés Island, Colombia

© The Author(s) 2025
J. E. Mancera Pineda et al. (eds.), *Climate Change Adaptation and Mitigation in the Seaflower Biosphere Reserve*, Disaster Risk Reduction,
https://doi.org/10.1007/978-981-97-6663-5_3

mitigate the intensity of Eta and Iota on the coast. This study encourages science-based decision-making and provides information for policymakers to consolidate risk assessments in vulnerable zones like the archipelago.

Keywords Tropical cyclones · Storm surge · Urban flooding · Coastal flooding · Archipelago of San Andrés · Providencia and Santa Catalina

1 Introduction

Tropical cyclones (TC) are atmosphere–ocean coupled systems that can generate high-energy winds and waves, coastal flooding from storm surge, and heavy rainfall. These phenomena cause significant human and economic losses, being among the most destructive meteorological events on Earth (Aon 2017). In the Caribbean, hurricanes are by far the most hazardous phenomena, leaving devastating effects on the ecosystems and coastal communities (Spencer and Urquhart 2018; Tanner et al. 1991; Walcker et al. 2019; Wiley and Wunderle 1993), as well as serious political, social, and economic consequences across the region (Ishizawa et al. 2019; Johnson 2015; Pielke et al. 2003; Watson and Johnson 2004).

Specifically, in Colombia, the Archipelago of San Andrés, Providencia, and Santa Catalina (hereafter, the archipelago) (Fig. 1) is the most susceptible region to the occurrence of hurricanes (Montoya et al. 2018; Ortiz-Royero 2012). Recently, Hurricanes Eta and Iota caused in Providencia the death of 3 people, and the complete or partial losses of approximately 98% of building infrastructure and 90% of the tropical dry forest. During Hurricane Iota, one death and the associated damage of around 90% of the houses were reported for San Andrés (UNGRD 2020). The effects of these hurricanes have cast doubt on the current ability of the population to face the possible damages generated by the passage of a hurricane and suggest that the coastal hazards during these events are poorly managed and understood.

Moreover, an increase in the occurrence of extreme wave events has been registered in the Caribbean Sea (Montoya et al. 2018) and a greater intensity of the most extreme TC is expected (Knutson et al. 2020; Seneviratne et al. 2021), so that hurricanes will have greater potential to affect the coast. As a result, all countries dealing with these extreme events each year must endeavor to develop an effective response to possible mass-casualty incidents. Due to the critical role of long- and short-term warnings on risk management, an accurate identification of the hazard is required in coastal populations exposed to hurricanes, as in the case of the archipelago (Committee on Homeland Security and Governmental Affairs 2006).

A TC hazard assessment implies the estimation of the hazards associated with extreme winds, coastal and urban flooding (Abtew 2019; Rezapour and Baldock 2014). Previous studies have evaluated the impact of hurricanes on different regions in the world, specifically through intensity, hurricane-induced waves, and storm surge modeling using historical and synthetic TC events (Kowaleski et al. 2020; Tian and Zhang 2021; Vickery et al. 2009; Yin et al. 2021). Marsooli and Lin (2020) showed

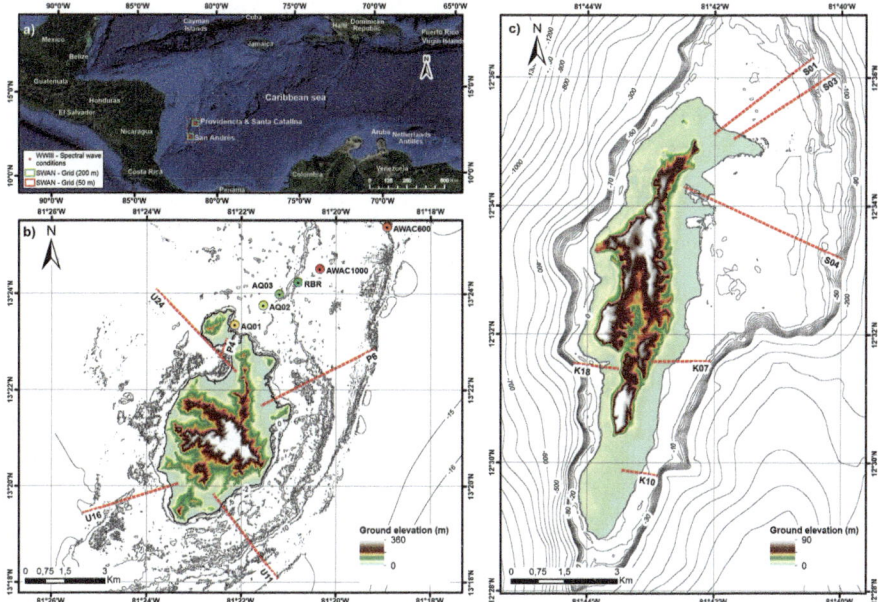

Fig. 1 a Location of the archipelago in the Caribbean Sea, including the location of the spectral wave conditions (red dots) obtained from WWIII and intermediate nested domains with a spatial resolution of 200 m (green rectangles) 50 m (red rectangles) implemented for wave propagation using SWAN model. **b** Providencia and Santa Catalina, indicating the location of wave/current sensors during the field campaign and selected profiles for coastal flooding, and **c** San Andrés, indicating the location of the coastal flooding profiles (dashed red lines)

the impact of the surface waves and the extent of coastal flooding induced by several selected hurricanes in Jamaica Bay, New York, while Lin et al. (2010) estimated the multiple hazards from a specific event, Hurricane Isabel, with an atmospheric and oceanic circulation model. Extreme waves and water levels related to the passage of TC on both the Pacific and Atlantic Mexican coasts were simulated by Meza-Padilla et al. (2015). They found that the Caribbean Sea and the northern coast of the Gulf of Mexico were the areas most exposed to the highest waves, and the northern part of the Yucatan Peninsula to the highest flood levels.

A recent study of the flooding caused by Hurricane Iota on the archipelago has shown that the storm surge and wave setup generated a flooded area corresponding to 3.7% of the total area of Providencia, with maximum storm surge values of 1.25 m at the east side of the island (Rey et al. 2021). Along with the east of San Andrés, this region in Providencia coincided with the areas most likely to be flooded in the archipelago. Furthermore, Hurricane Iota flood levels modeled with a 1D model evidenced the importance of including the wave contribution (wave setup and swash) to correctly estimate the seawater level during extreme conditions (Rey et al. 2021). Despite these results, an individual estimation of the hazard related to winds, waves, and flooding is still needed in the archipelago to obtain a comprehensive analysis of hazards associated with hurricanes in the archipelago.

This chapter aims to reconstruct the passage of Hurricanes Eta and Iota in terms of the intensity of winds and waves, and the associated coastal and urban flooding effects by using numerical modeling. These events and the study zone are briefly described in Sects. 2 and 3, respectively. The materials and methods are described in Sect. 4. The results are shown in Sect. 5. Finally, the discussion and main conclusions are presented in Sect. 6.

2 Hurricanes Eta and Iota (2020): An Overview

Eta and Iota were the two most powerful TCs of the 2020 Atlantic hurricane season (Blunden and Boyer 2021). Eta was a category 4 hurricane that passed close to the archipelago on November 2, 2020. Moreover, only two weeks after, the category 4 Hurricane Iota affected the archipelago on November 16. Although both systems were classified as category 4, Iota presented minimal central pressure and sustained wind speed higher than Eta. A brief description of the evolution of each TC is presented below.

2.1 Eta Patterns Description

Eta started as a tropical wave on October 22, 2020, which moved across the tropical Atlantic until it reached the Eastern Caribbean and became a tropical depression about 350 km southwest of the Dominican Republic at 18:00 UTC on October 31. Six hours later, the depression became a tropical storm and turned toward the west due to a high-pressure center located to its north. By 06:00 UTC on November 2, Eta was already a 36 m/s hurricane with its center located 500 km to the south of Grand Cayman. A continuous rapid intensification generated a category 4 hurricane with roughly 59 m/s winds at around 18:00 UTC that day, that is, over a period of just 12 h it underwent an increase of 23 m/s in its intensity. Additionally, this strengthening occurred while passing to the north of the archipelago. The intensity peak reached 69.5 m/s winds and passed about 101 km northeast of Nicaragua. Subsequently, Eta became a tropical depression during its landfall and passage over the Gulf of Honduras. A re-intensification kept Eta as a tropical storm due to the interaction with the Loop Current at around 12:00 UTC on November 11, until it transformed into an extratropical cyclone by 12:00 UTC on November 13 (Pasch et al. 2021). Although Eta's center passed away from the archipelago, sustained surface winds and gusts of wind of 10 m/s and 20 m/s, respectively, were measured at San Andrés airport at 09:00 UTC on November 3, while winds up to 16 m/s were recorded on Johnny Cay by a weather station operated by the Marine Research Institute of Colombia (INVEMAR).

2.2 Iota Patterns Description

On October 30, 2020, a tropical wave in front of the western coast of Africa started moving westward until it became a low-pressure system close to the southwest of the Dominican Republic. Subsequently, the convective structure improved, resulting in a tropical depression formed over the south-central Caribbean Sea at around 12:00 UTC on November 13. Only six hours later, when the system was located northwest of Aruba, a strengthening in the depression led it to become a tropical storm. During a 42-h period, between 18:00 UTC on November 14 and 12:00 UTC on November 16, Iota underwent a strong deepening with an 80 mb decrease in its central pressure and a rapid intensification with a 46 m/s (from 23 to 69 m/s) increase of the 10 min averaged maximum wind speed. The deepening rate of Iota has been the third largest in the Atlantic basin since 1965. When Iota reached its peak of intensity, it was located about 37 km northwest of Providencia and Santa Catalina, after which it gradually weakened due to passing over a cool wake created by Hurricane Eta a few weeks earlier, and it made landfall in less than 24 h along the eastern coast of Nicaragua. At 18:00 UTC on November 17, Iota became a tropical storm. Finally, it dissipated over western El Salvador 6 h later than its passage across the mountains of southern Honduras (Stewart 2021).

Although the center of Iota's eye did not cross Providencia and Santa Catalina, the southern eyewall directly impacted them, where it was estimated that sustained category 4 wind speeds of at least 59 m/s winds occurred, while hurricane-force winds persisted for approximately 7 h. In San Andrés, 10-min average wind speeds of 17 m/s and gusts of 22.5 m/s were measured, and tropical-storm-force winds occurred for at least 14 h (Stewart 2021). The Colombian Institute of Hydrology, Meteorology and Environmental Studies (IDEAM) also reported that hurricane winds of 50 m/s and 17.5–32.5 m/s were expected in Providencia and San Andrés, respectively and waves higher than 3 m (IDEAM 2020). Iota destroyed Providence's electric plant, and its 5,000 inhabitants were entirely left without communications and electricity for about 24 h (Stewart 2021). Three people died and six people were injured, about 80% of buildings were destroyed, while another 20% were severely damaged (UNGRD 2020). Meanwhile, in San Andrés, communications were lost during the storm, the high-intensity winds (greater than 25 m/s) caused damage to several homes and one person died. A flood level of around 0.15 m generated a temporary shutdown of the island's international airport, according to Stewart (2021).

3 Study Zone

The archipelago is located in the Caribbean Sea, and it is the largest island region in Colombia. It is composed of a group of islands, lesser islands, atolls, and cays (Fig. 1a). It is located 110 km east of Nicaragua, and roughly 720 km northwest of the

Caribbean coast of Colombia. San Andrés and Providencia are the two most populated and largest islands in the archipelago, with surface areas of 27 km^2 and 17km^2, respectively. The other lesser formations in the archipelago occupy 8.5 km^2, approximately. A mountain range crosses San Andrés from north to south, with maximum elevations up to 85 m above sea level, while Providencia has a steeper topography with a mountainous inner region with maximum elevations up to 360 m. Coastal ecosystems surround both islands (e.g., coral reefs, mangroves, and seagrasses) that foster a vast variety of marine flora and fauna (CORALINA-INVEMAR 2012) being one of the top tourist travel destinations in Colombia. The estimated population of the archipelago in 2022 was around 58,817 inhabitants (DANE 2020), and it is concentrated in a few flat areas associated with beaches. Previous studies (Ortiz-Royero 2012) have suggested that the archipelago is the zone most likely to be affected by storms in the Colombian Caribbean, highlighting the importance of a better understanding of the physical processes that take place during the passage of a TC in terms of coastal/urban flooding as well as preparing the community to face these events (Ortiz-Royero et al. 2015).

According to Ricaurte-Villota et al. (2017), the climate in the archipelago is strongly influenced by the displacement of the intertropical convergence zone (ITCZ), generating two seasons of maximum winds during December–January–February (DEF) and June–July–August (JJA), with values between 3.8 and 6 m/s, while periods of weak winds occur during March–April–May (MAM) and September–October–November (SON), ranging between 3 and 5 m/s. This climatic pattern in the Caribbean Sea has been reported by several authors who research wind speed and its connections with the El Niño Southern Oscillation (ENSO) (Alexander and Scott 2002; Alfaro 2002; Enfield and Mayer 1997; Giannini et al. 2000, 2001a, b, c; Ruiz-Ochoa and Bernal 2009).

Additionally, due to the archipelago being highly influenced by trade winds, the wind direction does not change significantly during the year, showing semi-permanent winds blowing from the northeast. Regarding the mean wave distribution in the archipelago, there is a dominance of the waves from the 0–90° direction throughout the year with some southward variation during the lesser winds season and the highest values of the significant wave height occurring during the first season (DEF) with values between 1.6 and 2.0 m (Osorio et al. 2016). The tidal regime in the archipelago is mixed diurnal and its range is 0.31 m (IDEAM 2017).

The average annual precipitation in San Andrés and Providencia shows a unimodal monthly distribution with a dry season from January to April with minimum values in March (23.1 mm in Providencia and 25.3 mm in San Andrés) and a wet season from May to December with maximum values in October (344.5 mm in Providencia and 315.5 mm in San Andrés). In general, the precipitation and wind speed regimes are higher in San Andrés than in Providencia (Ricaurte-Villota et al. 2017).

4 Materials and Methods

4.1 Field Measurements

A field campaign was conducted (March 6–19, 2021) to measure hourly wave conditions at six locations around Providencia (Fig. 1b). Outside the reef barrier, an acoustic Doppler current profiler (AWAC 600 kHz from Nortek instruments) was installed at the sea floor at 22.6 m, followed by an AWAC 1000 kHz at 9.1 m, a pressure sensor (RBRduo from RBR Lda.) at 5.1 m and three pressure sensors (AQUAlogger P520) from AQUATEC Ltd. (see AQ3, AQ2, AQ1 in Fig. 1b). The transect of instruments was heading the line north-east considering the main flow direction and wave characteristics during the field campaign in Providencia island. The information recorded was used as a boundary condition for the calibration of the XBeach model, which in turn was used for the local modeling of coastal flooding in the archipelago during the hurricane events.

The topo-bathymetric model used as an input for the coastal and urban flood-level modeling was composed of the Digital Elevation Model (DEM) and the Digital Surface Model (DSM) supplied by the government of San Andrés and Providencia in 2020 and obtained from aerial photographs, LIDAR, and base mapping of the archipelago. Additionally, beach profiles were measured using differential GPS in Real Time Kinematic mode (RTK) along 7 km of coastal beaches in San Andrés and 1.6 km in Providencia and Santa Catalina, these recordings were used as contour conditions for the coastal flooding simulations along several beach profiles. Moreover, a mapping of the drainage system was built based on earlier design reports and in-situ inspections of the hydraulic structures (e.g., box-culverts, channels, ditches, pipes) to be considered in the urban hydrological model.

4.2 Numerical Modeling

Figure 2 shows the workflow of the methodology, where wave and hydrodynamic models (WAVEWATCH IIITM and SWAN) were forced with a parametric model for hurricane winds to estimate storm surge together with wave setup and run-up, and complemented with urban flood using a storm water management model (SWMM). The following subsections provide more detailed information on each model.

4.2.1 Atmospheric Parametric Modeling

Several free wind field databases, such as the European Center for Medium-Range Weather Forecast Reanalysis (ECMWF, ERAinterim, and ERA5), the North American Regional Reanalysis Center (NARR), and the National Center for Environmental Prediction and Atmospheric Research reanalysis (NCEPR1), among others,

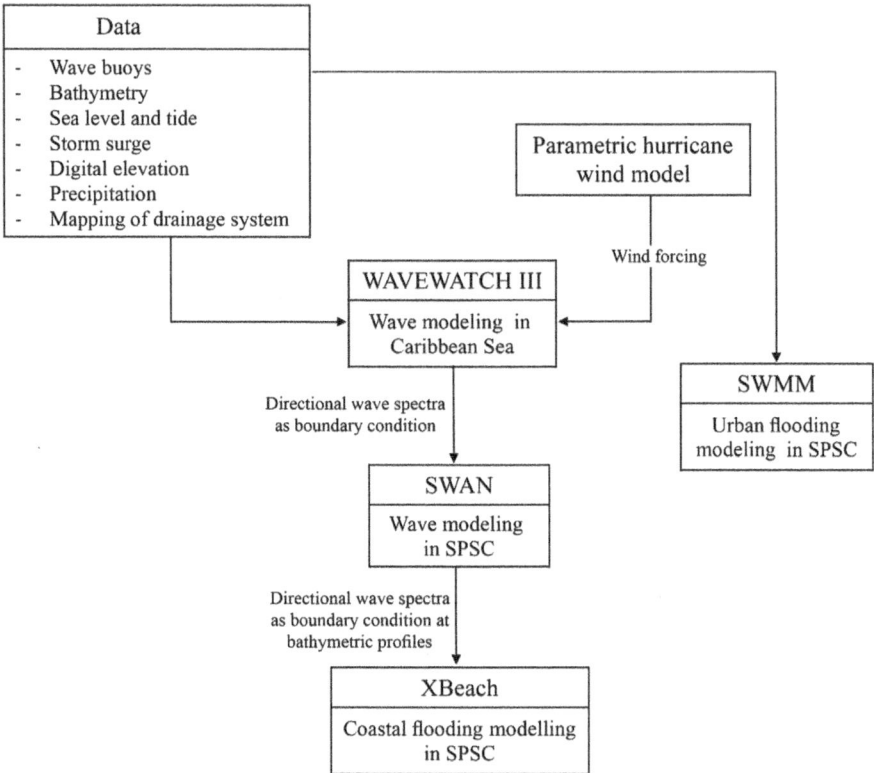

Fig. 2 Workflow methodology of the numerical modeling process

underestimate wind speed and do not adequately represent the spatial distribution of winds near the eye of the hurricane (Cavaleri and Sclavo 2006; Montoya et al. 2013; Ruti et al. 2008; Sharma and D'Sa 2008). These databases have low spatial and temporal resolution, which do not allow an accurate capture of the evolution of the phenomena next to the eye of the hurricane (maximum speed, asymmetry, maximum wind radius, and trajectory), even though they use the values of magnitude and wind direction of oceanographic buoys in their assimilation process. To improve this issue, different authors have proposed methodologies to calculate the wind field under extreme conditions (Lizano 1990; Visbal and Ortiz 2006; Willoughby and Rahn 2004). In this study, numerical wind field simulations were conducted considering the methodology proposed by Montoya et al. (2013), combining wind data from ERA5 reanalysis and a parametric model of hurricane winds developed by Roldán-Upegui et al. (2022), in order to better estimate the magnitude, asymmetry, and spatial distribution of the wind field under extreme conditions compared to the mentioned reanalysis databases.

4.2.2 Wave Modeling

The third-generation wave model WAVEWATCH III (WWIII) (WW3DG 2019) was used to propagate waves from deep water in the Caribbean Sea to intermediate and shallow water near the archipelago during the conditions of Hurricanes Eta and Iota. The WWIII model solves the random phase spectral action density balance equation and was set up to consider the source term package (ST2) (Tolman and Chalikov 1996), the Discrete Interaction Approximation (DIA) (Hasselmann et al. 1985) for nonlinear wave-wave interactions, the source term to model the bottom friction (Hasselmann et al. 1973) and the parameterized linear input (Cavaleri and Rizzoli 1981). Additionally, sea ice dissipation and reflection were disabled, and a third-order propagation scheme was used (Tolman 2002). The model was executed considering two nested domains with horizontal resolutions of 1/3° (37.1 km) and 1/12° (9.3 km) respectively. The frequency-direction space was discretized in 72 directions (5°) and 30 frequencies, varying from 0.042 to 0.65 Hz with an increment factor of 1.1. The bathymetric conditions were obtained from 1 arc-minute Gridded Global Elevation Data (ETOPO-1) available at https://www.ngdc.noaa.gov/mgg/global/. Available wave data from buoys 42,056, 42,058, and 42,060 from the NDBC (National Data Buoy Center: https://www.ndbc.noaa.gov/) was used to calibrate and validate the significant wave height (H_s) simulated by WWIII during Hurricanes Dean 2007 and Matthew 2016 (not shown here). Furthermore, considering the best configuration parameters and mesh details, the regional model was used to simulate the extreme wave conditions during Hurricanes Eta and Iota.

For shallow water modeling, the Simulating Waves Nearshore model (SWAN) (Booij et al. 1999) was used to represent the nearshore wave conditions around the archipelago. This model also solves the action density balance equation considering shallow processes that affect wave propagation. The wind wave growth parametrizations proposed by Komen et al. (1984) and Cavaleri and Rizzoli (1981) were used as well as the white capping source term proposed by Komen et al. (1984). Nonlinear quadruplet and nonlinear triad wave-wave interactions were modeled with the schemes of Hasselmann et al. (1985) and Eldeberky and Battjes (1984), respectively. Bottom friction (Hasselmann et al. 1973) and depth-induced breaking (Battjes and Janssen 1978) source terms were also included. Similar to WWIII, the wave direction was discretized in 72 directions (5°) and considered 30 frequencies, varying from an initial value of 0.042 Hz with an increment factor of 1.1. The SWAN model was executed using a nesting scheme, where the initial modeling domain was delimited by the red dots in Fig. 1a, which correspond to the location of the directional wave spectra obtained from WWIII and used as boundary conditions for local wave propagation. Subsequently, 2 additional meshes with a spatial resolution of 200 m and 50 m (green and red boxes in Fig. 1a, respectively) were nested to allow for improving the spatial wave resolution results around each island.

The bathymetry in the nearshore areas of the archipelago (Fig. 1b, c) was obtained through interpolations using the Inverse Distance Weighting (IDW) method by combining data from the topo-bathymetric survey of the beaches (as described in Sect. 4.1), using differential GPS (DGPS) with RTK (real-time kinematic) + PPK

(Post Processed Kinematic) techniques, photogrammetric height points, and contour lines every 2 m obtained from the Agustín Codazzi Geographic Institute (IGAC), nautical charts from the Center for Oceanographic and Hydrographic Research of the Caribbean (CIOH) and information available from the General Bathymetric Chart of the Oceans (GEBCO) (https://download.gebco.net/).

4.2.3 Coastal Flood Modeling

The coastal flood level due to waves was estimated considering the sea level anomalies, the astronomical tide (AT), the meteorological tide or storm surge (SS), and the wave run-up (R2), as described below.

Sea Level Anomalies and Astronomical Tide

The sea level anomalies were obtained from altimeter data from the Copernicus Marine Environment Monitoring Service (CMEMS) and the Copernicus Climate Change Service (C3S) (https://cds.climate.copernicus.eu/), considering a spatial resolution of 1/4° (27.8 km) and daily temporal resolution. The astronomical tide was obtained from the TPXO8 database (https://www.tpxo.net/regional). TPXO models allow the estimation of the amplitudes and phases for M2, S2, N2, K2, K1, O1, P1, Q1, M4, MS4, and MN4 harmonics.

Storm Surge

The storm surge was estimated from wind variations and atmospheric pressure at the sea surface. To determine the contribution of each component to the total flood level, empirical formulations were used (Benavente et al. 2006; Genes et al. 2021; Isobe 2013; Li et al. 2020). Sea level rise by winds $d\xi_w$ was calculated with the expression proposed by Bowden (1983):

$$d\xi = \frac{\rho_a C_d F}{\rho_w g h} U_{10}^2$$

where ρ_a is the air density (1.25 kg/m^3), ρ_w is the water density (1025 kg/m^3), g is the gravitational acceleration (9.81 m/s), h represents the wave relative depth, U_{10} is the wind speed at 10 m height obtained from ERA5 Reanalysis data and the parametric model of hurricane winds, the drag coefficient C_d was estimated for extreme wind conditions according to Peng and Li (2015), and the fetch (F) was obtained from Isobe (2013) considering the significant wave height H_s simulated by SWAN model.

The storm surge due to atmospheric pressure variations was estimated according to Benavente et al. (2006), as $d\xi_p = (\Delta P_a)/(\rho_w g)$, which considers that a decrease in atmospheric pressure by 1 mbar implies a 1 cm increase in sea level. The gradient ΔP_a represents the atmospheric pressure difference between the minimum pressure in the eye of the hurricane at time t, and the pressure at different points of interest around the archipelago. The minimum pressure data in the eye of the hurricane were obtained from the NHC database HURDAT2, and the pressure at the points of interest were obtained from the ERA5 database with an hourly temporal resolution and a spatial resolution of 1/4° (27.8 km).

Wave Run-Up

Wave run-up, defined as the maximum onshore elevation reached by waves, was calculated by subtracting the tidal components (obtained by the tidal component explained in the section above) from the free sea surface elevation simulated with the two-dimensional model Xbeach (Roelvink et al. 2010), originally designed to simulate the hydrodynamic and morphological processes and their impact on the coast. Xbeach was executed in non-hydrostatic and non-stationary mode due to all hydrodynamic processes involved in the non-linear shallow water equations. Directional wave spectrums simulated from SWAN were used as boundary conditions at six (6) bathymetric profiles placed around San Andrés (Fig. 1c) and five (5) at Providencia and Santa Catalina (Fig. 1b).

The calibration process was done by comparing the model results (significant wave height) with the wave data from Doppler (AWAC 600, AWAC1000) and pressure sensors (RBR, AQ3, AQ2, AQ1) recorded during 13 days at 6 points along a bathymetric profile in Providencia (Fig. 1a) and adjusting the non-dimensional wave friction coefficient (C_f), which has been used for model calibration under similar seafloor configurations (Roelvink et al. 2021). C_f values between 0.01 and 0.9 were obtained (not shown here) and were associated with the different bottom characteristics (sand, coral, coral rubble) to find the zones of the model domain where each C_f value applies. The calibrated C_f values were used to simulate the coastal flooding during the hurricane periods.

4.2.4 Urban Flood Modeling

The Storm Water Management Model (SWMM) (Rossman 2015) is a one-dimensional model that solves the Saint–Venant equations and it was used to simulate the urban flood level during Hurricanes Eta and Iota by analyzing storm sewer and other drainage systems. The main parameters required in the SWMM model for the sub-catchment elements are area, width, percentage slope, percentage of impervious areas, the manning coefficient for pervious and impervious areas, and depression storage for impervious and pervious areas. For San Andrés, the SWMM model was composed using two different configurations: (i) Areas with a storm sewer network obtained from the *Plan Maestro de Alcantarillado*, or Sewer Master Plan (Consorcio Plan Vial Caribe 2007) located along the road system and conceptualized as a dual drainage model, and (ii) Areas with roads without a storm sewer network and conceptualized as artificial channels generating flooding along them. For both cases, afferent areas to the storm sewer channels (rectangular in San Andrés) and streets for areas without the storm sewer system were estimated based on the Euclidean distance technique (Fig. 3a). The sub-catchments and their afferent areas in Providencia and Santa Catalina (Fig. 3b) were estimated based on the box culverts located along the

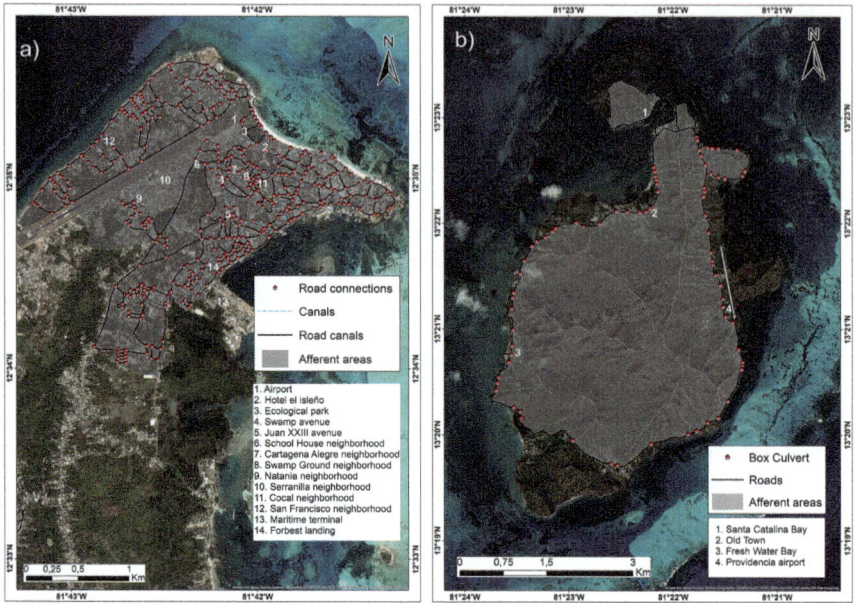

Fig. 3 **a** Storm sewer system and road system with afferent areas for San Andrés and **b** Perimetral road configuration and afferent areas for Providencia. Information available from Plan Maestro de Alcantarillado (Consorcio Plan Vial Caribe 2007)

perimeter of the islands and the rural basins between each of them. The main sub-catchments are conceptualized in the model as rural catchments draining to the box culverts.

The main parameters of the sub-catchments were estimated as follows: The *percentage of impervious areas* was obtained from the maps of land use and vegetation cover for the different sub-basins, the *depression store* was obtained as an initial approximation using the values proposed in the SWMM model user manual (Rossman 2015) according to the land uses and vegetation cover for San Andrés. For Providencia and Santa Catalina, given their high proportion of rural areas, constant values were selected for the entire area (5 mm for permeable areas corresponding to grass), the *Slope* was obtained automatically for each sub-catchment from the approximate surface model (MDS) that accounts for all the natural terrain modifications due to the construction of urban infrastructure, and from the DEM for those rural basins interacting with urban drainage (a filter was employed for slopes greater than 100% to avoid unrealistic slopes). The manning coefficients for the different sub-basins in San Andrés were obtained for impermeable and permeable urban areas based on land use and vegetation cover. The values were selected from the SWMM manual (Rossman 2015). For Providencia, an average value for dense grass of 0.27 was selected.

The *width* parameter was obtained using the approximated formulation based on geomorphology proposed by Babaei et al. (2018) for higher values $C = 1.28$. This formulation was employed for both islands. Finally, the curve number CN, from the Soil Conservation Service (SCS) method for runoff abstractions, proposed by the Soil Conservation Service (1972) was employed. This CN value was obtained from Chow et al. (1994) based on the conditions for the influential variables and information on land use and soil type for the archipelago.

For the storm sewer system and the associated joints in San Andrés, the main parameters such as length, bottom elevation, hydraulic section, and roughness, among others, were obtained from information provided by CORALINA (Consorcio Plan Vial Caribe 2007). The details of the streets located in areas with and without storm drainage systems acting in the SWMM model as channels during extreme rain events were estimated from the DSM.

Dual Drainage System

For the dual model in San Andrés, the runoff is transferred from the streets (main system) toward the secondary drainage system corresponding to the storm sewer system. For areas without the existence of a storm sewer system, the drainage is composed of areas related to the road sections and its direct discharge towards the representative nodes of road intersections. The flow is discharged directly towards the roads represented in the SWMM model by channels with a defined cross section. To determine the cross sections on the roads, a review of the field campaign information and the orthophoto was performed. A typical street cross section of 8 m roadway width, 2 m sidewalk, 0.2 m elevation above the roadway, and outer wall edifications of 3 m were defined for the street systems of San Andrés. For Providencia and Santa Catalina, only the roadway width was modified to 7 m.

For discharges from the main drainage system (streets or roads) toward the secondary system (pluvial sewer system), the equations for INOS gratings on slope type I and type II presented by Rincón and Muñoz (2013) were used. For the roads of San Andrés and Providencia, a transverse slope of the road equal to zero is assumed from what was observed in the field campaigns. The Manning roughness coefficient used was 0.018 for concrete.

Extreme Storm Rainfall During Eta and Iota

To determine the precipitation in each sub-catchment of the archipelago, rain information from the Integrated Multi-satellitE Retrievals for GPM (IMERG) (Huffman et al. 2020) was downloaded (with an original spatial resolution of $0.1° \times 0.1°$) and modified to a higher resolution (300–500 m) through a downscaling method. Those events with the highest rainfall intensity for each of the islands were associated with high water elevations over urban roads and with the highest potential damage across the islands. Figure 4 shows the rainfall intensity during the storm duration (black line) and the range of variability associated with all the main sub-catchments (gray shaded area) in San Andrés (22 drainage sub-catchments) for Hurricanes Eta, while for Providencia and Santa Catalina, just one integrated catchment was assumed as representative since not significant variation between sub-catchments was obtained.

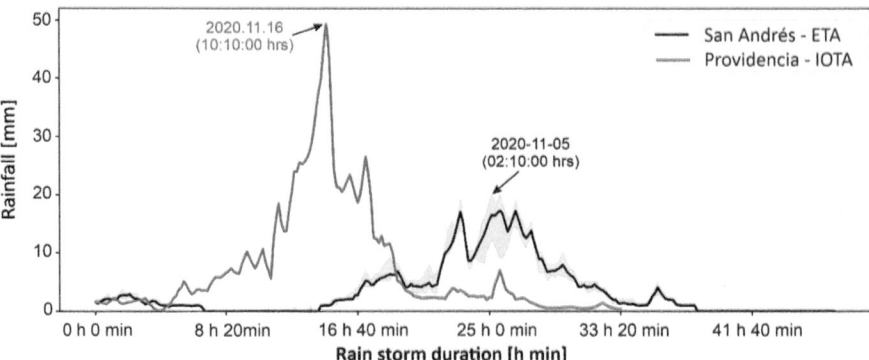

Fig. 4 Rainfall intensity during the storm duration occurrence of Hurricane Eta for San Andrés (starting at 2020.11.04—00:30:00 UTC) and Iota for Providencia and Santa Catalina (starting at 2020.11.15—19:30:00 UTC), including the storm peak date. The gray-shaded area around the black solid line represents the range of rain variability for all the main sub-catchments in San Andrés

Hence, the flood modeling was carried out considering hurricane Eta as the most intense event for San Andrés and hurricane Iota for Providencia and Santa Catalina (see Fig. 4).

It is observed that the accumulated precipitation in Providencia during the passage of Iota was around 247 mm, and maximum intensities of almost 49.24 mm/h were reached. For San Andrés, significantly lower accumulation was reached for Iota (maximum intensities of around 19 mm/h and accumulated precipitation of 175.5 mm). During Hurricane Eta, around 159.2 mm of accumulated rain was obtained, with maximum intensity values of up to 17.2 mm/h on average for all the urban areas.

5 Results

For Hurricanes Eta and Iota, Fig. 5 shows the spatial patterns occurring during the most extreme wind speed near San Andrés and Providencia and the time series of wind speed for points located in the middle of both islands. The specific time for each hurricane was selected based on the approximate distance between the hurricane eye and the center of the respective island. For Hurricane Eta, the maximum and nearest wind speed spatial pattern (Fig. 5a) occurred during 12:00 UTC on November 2, with a maximum wind speed of around 40 m/s and an average distance of 162 km for Providencia and 257 km for San Andrés. Time series of wind speed in the middle of both islands show maximum wind speed reaching values of around 16.5 m/s during 18:00 UTC on November 2 for Providencia, and slightly lower values of around 15.4 m/s for San Andrés.

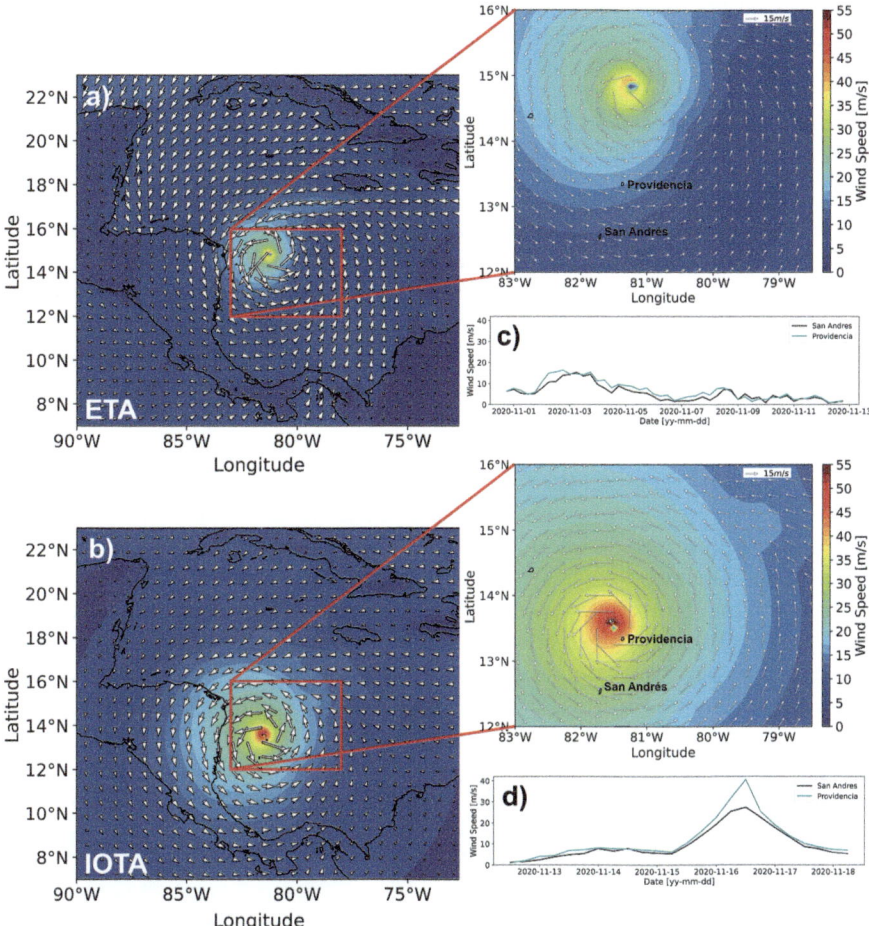

Fig. 5 Spatial patterns **a**, **b** and time series **c**, **d** of the wind speed during the passage of Hurricanes Eta and Iota, respectively obtained using the methodology proposed by Roldan et al. (In preparation)

For Hurricane Iota the maximum and nearest wind speed spatial pattern (Fig. 5b) occurred at 12:00 UTC on November 16 with a maximum wind speed of around 52 m/s. The average distance between the eye of the hurricane and the centers of Providencia and San Andrés are around 22.32 km and 109.2 km, respectively. The time series of wind speed in both islands shows clearly how the maximum wind speed during the most intensive occurrence was obtained for Providencia and Santa Catalina with maximum winds of around 40 m/s when compared to San Andrés with maximum values around 28 m/s. These results agree with the damage reported by several governmental agencies and newspapers, among others, showing the most catastrophic situation for Providencia and Santa Catalina with infrastructure widely affected.

According to the evaluations, there were approximately 6,300 people affected in Providencia and at least 700 families in San Andrés.

5.1 Wave Modeling and Coastal Flooding

Figure 6 shows the spatial patterns of H_s estimated by WWIII during the passage of Hurricane Iota (the H_s values for Hurricane Eta are not presented here since their magnitudes were smaller than Iota). Before the rapid intensification of Iota, leading it to become a category 4 hurricane, the maximum values of H_s were between 7 and 8 m, while from 07:00 UTC on November 16, when the cyclone was close to Providencia, they reached values of up to 10 m in the right forward quadrant (with respect to the direction of the hurricane translation). Even after passing Providencia, these H_s values were maintained near the eastern coast of Nicaragua.

The highest H_s values seem to be located under the hurricane eyewall, suggesting that they are directly generated by the wind hurricane action and are not advected from other regions outside the simulation domain. From the temporal evolution, it can be noticed that the higher waves are in the right forward quadrant as expected. In this quadrant, the forward motion of the hurricane increases the wind speed and wave-growth processes, together with a partial resonance effect (Wright et al. 2001). In general, the temporal changes, the asymmetry, and the spatial distribution of H_s during these extreme conditions were well represented by the regional wave model.

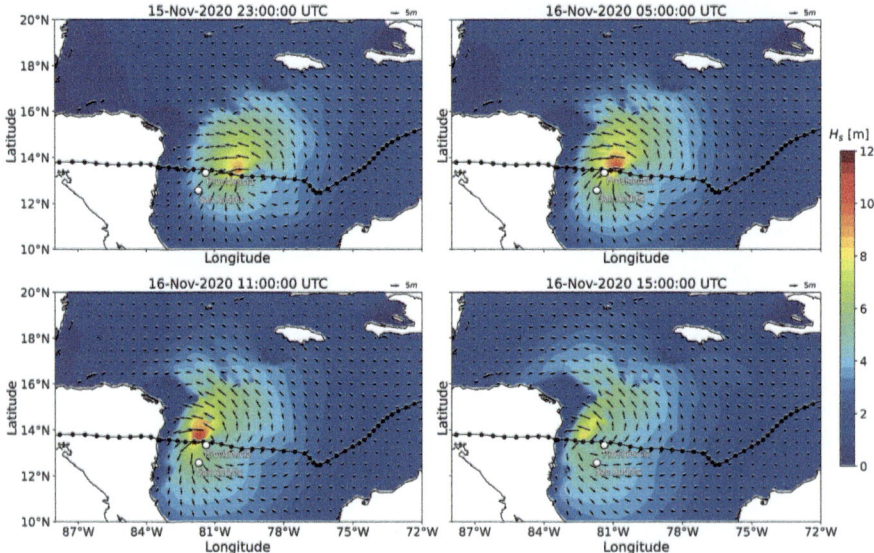

Fig. 6 Evolution of significant wave height simulated by WWIII during Iota. Arrows indicate the wave direction. The dotted line represents the track of Iota

The directional wave spectrum across the whole domain boundary was provided by WWIII as boundary conditions (red dots in Fig. 1a) to simulate the spatial distribution of local waves (H_s) with the SWAN model around the archipelago (Fig. 7).

During the approach and passage of Iota around the archipelago, the islands were affected by the left forward quadrant of the hurricane. In San Andrés, even though waves lower than 2 m were estimated numerically in a large part of the island at

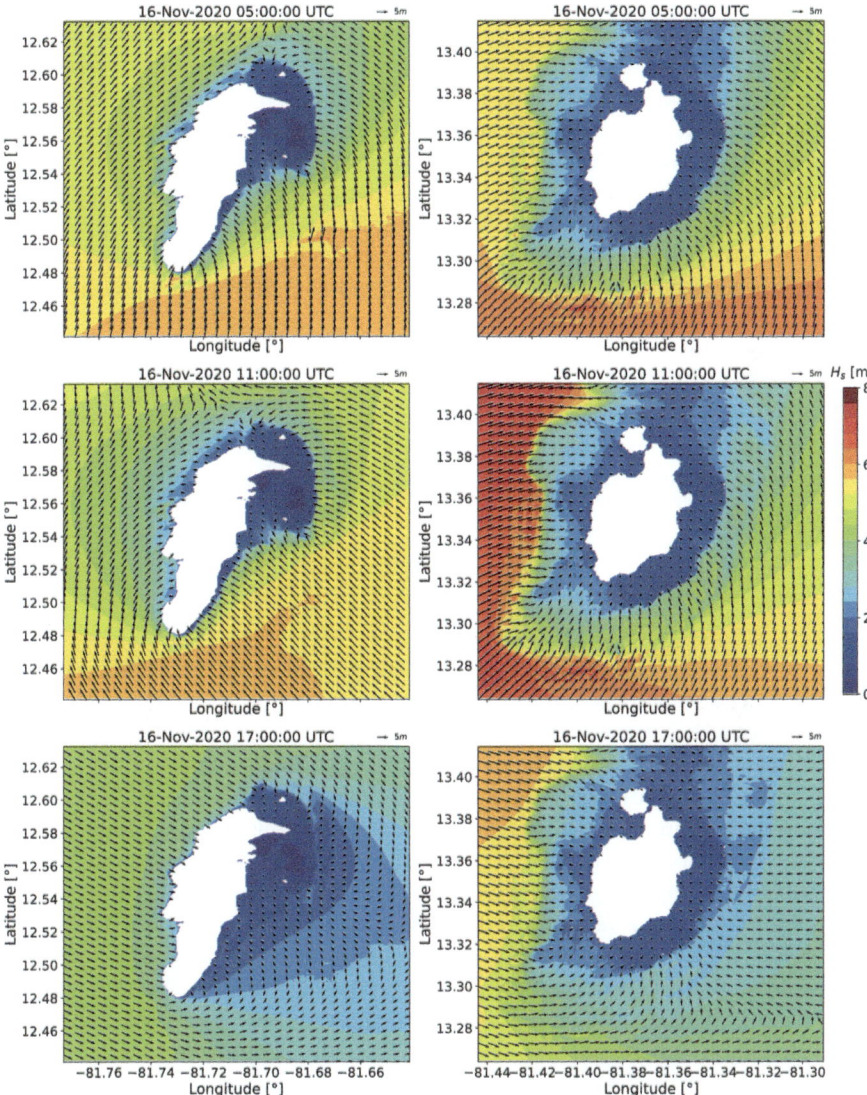

Fig. 7 Significant wave height and wave direction during the passage of Iota over the archipelago between 05:00 and 17:00 UTC on November 16. Black arrows indicate the wave direction

05:00 UTC on November 16, waves roughly 4 m may have hit the west and south. At 11:00 UTC on November 16, Iota generated waves up to 6 m in the south of San Andrés. It can also be noted that the coral reef attenuated the waves from the north to northeast of San Andrés, where the H_s values decreased significantly from 5 m to 1–2 m. Regions like the southeast and west were impacted with waves of 4–5 m, approximately. After the passage of Iota, there are no significant changes on the east coast of San Andrés, although the waves far from the coast no longer exceed heights of 2–3 m. The west and south of San Andrés show the highest values (less than 4 m).

In Providencia, waves about 6 m far from the coast were estimated before the major impact of Iota, especially over the south of the island. A H_s change is evidenced close to shore, where wave heights are lower than 2–3 m. When Iota directly hit Providencia, waves of 7–8 m were simulated outside the coral reef that surrounds Providencia to the west. Despite the enormous decrease in height (~6 m), waves of 2.5 m impacted Providencia, especially in the west. The H_s estimated in the east, south, and north were around 1 m. At 17:00 UTC on November 16, wave heights outside the barrier reef continued to be higher in the west of the island (6 m) than in the rest of the island (2–4 m), while H_s values of 2–3 m were still present on the coast.

The values estimated are coherent and show good agreement with warning reports issued by IDEAM, so waves higher than 3–4 m were expected on November 15 and 16 (IDEAM 2020). However, the results are presented as indicative of the possible physical values sensed in the archipelago due to the lack of available information to corroborate the estimations.

5.2 Coastal Flooding

Figure 8 shows the correlation coefficient (*Corr.*), the Root Mean Square Error (*RMSE*), and de Mean Bias Error (MBE) between the XBeach model and H_s values from the AWAC1000 and RBR locations, where H_s values from AWAC600 were used as boundary conditions. The calibration was carried out considering the first 9 days of the 13 days of measurements and the best input parameters (e.g., bottom roughness and frictional coefficients) were later used to simulate the coastal flooding during the hurricane periods. These results show the high performance of XBeach in reproducing the hydrodynamic processes in the surf zone.

The spatial–temporal variation of the maximum flood level or run-up (R_{high} = AT + SS + R2) and its components (AT = Astronomical Tide, SS = Storm Surge, and R2 = wave run-up) during Hurricane Iota are presented in Figs. 9 and 10 for San Andrés and Providencia (including Santa Catalina), respectively. From the run-up time series, the 98% percentile (R_{high2}) was estimated every hour during a simulation period of 72 h (36 h before and 36 h after the maximum flood peak recorded by each profile). Additionally, the time series of the meteorological tide or Storm Surge (SS) is presented, as well as the contribution to the total flood level given by its pressure (SP) and wind (SV) components.

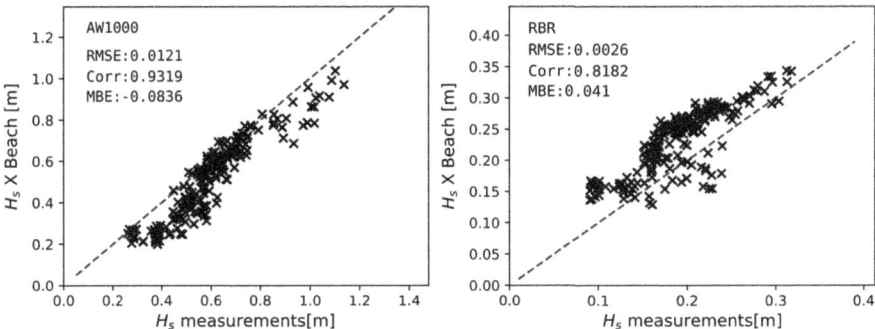

Fig. 8 H_s comparison between the XBeach model and field measurements in Providencia. Data from the AWAC100 and RBR devices are shown on the left and right plots, respectively. The locations of these measurements are indicated in Fig. 1b

A significant contribution (>62%) to the total flood level due to Pressure (SP) is observed in San Andrés (Fig. 9), especially at the northeast profiles (S01, S03) and between 45–52% for the rest of the island (S04, K07, K10, K18). On the other hand, the contribution of wave run-up (R2) to the total flood level is more noticeable towards the area of Sound Bay (K07) with a percentage around 54% with flood levels around 2 m, reaching more than 200 m of flooding towards the coast. In the area near the Old Point mangrove (S04), given the topo-bathymetric configuration and the low slope of the terrain, a 901 m advance of the seawater towards the coast and a maximum flood level, which exceeds 2 m during 5 consecutive hours, can be observed during the peak of the hurricane according to the numerical model.

In Providencia and Santa Catalina (Fig. 10), the contribution of the pressure component (SP) from the storm surge to the total flood level is dominant in P04 (Santa Catalina) and P06 (McBean Lagoon) with contributions of 65.9% and 71.3% respectively, reaching flood levels around 1.5 m and 472 m of coastal flooding at the P06 profile. The maximum run-up was obtained at U24 (profile located at the northwest of Providencia) where the wave run-up (R2) represented 70.8%, exceeding the effect generated by the storm surge (SS) and reaching a maximum flood level of 5 m above the mean sea level. In general, the contributions of the astronomical tide (AT) are not significant compared to the contributions of SP and R2 in the archipelago.

5.3 Urban Flooding

Figure 11 shows the SWMM maximum nodal levels for Hurricane Eta for San Andrés, and Hurricane Iota for Providencia and Santa Catalina.

The results obtained from the SWMM simulations for San Andrés show that the most critical values of maximum flood levels occur toward the north and center of the urban area near the airport. The most critical region is located in the vicinity

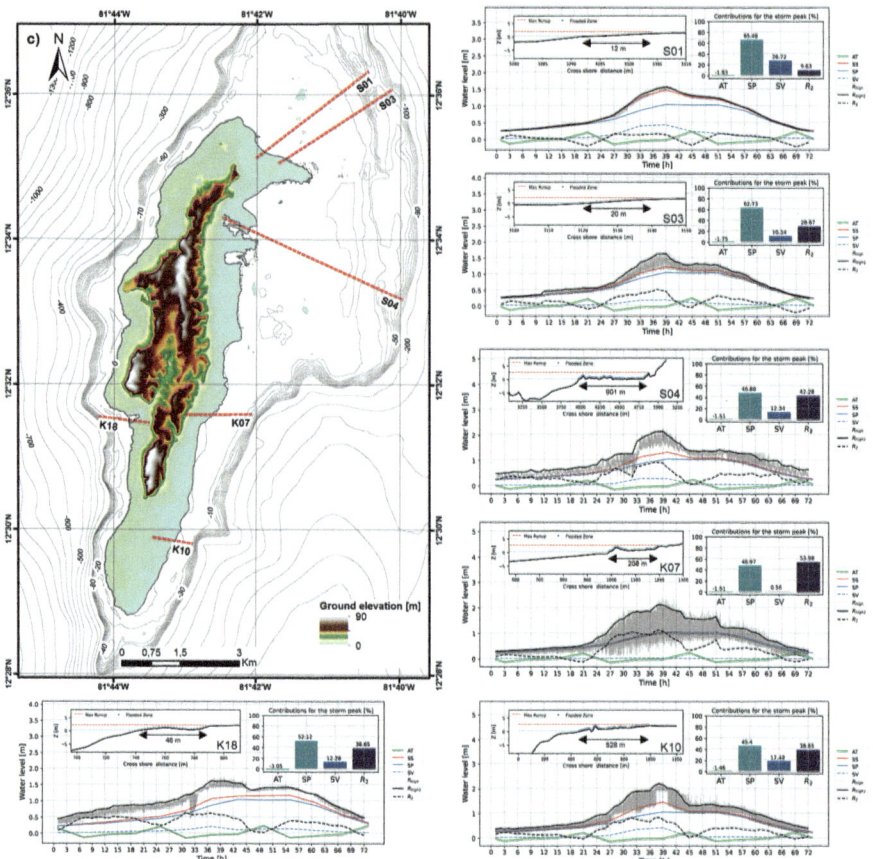

Fig. 9 Spatial distribution of the total coastal flooding during the passage of Iota in San Andrés including percentage components and time series during the storm peak (72 h) of Astronomical Tide (AT), Storm Surge due to pressure (SP), Storm Surge due to wind (SV), wave run-up (R_{high}), 98th percentile of wave run-up (R_{high2}), and the run-up component subtracting sea level component (R2)

of the El Isleño hotel, around the ecological park of San Andrés near the entrance to the airport runway. The critical zone extends around the airport toward the south through Swamp and Juan XXIII Avenues in the area of influence of the School House neighborhood. An area of high flooding is also observed near the Cartagena Alegre and Swamp Ground neighborhoods. These results are consistent with the flood sectors provided by CORALINA, historical photographic records, citizen reports, and even news records generated during the passage of Hurricane Eta (in early November 2020), showing strong floods in neighborhoods such as Natania, Serranilla, School House, Cocal, Santana, and Juan Avenue XXIII, areas of high population density in San Andrés (see Fig. 3a for neighborhoods).

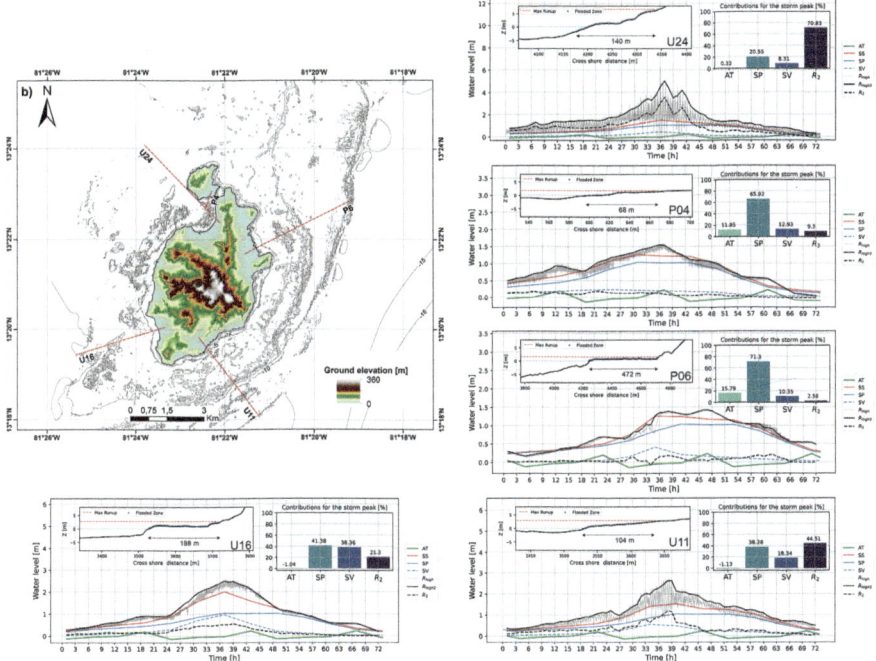

Fig. 10 Spatial distribution of total coastal flooding during the passage of Iota in Providencia and Santa Catalina, including percentage components and time series during the storm peak (72 h) of Astronomical Tide (AT), Storm Surge due to pressure (SP), Storm Surge due to wind (SV), wave run-up (R_{high}), 98th percentile of wave run-up (R_{high2}), and the run-up component subtracting sea level component (R2)

Other regions with extreme values observed are found towards the north of the island in the San Francisco de Asís neighborhood. This area is not mentioned in the vulnerable areas reported in the information provided by CORALINA. Towards the southeast of the urban area of San Andrés, throughout the eastern coastal zone, areas with maximum flood values are observed near the maritime terminal and Forbes Landing, further to the northeast, Colombia Avenue near the Aquarium hotel and Las Américas Avenue are also reported as areas of high recurrence of flood events by CORALINA (see Fig. 3a for neighborhoods).

For Providencia, the results for the approximate drainage conceptualization show how the areas of greatest flooding are located toward the north of Santa Catalina Bay in the area known as Old Town, to the west in the region of the Agua Dulce (Fresh Water Bay) and to the east in the area surrounding the Providencia airport (see Fig. 3b for neighborhoods).

Since the SWMM model is a unidimensional model with several approximations and despite the fact that water flooding elevation was tested comparing with real values observed by the local community in some specific points, and that the SWMM model results are consistent considering the frequent flooding areas in San Andrés

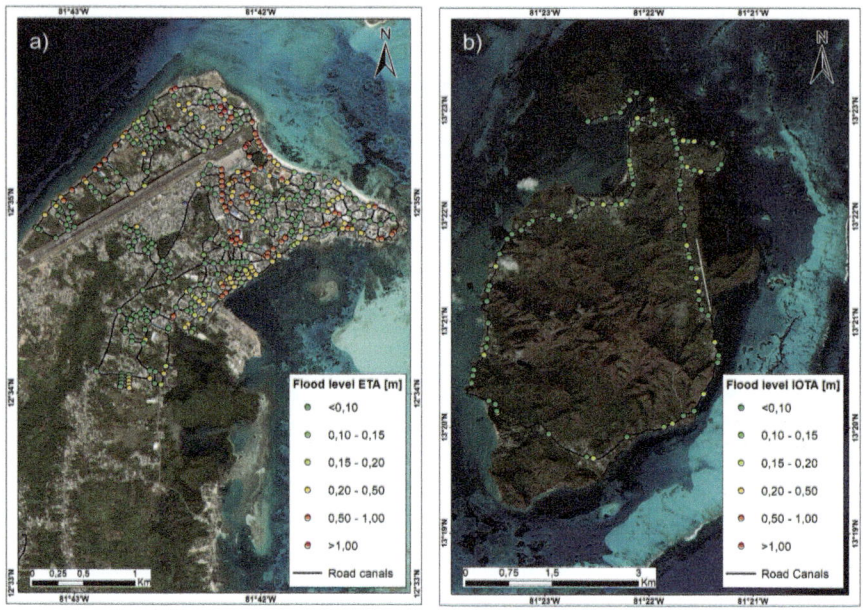

Fig. 11 **a** Maximum extreme water levels for San Andrés and **b** Providencia and Santa Catalina during the most extreme hurricane events

island reported by the Plan Maestro de Alcantarillado (Consorcio Plan Vial Caribe 2007), the flooding results during hurricanes Eta and Iota must be considered as indicatives.

6 Discussion and Conclusions

The results show the potential of mathematical modeling and field measurements to explain the behavior of the atmosphere-ocean-land interaction, particularly extreme hurricane events, and their impacts on coastal areas. The passage of Hurricanes Eta and Iota through the archipelago showed the differentiated contribution of each hazard (rain, wind, and coastal flooding) on physical infrastructure, coastal ecosystems, and population. Particularly, for Iota the magnitudes of winds are influenced by the distance to the coast, thus the impact was not the same for San Andrés (around 103 km from the hurricane's eye) as for Providencia (approximately 20 km) during the maximum development of the cyclone.

The wave results were validated (not shown here) for deep water with other hurricanes (Dean and Matthew) that have been reported in the literature using in-situ data available from the National Data Buoy Center (NDBC). In shallow waters for waves and storm surge levels, rainfall, drainage, among others, a direct validation was not possible since there is no monitoring system at the institutional level that permanently records these variables. Therefore, it is recommended to complement the information obtained with in-situ operational measurement systems (coastal buoys, water level sensors, and weather stations) and remote data provided by video camera systems, among others. All these measurements and modeling elements are part of the recommendations of the Sustainable Development Goals (SDGs) and the Decade of Ocean Sciences declared by the United Nations (UN).

For coastal flooding, the contributions of the storm surge (contributions of pressure and wind forcing) were larger for San Andrés, while in some areas of Providencia (e.g., the central area of Pueblo Viejo) the effect of the wave run-up was four times higher than the storm surge effect. The wave attenuation in distinct locations of the archipelago might suggest the importance of coastal ecosystems (e.g., coral reefs, mangroves, seagrasses, or beaches) for protecting coastal communities. In San Andrés, the profiles S01 and S03 showed the lowest contribution of wave run-up, probably due to the wave damping provided by the barrier reef at the northeast side of San Andrés, while in Providencia, the lowest wave run-up contribution was obtained along the main mangrove area of the McBean Lagoon National Natural Park.

Regarding urban flooding modeling for the archipelago, it is suggested to involve all possible existing hydraulic elements in the complex network of the archipelago in order to achieve a more comprehensive representation of the urban drainage process. Although the dynamic wave method to solve physical processes of urban drainage was used, the conceptualization carried out for the archipelago represents a simplification of reality and does not consider all the existing hydraulic elements in such complex networks. Considering the lack of information, further implementation of an automatic flow and level monitoring system is also recommended for the storm sewer network or other hydraulic elements (e.g., gutters and natural channels). Although the results were consistently adjusted to the visual perception of the inhabitants, it is recommended to keep an accurate record of the water sheet through a community monitoring system, including flood level and photographic records, among others. Such data could complement the information supplied by other entities in charge of risk management in the archipelago. The urban flood modeling showed the enormous potential to incorporate the results into urban planning schemes and the development of sustainable drainage systems.

Regarding atmospheric modeling, simple parametric methodologies were used to define flooding components such as storm surge (effect of wind and pressure). Although the results are robust based on historical reports, newspapers, and information provided by inhabitants, there is a need to employ methodologies that consider the coupling of physical-based numerical models such as the Regional Ocean Modeling System (ROMS) and Delft 3D, among others. This will allow for improvements in the coastal flooding representation, given that storm surge represents one of the most important components, even exceeding other essential components like sea level anomalies, astronomical tide, or wave run-up in the archipelago.

Topo-bathymetric changes are among the most relevant factors for wave propagation in shallow waters and coastal flooding. The results obtained from XBeach are strongly dependent on the available bathymetric information and bottom roughness. Hence, the influence of coral cover, grasses, mangroves, and dune vegetation, among others, should be incorporated into the models and strategies for planning Nature-based Solutions (Osorio-Cano et al. 2019). As such, it is recommended to apply satellite, drone, Remotely Operated Vehicle (ROV), and other techniques to make more robust digital terrain models.

These effects related to physical processes due to hurricanes and extreme events must be incorporated into territorial planning and decision-making regarding vulnerability and risk. Likewise, the reconstruction of infrastructure must be guided by the accurate diagnosis offered by numerical tools to predict the wind/wave climate and the interaction with the insular zones. The maximum winds determine the building material to be used and/or the ecosystems that can provide urban protection as an ecosystem service. The level of coastal flooding brings elements to define the retreat and/or shelter zones for evacuation, as well as the type of infrastructure that can be considered in these vulnerable zones. Hence, strategic coastal ecosystems (e.g., mangroves, corals, and seagrasses) are highlighted for providing coastal protection. The level of urban flooding allows the re-dimensioning of drainage systems and thinking about territorial planning solutions that include Sustainable Urban Drainage Systems (SUDS).

Nature shows us the way for future planning, enhancing the chance of Building with Nature (BwN) and not against it, in order to achieve protection against flooding and coastal erosion. Although the archipelago can be considered a region with scarce historical hurricane events, knowing the more susceptible areas in this region is useful information for policymakers as it encourages science-based decision-making and constitutes a key step towards the consolidation of a risk assessment in the archipelago.

Acknowledgements The authors want to thank CORALINA for providing data and financial support. Thanks also to the collaborators Victor Saavedra, Jose Daniel Rios, Juan Pablo Ramirez, Alejandro Álvarez, Jhayron Pérez, Mauricio Zapata, Victor Rua, and Mariana Roldán for the technical support and numerical modeling. Thanks also to Alejandro Henao, David Quintero, Simon Acevedo, and Margarita López during field instrumentation and data processing. Thanks also to IDEAM, IGAC, and NOAA for allowing open access data. Finally, to CEMarin, which supports the publication of this chapter.

References

Abtew W (2019) Hurricanes: wind, rain, and storm surge. In: Melesse AM, Abtew W, Senay G (eds) Extreme hydrology and climate variability. Elsevier, pp 367–378. https://doi.org/10.1016/B978-0-12-815998-9.00028-2

Alexander M, Scott J (2002) The influence of ENSO on air-sea interaction in the Atlantic. Geophys Res Lett 29(14):1–4. https://doi.org/10.1029/2001GL014347

Alfaro EJ (2002) Response of air surface temperatures over Central America to oceanic climate variability indices. Investig Mar 30(1):63–72. https://doi.org/10.4067/s0717-71782002030100006

Aon (2017) Weather, climate & catastrophe insight. In: 2017 annual report. https://www.aon.com/global-weather-catastrophe-natural-disasters-costs-climate-change-2020-annual-report/index.html?utm_source=prnewswire&utm_medium=mediarelease&utm_campaign=natcat21

Babaei S, Ghazavi R, Erfanian M (2018) Urban flood simulation and prioritization of critical urban sub-catchments using SWMM model and PROMETHEE II approach. Phys Chem Earth, Parts A/B/C 105:3–11. https://doi.org/10.1016/J.PCE.2018.02.002

Battjes JA, Janssen JPFM (1978) Energy loss and wave set-up due to breaking of random waves. In: 16th international conference on coastal engineering, pp 569–587. https://doi.org/10.1061/9780872621909.034

Benavente J, del Río L, Gracia FJ et al (2006) Coastal flooding hazard related to storms and coastal evolution in Valdelagrana spit (Cadiz Bay Natural Park, SW Spain). Continent Shelf Res 26(9):1061–1076. https://doi.org/10.1016/J.CSR.2005.12.015

Blunden J, Boyer T (2021) State of the climate in 2020. Bull Amer Meteor Soc 102(8):S1–S475. https://doi.org/10.1175/2021BAMSSTATEOFTHECLIMATE.1

Booij N, Ris RC, Holthuijsen LH (1999) A third-generation wave model for coastal regions 1. Model description and validation. J Geophys Res 104(C4):7649–7666. https://doi.org/10.1029/98JC02622

Bowden KF (1983) Physical oceanography of coastal waters. E. Horwood; Halsted Press Chichester [West Sussex], New York

Cavaleri L, Rizzoli PM (1981) Wind wave prediction in shallow water: theory and applications. J Geophys Res 86(1):10961–10973. https://doi.org/10.1029/JC086iC11p10961

Cavaleri L, Sclavo M (2006) The calibration of wind and wave model data in the Mediterranean Sea. Coast Eng 53(7):613–627. https://doi.org/10.1016/J.COASTALENG.2005.12.006

Chow VT, Maidment DR, Mays LW (1994) Hidrología Aplicada. McGraw-Hill Interamericana

Committee on Homeland Security and Governmental Affairs (2006) Hurricane Katrina: a nation still unprepared. S. Rept., pp 109–322

Consorcio Plan Vial Caribe (2007) Elaboración del plan vial y de transporte para la isla de San Andrés y plan maestro de alcantarillado pluvial en el sector North End ajustado al RAS 2000

CORALINA-INVEMAR (2012) Atlas de la Reserva de Biósfera Seaflower. Archipiélago de San Andrés, Providencia y Santa Catalina. In: Gómez-López DI, Segura-Quintero C, Sierra-Correa PC et al (eds) Serie de publicaciones especiales de Invemar No. 28. Instituto de Investigaciones Marinas y Costeras "José Benito Vives De Andréis" -INVEMAR- y Corporación para el Desarrollo Sostenible del Archipiélago de San Andrés, Providencia y Santa Catalina, CORALINA. http://www.invemar.org.co/redcostera1/invemar/docs/10447AtlasSAISeaflower.pdf

DANE (2020) Encuesta de hábitat y usos socioeconómicos, 2019 - Archipiélago de San Andrés, Providencia y Santa Catalina. https://www.dane.gov.co/index.php/estadisticas-por-tema/inf ormacion-regional/encuesta-de-habitat-y-usos-socioeconomicos-2019-archipielago-de-san-andres-providencia-y-santa-catalina

Eldeberky Y, Battjes JA (1984) Nonlinear coupling in waves propagating over a bar. In: 24th international conference on coastal engineering, pp 157–167

Enfield DB, Mayer DA (1997) Tropical Atlantic sea surface temperature variability and its relation to El Niño-southern oscillation. J Geophys Res 102(C1):929–945. https://doi.org/10.1029/96jc03296

Genes LS, Montoya RD, Osorio AF (2021) Coastal sea level variability and extreme events in Moñitos, Cordoba, Colombian Caribbean Sea. Continent Shelf Res 228:104489. https://doi.org/10.1016/j.csr.2021.104489

Giannini A, Cane MA, Kushnir Y (2001) Interdecadal changes in the ENSO teleconnection to the Caribbean region and the north Atlantic oscillation. J Clim 14(13):2867–2879. https://doi.org/10.1175/1520-0442(2001)014%3c2867:ICITET%3e2.0.CO;2

Giannini A, Chiang JCH, Cane MA et al (2001) The ENSO teleconnection to the tropical Atlantic ocean: contributions of the remote and local SSTs to rainfall variability in the tropical Americas. J Clim 14(24):4530–4544. https://doi.org/10.1175/1520-0442(2001)014%3c4530:TETTTT%3e2.0.CO;2

Giannini A, Kushnir Y, Cane MA (2000) Interannual variability of Caribbean rainfall, ENSO, and the Atlantic Ocean. J Clim 13(2):297–311. https://doi.org/10.1175/1520-0442(2000)013%3c0 297:IVOCRE%3e2.0.CO;2

Giannini A, Kushnir Y, Cane MA (2001) Seasonality in the impact of ENSO and the north Atlantic high on Caribbean rainfall. Phys Chem Earth, Part B 26(2):143–147. https://doi.org/10.1016/S1464-1909(00)00231-8

Hasselmann K, Barnett TP, Bouws E et al (1973) Measurements of wind-wave growth and swell decay during the joint North Sea wave project (JONSWAP). Suppl. A, Deutsches Hydrographisches Institut

Hasselmann S, Hasselmann K, Allender JH et al (1985) Computations and parametrizations of the nonlinear energy transfer in a gravity-wave spectrum. Part II: parametrizations of the nonlinear energy transfer for application in wave models. J Phys Oceanogr 15:1378–1391. https://doi.org/10.1175/1520-0485(1985)015<1378:CAPOTN>2.0.CO;2

Huffman GJ, Bolvin DT, Braithwaite D et al (2020) Integrated multi-satellite retrievals for the global precipitation measurement (GPM) mission (IMERG). In: Levizzani V, Kidd C, Kirschbaum DB, Kummerow CD, Nakamura K, Turk FJ (eds) Satellite precipitation measurement, vol 1. Springer International Publishing, pp 343–353. https://doi.org/10.1007/978-3-030-24568-9_19

IDEAM (2017) Pronóstico de pleamares y bajamares en la costa Caribe Colombiana año 2018. IDEAM

IDEAM (2020) Comunicados especiales N° 132–135. Comunicados Especiales Noviembre 2020. IDEAM

Ishizawa OA, Miranda JJ, Strobl E (2019) The impact of hurricane strikes on short-term local economic activity: evidence from nightlight images in the Dominican Republic. Int J Disaster Risk Sci 10(3):362–370. https://doi.org/10.1007/s13753-019-00226-0

Isobe M (2013) Impact of global warming on coastal structures in shallow water. Ocean Eng 71:51–57. https://doi.org/10.1016/J.OCEANENG.2012.12.032

Johnson S (2015) The history and science of hurricanes in the Greater Caribbean. Oxford Research Encyclopedia of Latin American History, pp 1–23. https://doi.org/10.1093/acrefore/9780199366439.013.57

Knutson T, Camargo SJ, Chan JCL et al (2020) Tropical cyclones and climate change assessment Part II: projected response to anthropogenic warming. Bull Amer Meteorol Soc 101(3):E303–E322. https://doi.org/10.1175/BAMS-D-18-0194.1

Komen GJ, Hasselmann S, Hasselmann K (1984) On the existence of a fully developed wind-sea spectrum. J Phys Oceanogr 14:1271–1285. https://doi.org/10.1175/1520-0485(1984)014<1271:OTEOAF>2.0.CO;2

Kowaleski AM, Morss RE, Ahijevych D et al (2020) Using a WRF-ADCIRC ensemble and track clustering to investigate storm surge hazards and inundation scenarios associated with hurricane Irma. Wea Forecast 35(4):1289–1315. https://doi.org/10.1175/WAF-D-19-0169.1

Li J, Tan W, Chen M et al (2020) An extreme sea level event along the northwest coast of the south China sea in 2011–2012. Continent Shelf Res 196:104073. https://doi.org/10.1016/j.csr.2020.104073

Lin N, Smith JA, Villarini G et al (2010) Modeling extreme rainfall, winds, and surge from hurricane Isabel (2003). Wea Forecast 25(5):1342–1361. https://doi.org/10.1175/2010WAF2222349.1

Lizano OG (1990) Un modelo de viento ajustado a modelo de generación de olas para pronóstico de oleaje durante huracanes. Geofísica 33:75–103

Marsooli R, Lin N (2020) Impacts of climate change on hurricane flood hazards in Jamaica Bay, New York. Clim Change 163:2153–2171. https://doi.org/10.1007/s10584-020-02932-x

Meza-Padilla R, Appendini CM, Pedrozo-Acuña A (2015) Hurricane-induced waves and storm surge modeling for the Mexican coast. Ocean Dyn 65:1199–1211. https://doi.org/10.1007/s10236-015-0861-7

Montoya RD, Arias AO, Royero JCO et al (2013) A wave parameters and directional spectrum analysis for extreme winds. Ocean Eng 67:100–118. https://doi.org/10.1016/j.oceaneng.2013.04.016

Montoya RD, Menendez M, Osorio AF (2018) Exploring changes in Caribbean hurricane-induced wave heights. Ocean Eng 163:126–135. https://doi.org/10.1016/j.oceaneng.2018.05.032

Ortiz-Royero JC (2012) Exposure of the Colombian Caribbean coast, including San Andrés Island, to tropical storms and hurricanes, 1900–2010. Nat Hazards 61(2):815–827. https://doi.org/10.1007/s11069-011-0069-1

Ortiz-Royero JC, Plazas-Moreno JM, Lizano O (2015) Evaluation of extreme waves associated with cyclonic activity on San Andrés Island in the Caribbean Sea since 1900. J Coast Res 31(3):557–568. https://doi.org/10.2112/JCOASTRES-D-14-00072.1

Osorio AF, Montoya RD, Ortiz JC et al (2016) Construction of synthetic ocean wave series along the Colombian Caribbean coast: a wave climate analysis. Appl Ocean Res 56:119–131. https://doi.org/10.1016/j.apor.2016.01.004

Osorio-Cano JD, Osorio AF, Peláez-Zapata DS (2019) Ecosystem management tools to study natural habitats as wave damping structures and coastal protection mechanisms. Ecol Eng 130:282–295. https://doi.org/10.1016/j.ecoleng.2017.07.015

Pasch RJ, Reinhart BJ, Berg R et al (2021) Tropical cyclone report: Hurricane Eta (AL292020). https://www.nhc.noaa.gov/data/tcr/AL292020_Eta.pdf

Peng S, Li Y (2015) A parabolic model of drag coefficient for storm surge simulation in the south China Sea. Sci Rep 5(15496):1–6. https://doi.org/10.1038/srep15496

Pielke RA, Rubiera J, Landsea C et al (2003) Hurricane vulnerability in latin America and the Caribbean: normalized damage and loss potentials. Nat Hazards Rev 4(3):101–114. https://doi.org/10.1061/(asce)1527-6988(2003)4:3(101)

Rey W, Ruiz-Salcines P, Salles P et al (2021) Hurricane flood hazard assessment for the Archipelago of San Andres, Providencia and Santa Catalina, Colombia. Front Mar Sci 8:1–18. https://doi.org/10.3389/fmars.2021.766258

Rezapour M, Baldock TE (2014) Classification of hurricane hazards: the importance of rainfall. Wea Forecast 29(6):1319–1331. https://doi.org/10.1175/WAF-D-14-00014.1

Ricaurte-Villota C, Murcia-Riaño M, Ordoñez-Zuñiga SA (2017) Región 3: Insular. In: Ricaurte-Villota C, Bastidas-Salamanca ML (eds) Regionalización oceanográfica, una visión dinámica del Caribe. Instituto de Investigaciones Marinas y Costeras José Benito Vives De Andréis (INVEMAR), pp 62–81

Rincón JC, Muñoz MF (2013) Hydraulic design of dual drainage systems using the SWMM model. RIHA 34(2):103–117. http://scielo.sld.cu/scielo.php?script=sci_arttext&pid=S1680-03382013000200009&lng=es&tlng=es

Roelvink D, Reniers A, van Dongeren A et al (2010) XBeach model description and manual

Roelvink FE, Storlazzi CD, van Dongeren AR et al (2021) Coral reef restorations can be optimized to reduce coastal flooding hazards. Front Mar Sci 8:1–11. https://doi.org/10.3389/fmars.2021.653945

Roldán-Upegui M, Montoya RD, Osorio AF et al (2022) Modified parametric hurricane wind model to improve asymmetry representation in the region of maximum winds: study case for the Caribbean Sea. Manuscript in Preparation

Rossman LA (2015) Storm water management model (SWMM) user's manual version 5.1 (Issue September)

Ruiz-Ochoa MA, Bernal G (2009) Variabilidad estacional e interanual en los datos del reanálisis NCEP/NCAR en la cuenca Colombia, mar Caribe. Avances en Recursos Hidráulicos 20:7–20. https://revistas.unal.edu.co/index.php/arh/article/view/14328

Ruti PM, Marullo S, D'Ortenzio F et al (2008) Comparison of analyzed and measured wind speeds in the perspective of oceanic simulations over the Mediterranean basin: analyses, QuikSCAT and buoy data. J Mar Syst 70(1–2):33–48. https://doi.org/10.1016/J.JMARSYS.2007.02.026

Sharma N, D'Sa E (2008) Assessment and analysis of QuikSCAT vector wind products for the Gulf of Mexico: a long-term and hurricane analysis. Sensors 8(3):1927–1949. https://doi.org/10.3390/S8031927

Seneviratne SI, Zhang X, Adnan M et al (2021) Weather and climate extreme events in a changing climate. In: Climate change 2021: the physical science basis. Contribution of working group

I to the sixth assessment report of the intergovernmental panel on climate change. Cambridge University Press, pp 1513–1766. https://doi.org/10.1017/9781009157896.013

Soil Conservation Service (1972) National engineering handbook. Section 4. Hydrology. US Department of Agriculture

Spencer N, Urquhart MA (2018) Hurricane strikes and migration: evidence from storms in Central America and the Caribbean. Wea Climate Soc 10(3):569–577. https://doi.org/10.1175/wcas-d-17-0057.1

Stewart SR (2021) Tropical cyclone report: Hurricane Iota (AL312020). https://www.nhc.noaa.gov/data/tcr/AL312020_Iota.pdf

Tanner EVJ, Kapos V, Healey JR (1991) Hurricane effects on forest ecosystems in the Caribbean. Biotropica 23(4):513. https://doi.org/10.2307/2388274

Tian Z, Zhang Y (2021) Numerical estimation of the typhoon-induced wind and wave fields in Taiwan Strait. Ocean Eng 239(June):109803. https://doi.org/10.1016/j.oceaneng.2021.109803

Tolman HL (2002) Alleviating the garden sprinkler effect in wind wave models. Ocean Model 4(3–4):269–289. https://doi.org/10.1016/S1463-5003(02)00004-5

Tolman HL, Chalikov D (1996) Source terms in a third-generation wind wave model. J Phys Oceanogr 26:2497–2518. https://doi.org/10.1175/1520-0485(1996)026%3c2497:STIATG%3e2.0.CO;2

UNGRD (2020) Boletín N°198: UNGRD Entregó el primer balance de la evaluación de daños y análisis de necesidades, tras emergencia generada por el paso del huracán Iota en el archipiélago. In: Boletín informativo: Unidad Nacional para la Gestión del Riesgo de Desastres

Vickery PJ, Masters FJ, Powell MD et al (2009) Hurricane hazard modeling: the past, present, and future. J Wind Eng Ind Aerodyn 97(7–8):392–405. https://doi.org/10.1016/j.jweia.2009.05.005

Visbal J, Ortiz J (2006) Simulación de huracanes bajo el lenguaje Java a partir del modelo HURWIN (HURricane WINd model) para su aplicación sobre la costa Caribe Colombiana. Undergraduate Thesis in Systems Engineering, Universidad Del Norte

Walcker R, Laplanche C, Herteman M et al (2019) Damages caused by hurricane Irma in the human-degraded mangroves of Saint Martin (Caribbean). Sci Rep 9(1):18971. https://doi.org/10.1038/s41598-019-55393-3

Watson CC, Johnson ME (2004) Hurricane loss estimation models: opportunities for improving the state of the art. Bull Amer Meteor Soc 85(11):1713–1726. https://doi.org/10.1175/BAMS-85-11-1713

Wiley JW, Wunderle JM (1993) The effects of hurricanes on birds with special reference to Caribbean islands. Bird Conserv Int 3(4):319–349. https://doi.org/10.1017/S0959270900002598

Willoughby HE, Rahn ME (2004) Parametric representation of the primary Hurricane Vortex. Part I: observations and evaluation of the Holland (1980) Model. Mon Wea Rev 132(12):3033–3048. https://doi.org/10.1175/MWR2831.1

Wright CW, Walsh EJ, Vandemark D et al (2001) Hurricane directional wave spectrum spatial variation in the open ocean. J Phys Oceanogr 31:2472–2488. https://doi.org/10.1175/1520-0485(2001)031%3c2472:hdwssv%3e2.0.co;2

WW3DG (2019) User manual and system documentation of WAVEWATCH III version 6.07. In: Technical note 333

Yin J, Lin N, Yang Y et al (2021) Hazard assessment for typhoon-induced coastal flooding and inundation in Shanghai, China. J Geophys Res Oceans 126(7):e2021JC017319. https://doi.org/10.1029/2021JC017319

Open Access This chapter is licensed under the terms of the Creative Commons Attribution 4.0 International License (http://creativecommons.org/licenses/by/4.0/), which permits use, sharing, adaptation, distribution and reproduction in any medium or format, as long as you give appropriate credit to the original author(s) and the source, provide a link to the Creative Commons license and indicate if changes were made.

The images or other third party material in this chapter are included in the chapter's Creative Commons license, unless indicated otherwise in a credit line to the material. If material is not included in the chapter's Creative Commons license and your intended use is not permitted by statutory regulation or exceeds the permitted use, you will need to obtain permission directly from the copyright holder.

Rapid Remote Sensing Assessment of Impacts from Hurricane Iota on the Coral Reef Geomorphic Zonation in Providencia

Hernando Hernández-Hamón⦿, Paula A. Zapata-Ramírez⦿, Rafael E. Vásquez⦿, Carlos A. Zuluaga, Juan David Santana Mejía, and Marcela Cano⦿

Abstract This study assesses Hurricane Iota's impact on Providencia island's reef environments, using Google Earth Engine, Satellite Derived Bathymetry, and machine learning to calculate a supervised classification process to delineate six geomorphic reef units. Results reveal dynamic changes, including erosion in the Lagoon unit (4.47% pre-Iota, 2.27% post-Iota), loss on the Back Reef (38.14%), and Rock Terrace (6.15%). Reef Ridge showed minimal change, acting as an effective wave barrier. Back Reef and the deep Rock Terrace experienced significant erosion (-3 to -14 m) to the northeast, with sedimentary dynamics observed in deeper units (up to 22 m). The high thematic accuracies found (Kappa 99%) illustrate the effectiveness of the assessment to (i) map the reef rapidly, (ii) provide tools for long-term monitoring of changes over time and (iii) improve management strategies and decision-making.

H. Hernández-Hamón · P. A. Zapata-Ramírez (✉) · R. E. Vásquez · C. A. Zuluaga
School of Engineering, Universidad Pontificia Bolivariana, Medellín, Colombia
e-mail: paula.zapataramirez@upb.edu.co

H. Hernández-Hamón
e-mail: hernando.hernandezh@upb.edu.co

R. E. Vásquez
e-mail: rafael.vasquez@upb.edu.co

C. A. Zuluaga
e-mail: carlos.zuluaga@upb.edu.co

J. D. S. Mejía
Center for Oceanographic and Hydrographic Research of the Caribbean, Cartagena, Colombia
e-mail: jsantana@dimar.mil.co

M. Cano
National Natural Parks of Colombia, Providencia, Colombia
e-mail: marcela.cano@parquesnacionales.gov.co

© The Author(s) 2025
J. E. Mancera Pineda et al. (eds.), *Climate Change Adaptation and Mitigation in the Seaflower Biosphere Reserve*, Disaster Risk Reduction,
https://doi.org/10.1007/978-981-97-6663-5_4

Keywords Remote sensing · Google earth engine · Hurricane impacts · Geomorphic changes · Providencia Island

1 Introduction

Shallow water coral reefs provide valuable protection to coastal infrastructure from storm surge and waves. However, the same reefs that provide protection can also be damaged by storm-related wave energy. As the environment continues to change through both natural and human-influenced means, catastrophic events such as hurricanes are projected to have much larger devastating effects on small island territories. The intensity, frequency, and duration of hydrometeorological events such as hurricanes increase the risk of a grave danger by storm surge and waves substrate erosion, or sediment deposition (Verfaillie et al. 2009; Goes et al. 2019; Kumar et al. 2021). In addition to the socioeconomic risk, the spatiotemporal disturbance affects the habitat suitability for benthic flora and fauna (Post 2008; Lecours et al. 2015).

A novel approach introduces Nature-based Solutions (NBS), inspired by nature, and more efficient cost-effective management strategies to mitigate the increasing risk from hydrometeorological hazards (HMHs). For marine spaces, the NBS strategy focuses on developing large-scale bathymetry and geomorphological preliminary approaches to monitor the complexity of seafloor changes caused by HMHs that may generate socioeconomic and environmental losses (UNISDR 2009). Although the technologies developed in the field of remote sensing for marine environments have included mechanisms for geomorphic mapping and classification using strong computational approaches (Kennedy et al. 2021), machine learning classification and cloud computing processing offer a unique opportunity to access petabytes of free-access data generated from moderate-resolution satellites such as Sentinel 2 (Gorelick et al. 2017).

Hurricane Iota made landfall on the island of Providencia as a strong category 5 storm on November 17, 2020, between 4:00 and 7:00 a.m., passing within 10 km north of Providencia with sustained wind speeds >250 km/h and gusts of 270 km/h (INVEMAR 2021). Hurricane impacts on coral reefs can come in many forms, from broken pieces missing from branching coral species to entire colonies dislodged, cracked, or shattered into multiple pieces or fragments. They can also strip off the superficial reef framework, deposit loosened material onto beaches or cays above sea level, or propel them into deeper sub-reef environments. However, the assessment of these impacts by traditional field surveys is time-consuming and expensive to investigate the full extent and magnitude of the reef geomorphic changes, in order to support effective post-hurricane management approaches such as the emerging areas of NBS and climate change adaptation. Thus, this powerful and rare weather event provides an opportunity to examine the effects of extreme physical forces on coral reefs and their impact on reef geomorphology.

Accordingly, this study is focused on a rapid assessment of Hurricane Iota's impact on the coral reef geomorphology in Providencia, located in the Seaflower

Biosphere Reserve that offers the ideal conditions to generate accurate maps using remote sensing techniques, due to its clear water conditions and shallow depth platforms. Calibrated and corrected Sentinel image composites for the entire island were generated using Google Earth Engine (GEE) for a comparable pre-Iota and post-Iota timeframe that accounted for reef geomorphic zonation. Our results demonstrate how open-source satellite imagery (Sentinel 2) enables efficient analytic processes of changes during the investigated event. Results show dynamic variation in the geomorphic units following a major hurricane event and could lead to improved management strategies such as (i) restoration efforts, (ii) monitoring activities for the analysis of the timing and nature of recovery initiation after impact, (iii) the spatial prioritization of conservation activities such as the reefs with the best chance of survival with the increasing frequency of extreme events, and (iv) to improve the implementation of NBS approaches and climate change adaptation strategies.

1.1 Study Area

In the Archipelago of San Andrés, Providencia and Santa Catalina, among the oceanic islands, atolls and banks, the island of Providencia—also known as Old Providence—is found, which extends 7.2 km across from north to south. The barrier reef—the second largest in the Southeastern Caribbean—is an extensive calcareous platform stretching over 32 km (Diaz et al. 2000; Sanchez et al. 1998). Situated at 25 m deep, there is a submerged elongated ridge in a shelf-margin position, which may be a drowned shelf-edge barrier reef. Geister (1992) and Geister and Diaz (2007) clearly describe the reef complex's geomorphology: the lagoon platform is occupied by extensive semi-closed and gently sloped terraces up to 14 m deep with areas 2–6 m wide that are occupied by an extensive shallow lagoonal terrace. Front Reef is a fore-reef terrace in front of the shallow peripheral reef, it is up to several meters wide and slopes gently to the Rock Terrace. A significant part of the barrier reef is formed by a wide belt made up of numerous patch reefs, mostly of the pinnacle type, which rise from the seafloor at -6 to -8 m and reach the low-tide level. Occasional storms with westerly or northwesterly winds reaching speeds over 20 m/s do occur, mostly in the second half of the year (Geister and Diaz 2007). The mean annual air temperature is 27 °C, with a 1 °C range between monthly values, while rainfall is irregular and varies greatly from year to year. According to Geister and Diaz (2007), sedimentation processes in the area are controlled by the surface persistent northward flow of the Caribbean Current through large gaps and narrow open seaways across the top of the Nicaraguan Rise.

Table 1 Sentinel 2 MSI scenes in Providencia

Date	Image ID	Cloud cover (%)
2019-09-10 to 2019-09-20	20190910T155752_T17PMQ-PMR 20190920T155836_T17PMQ-PMR	<10
2020-09-08 to 2020-09-19	20200909T155527_T17PMQ-PMR 20200919T155527_T17PMQ-PMR 200914T155529_T17PMQ-PMR	<10
2021-01-27 to 2021-02-01	20210201T155525_T17PMR-PRM 20210201T155525_T17PMQ-PMR	<10

2 Methodology

2.1 Image Pre-processing

Moderate-resolution satellite imagery from GEE JavaScript-API was selected from Optical Sentinel 2-MSI Multispectral imaging bands and 10 m of spatial resolution, as shown in Table 1. Using image metadata (cloud < 10%) and reducing filtering in GEE libraries, we selected atmospherically corrected scenes from "COPERNICUS/S2" level-2A of surface reflectance before and after the Iota HMHs event (November 17, 2020). Scenes of the shallow and clear sea bottom in Providencia were reduced to annual averaged mosaics with few clouds between 2019 and 2021 (Table 1). We selected all bands used in the visible regions: red, green, and blue (0.45–0.68 μm) in the image, because they provide the spectral attenuation differences in seafloor structures at different depths as a function of wavelength. In addition, the near-infrared band (0.78–0.90 μm) was included to reduce the solar brightness on the sea surface, to mask the land and the wave crest (Fig. 1).

2.2 Image Processing

2.2.1 Deglint Correction

Water outflow reflectance can be difficult to observe due to the reflection of direct sunlight at the air–water interface in the satellite direction. In addition, specular reflection of the incident radiation occludes the benthic component in optical remote sensing and confounds the visual identification of the bottom feature that could influence the image classification. As a result, brightness in the surface water pixels was removed using the algorithm proposed by Hedley et al. (2005) (Table 2 and Fig. 1). To do so, a regression was performed in GEE between the near infrared (NIR) brightness and the visible bands using open-water pixels free of the NIR bottom reflectance and following the indications of ESA (2019). The outcome of this correction is illustrated in Fig. 2.

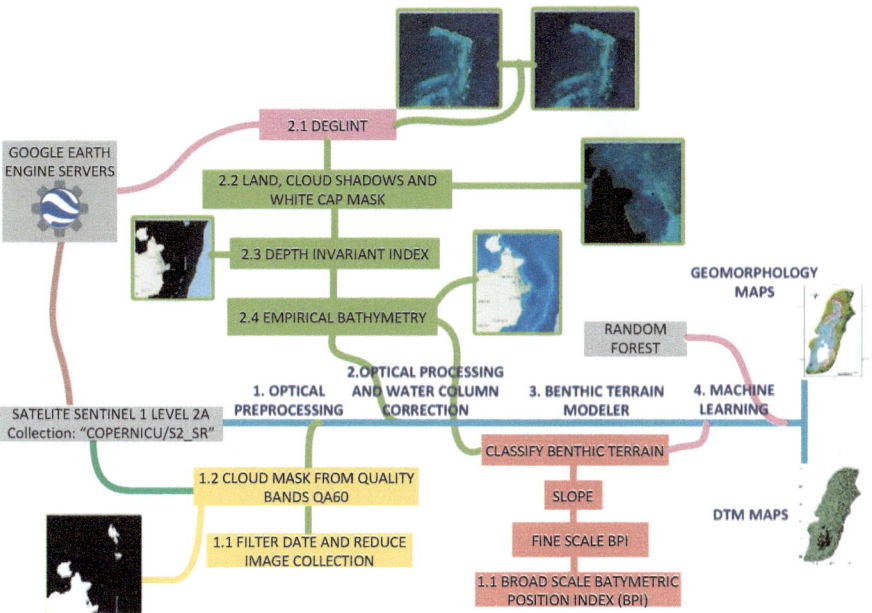

Fig. 1 Methodological phases of optical preprocessing, water column correction and machine learning in Sentinel 2

2.2.2 Land, Wave Crest, and Cloud Shadow Masking

Masking the land, white wave crest, and cloud shadow is an essential processing step to separate bright features that can be identified by high reflectance in the NIR. These areas interfere with the spectral response in the classification; therefore, it is important to mask them. NIR wavelengths do not penetrate the water, so after deglint, clear areas of water appear very dark. To solve this, an operation threshold was applied in NIR < 0.05 over the deglinted image (Table 1). The resulting mask was vectorized and the image clipped from the shallow bottom to the intermediate depth up 20 m. In addition, the dark shadows of clouds were also clipped using a threshold (B2 > 0.01) in the blue band.

2.2.3 Depth Invariant Index

To remove the confusing influence of variable water depth we applied the method provided by Lyzenga (1978) that compensates for the variable effect of depth when mapping bottom features. The first processing linearizes the effect of depth on reflectance with a natural logarithm (Table 2). To establish the depth-invariant indices, we compute the ratio between Sentinel-2 bands 1–2 in GEE over the same bottom type at different depths following the indications provided by ESA (2019).

Table 2 Column water correction and image processing algorithms

Method	References	Algorithm	Notes
Deglint	Hedley et al. (2005)	$R'_i = R_i - b_i(R_{NIR} - Min_{NIR})$	R'_i is the deglinted pixel in band i. R_i is the reflectance from a visible band i. b_i is the regression slope. R_{NIR} is the NIR band value. Min_{NIR} is the minimum NIR value of the sample
Shadow cloud and terrain mask	ESA (2019)	$NIR < 0.05$ and $BLUE > 0.02$	Threshold NIR < 0.05 and Blue > 0.02 bands value
Invariant index	Lyzenga (1978)	$Ln(B1) = P + Q \cdot Ln(B2)$ $P = Ln(B1) - Q \cdot Ln(B2)$	P is the Invariant index or y-intercept, Q is the gradient of the regression of $Ln(B1)$ on $Ln(B2)$
SDB Satellite-derived bathymetry	Stumpf et al. (2003)	$Z = m_1 \frac{Ln(nR_w(\mu))}{Ln(nR_w(\mu))} + m_0$	Z is the derivate depth. m_1 is a tunable constant to scale the ratio to depth, n is a fixed constant for all areas, and m_0 is the offset for a depth of 0 m
Slope	Horn (1981)	$\arctan = \sqrt{\frac{dz^2}{dx} + \frac{dz^2}{dy}}$	Least-squares fitting of the curvature calculations, including that used by the geodesic slope computation
BPI	Goes et al. (2019)	$Zxy - Zannulus$	Zannulus is the mean elevation value of all cells within an annulus-shaped neighborhood

Fig. 2 Glint (**a**) and Deglint (**b**) correction in east Providencia shallow water setting

a
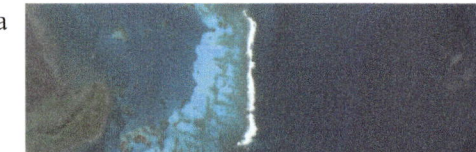

b

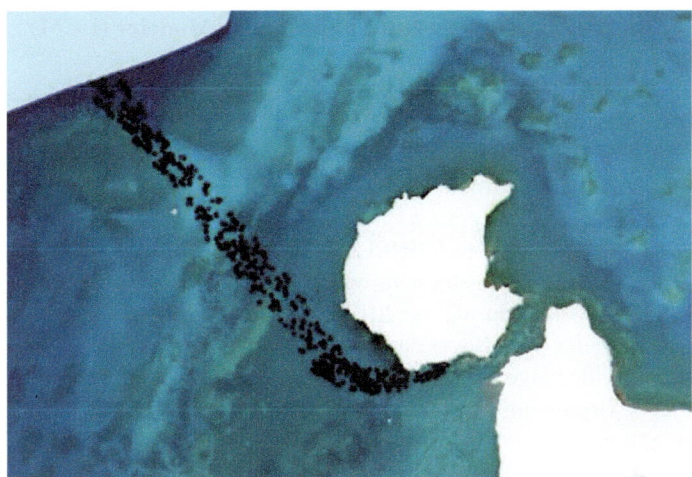

Fig. 3 Bathymetry points in Providencia collected by *Consorcio Dragado Providencia* and provided by *Aqua & Terra*. The data were collected with a multi-beam High Frequency (200 kHz) MBI ODOM echosounder system

2.2.4 Satellite-Derived Bathymetry (SDB)

The satellite-derived bathymetry (SDB) implemented was based on the principle that the water column attenuation coefficients differ between spectral bands and that the ratio between the two visible bands will change with depth (Stumpf et al. 2003). To determine the ratio between the two bands, we used GEE to apply the empirical regression method for mapping shallow waters using the algorithm proposed by Stumpf et al. (2003) (Table 2). The Z data comes from *in-situ* bathymetric data collected by *Consorcio Dragado Providencia* and provided by *Aqua & Terra S.A.S* (Fig. 3). The data were collected from a multi-beam High Frequency (200 kHz) MBI ODOM echosounder. Finally, to better observe the geomorphic features of the reef, we built a Digital Terrain Model (DTM) in Surfer 17.1.288.

2.2.5 Benthic Terrain Modeler (BTM)

We employed the derived factors such as the Bathymetric Position Index (BPI) broad and fine, and the slope calculated by the Benthic Terrain Modeler (ArcGIS 10.8 toolbox) and the satellite-derived Bathymetry (SDB) as input layers for the machine learning processes.

2.2.6 Machine Learning (ML) and Land Change Modeler (LCM)

We selected the Random Forest (RF) algorithm in GEE to perform a supervised classification image of Providencia reef geomorphic zones. As indicated above, we employed the 2.3.5 BTM data, plus the visible bands (Blue, Green, and Red) of the Sentinel 2 image. In addition, we directly photo-interpreted the corrected image in GEE to produce the polygons as training data. The resulting supervised classification was then imported to IDRISI Land Change Modeler (LCM) toolbox, to perform a multitemporal analysis with which to evaluate the spatial changes in the reef geomorphic zonation during the 2019–2020 and 2020–2021 study periods. The Land Change Modeler (LCM) was developed (Eastman 2006) as a change projection tool to support a wide range of planning activities. This modeler has been designed for REED projects. (Reducing Emissions from Deforestation and forest Degradation), but its applications can be observed in different earth change modeling investigations using CA-Markov (Areendran et al. 2013; Halmy et al. 2015; Eastman et al. 2018), modeler the Earth Trends (Fuller et al. 2012), Forestry changes (Holmer et al. 2001; Hill et al. 2003) and biodiversity and Habitats (Poirazidis et al. 2006; Bino et al. 2008). LCM examines each of the historical classification's pixel transitions, change pixels and number of persistence pixels. The resulting models show the units of change without subjective intervention and are an alternative to geo-statistical techniques.

3 Results

3.1 *Geomorphic Classification*

Using the optical properties of Sentinel 2, the SBD bathymetry, and the factors derived from the BTM, the Bathymetric Position Index (BPI) and the slope, we delineated six geomorphic units (GU) occupying different percentages of surface in shallow waters (<25 m): Lagoon (44.2%), Rock Terrace (40.15%), Back Reef (6.3%), Reef Crest (4.4%), Sand Terrace (3.3%), and Front Reef (1.5%).

The 3D Digital Terrain Model (Fig. 4) of the Satellite-Derived Bathymetry (SDB) shows a heterogeneous and discontinuous, 26 km long stretch of reef complex in the eastern and western part of the island with an extensive Reef Crest and pinnacles segments to the north (4–10 km) and to the windward east (10 km). An extensive lagoon area towards the north of the island showed the presence of irregularly distributed patch reefs and a broad shallow marine terrace covered by fine sediments. In the deepest areas, satellite information allowed us to identify the outer reef Rock Terrace (18 m) and the Sand Terrace up to 20 m deep, demonstrating the capabilities of the sensor reflectance response in the deepest zone of oligotrophic and ocean waters.

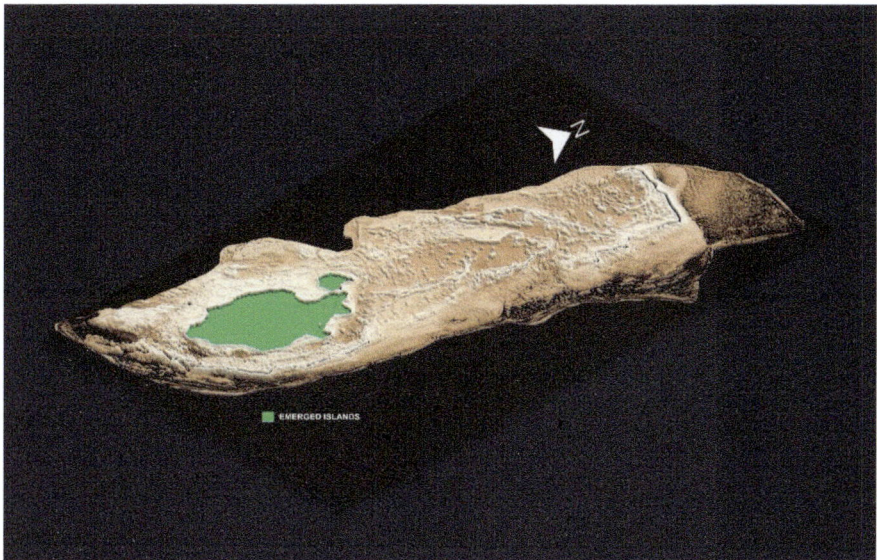

Fig. 4 Digital terrain model (DTM) showing the reef complexity and the reef extension at Providencia

The slope layer showed very steep slopes (between 6° and 7°) at the Rocky Terrace and the Sand Terrace corresponding to the Fore Reef and the outer slope at depths between 11 and 15 m. In shallow areas, the Lagoon, the Back Reef, and the Reef Crest were the most gently sloped areas with 0.1°–1.8° between 1 and 10 m deep (Table 3). The BPI and the slope allowed to distinguish the reef landscape structures (e.g., plains and barriers). In particular, the slope position accounted over two scales for: (1) negative and near to zero values in the Lagoon, the Rock Terrace, and Sand Terrace, and (2) positive or positive to negative values in the Reef Crest, the Back Reef, and the Front Reef (Table 3). These differences of the BPIs layers (broad and fine), the slope, and the SBD served as data inputs to run the Machine Learning process and to delineate the GU in the reef as follows:

Reef Crest: Covers 11.2 km^2 of the survey area; this intertidal zone represents the shallowest or emerged part of the reef due to the presence of live coral mounds or pinnacles that reach the surface with a steep slope. The SBD-derived model showed extensive breaker zones located to the north and east with interruptions up to the main reef ridge that is slightly more continuous and curved to the east up to 4 m in depth (Fig. 5).

Lagoon: This zone was the most extended class (112.6 km^2) with highly depositional environments in the flat plains between -8 and -10 m. These areas included patch reefs observed in the images with darker spots and are easy to differentiate from sandy bottoms with higher reflectance. Additionally, the lagoon shows large coral

Table 3 Upper and lower factors resulting from the BTM processing 2019–2021 in Providencia

Class	Zone	BroadBPI_Lower	BroadBPI_Upper	FineBPI_Lower	FineBPI_Upper	Slope_Upper	Depth_Lower	Depth_Upper
1	Reef Crest	235	413	372	426	1.2	−4	−1
2	Lagoon	−120	−31	−35	118	0.5	−10	−8
3	Back Reef	57	324	−35	118	1.4	−6	−3
4	Front Reef	225	146	−35	118	2.6	−8	−6
5	Rock Terrace	−35	−31	−120	−35	6	−14	−11
6	Sand Terrace	−120	−31	−35	118	7	−15	−13

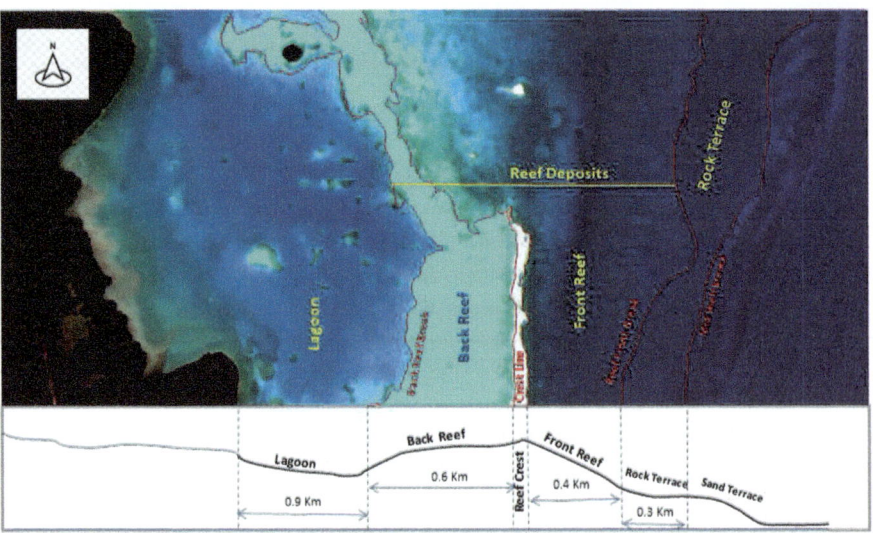

Fig. 5 Geomorphic units (GU) in the eastern reef of Providencia. Sentinel 2 corrected image date February 1, 2021

heads and pinnacles in shallow areas (Fig. 5). The lagoon is variable in extension from the east coast between 0.2 to 2 km to 10 km in the flat plain to the North (Fig. 6).

Back Reef: This unit is located behind the Reef Crest, covers around 16 km^2 and shows a gently sloping surface adjacent to a Reef Flat. The area is sheltered by sediments dominated by coral rubble and broken reef material over a variable bottom depth (6–3 m). The extent of the Back Reef to leeward is greater on the exposed eastern side of the island, averaging one kilometer in width (Fig. 5). The presence of debris detected in the image is different from the sand which shows more brightness

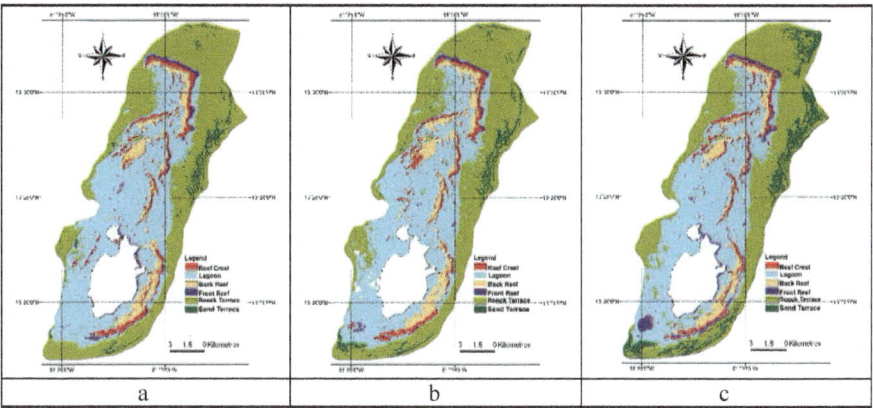

Fig. 6 Geomorphic units pre-Iota 2019 (**a**), 2020 (**b**), and post-Iota 2021 (**c**) at Providencia

reflectance. The debris area is located next to the Reef Crest, while the sand area is located towards the Lagoon. In both cases, the depositional material comes from the Reef Crest and the Front Reef.

Front Reef: This is a narrow and sloping area extending from the Reef Crest at the windward margin towards the Rock Terrace. This area covers 3.8 km^2. It is characterized by high wave exposure (Fig. 5).

Rock Terrace: This feature is the second most extensive GU in the area, covering 102.3 km^2, and represents a division between the Front Reef and the deeper Fore Reef in a sloping shallow area between 11 and 14 m. It shows a dark contrast in the satellite images probably due to the attenuation factor of the spectral response of the visible bands with increasing depth. Between the Rock Terrace and the Sand Terrace, the Mid-Shelf Break is a widespread slope break delimiting the edge of the inner Front Reef shelf and occurs in depths between 10 and 15 m.

Sand Terrace: The sandy surface contrasts with the Rock Terrace in some sectors of the survey area between 13 and 15 m and occupies an area of 8.42 km^2. The sandy terrace forms channels parallel to the reef crest about 2 km offshore (Fig. 5). The sandy formations are at the detection limit of the satellites, where the bathymetric models lose accuracy, and it is difficult to interpret the geomorphology of these structures.

3.2 Cartography Accuracy

The evaluation of the thematic accuracy in the models generated from the ML process resulted in high global accuracy and kappa index metrics in all years of the multi-temporal analysis as follows: The producer's accuracy ranged from 99.64 to 100%

Table 4 Confusion matrix, overall accuracy, and Kappa index for 2019, 2020, and 2021

Coverage classes	Reef Crest	Lagoon	Years 2019/2020/2021		Rock Terrace	Sand Terrace	User accuracy (%)
			Back Reef	Front Reef			
Reef Crest	688/493/835	0/0/7	3/0/0	9/0/0	0/0/0	0/0/0	98.3/100/99.2
Lagoon	3/0/3	10,136/6069/6569	0/0/0	0/0/0	0/0/0	0/0/0	99.9/100/99.9
Back Reef	0/1/0	0/0/0	2645/844/2595	0/0/0	0/0/0	0/0/0	100/99/100
Front Reef	1/0/0	15/0/0	0/0/0	799/155/502	0/0/0	0/0/0	98/100/100
Rock Terrace	0/0/0	0	0/0/0	0/0/0	1281/1454/1041	0/0/0	100/100/100
Sand Terrace	0/0/0	0	0/0/0	0/0/0	0/0/0	120/188/308	100/100/100
Producer accuracy (%)	99.4/99.8/99.6	99.8/100/99.9	99.9/100/100	98.9/100/100	100/100/100	100/100/100	
		Global accuracy	0.99/0.99/0.99	Kappa index	0.99/0.99/0.99		

accuracy, which evidences the suitability of the Random Forest algorithm to separate the pixels between the GU's using the BPI bands and the Sentinel 2 visible bands (Table 4). The user's accuracy showed high percentages between 98.04 and 100% representing an adequate selection of training samples associated with the established polygons in each of the GU assessed.

3.3 Geomorphic Cover Changes

3.3.1 Pre-Iota 2019–2020 Cover Changes

Based on the ML processing with high precision cartography, we detected conspicuous changes comparing the pre-Iota 2019–2020 and post-Iota 2020–2021 GU classifications (Fig. 6). The multi-temporal analysis assessed the biannual changes under typical climatic conditions and, without the presence of extreme weather events (pre-Iota). Results showed growths of the deposited material (18.55%) around the Back Reef covering a new area of 3.65 km^2. The Sand Terrace showed an increase of 22.44% covering a new area of 1.89 km^2. Finally, the area showing the largest increases was the Front Reef, which increased 88.5% covering an estimated area of

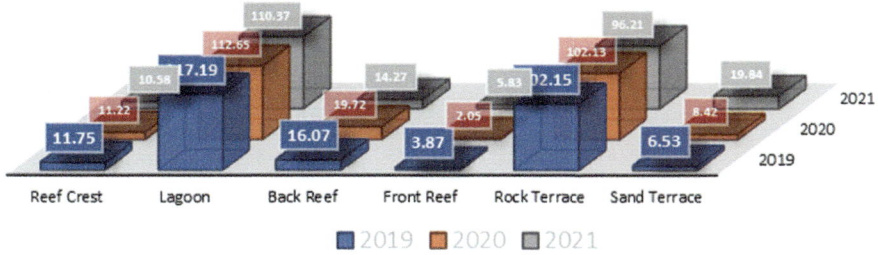

Fig. 7 Geomorphic changes in km^2, pre-Iota 2019–2020, and post-Iota 2020–2021, at Providencia

1.81 km^2. Finally, the GU that showed slight tendencies to lose material (0.01–4.74%) were the Reef Crest, the Lagoon, and the Back Reef (Fig. 7).

The multi-temporal analysis assessed the biannual changes under typical climatic conditions and, without the presence of extreme weather events (pre-Iota), showed deposited material (18.55%) around the Back Reef covering a new area of 3.65 km^2. The Sand Terrace showed a 22.44% covering a new area of 1.89 km^2. Finally, the area showing growth was the Front Reef, with 88.5% covering an estimated area of 1.81 km^2. Finally, the GU that showed slight tendencies to lose material (0.01–4.74%) were the Reef Crest, the Lagoon, and the Back Reef (Fig. 7).

3.3.2 Post-Iota 2020–2021 Cover Changes

After the passage of Hurricane Iota, between November 2020 and the first months of 2021 (Fig. 7), an opposite behavior was observed with a loss of coverage in the Back Reef of 5.4 km^2 (38.14%) and a decrease in the Front Reef with 3.77 km^2 (64.75%). Furthermore, there was a noteworthy increase in the area covered by the Sandy Terrace, 11.41 km^2 (57.56%).

A detailed analysis of the spatial distribution of changes calculated with the LCM for persistence, loss, and gain of the area in the pre- and post-Iota periods, showed different responses: The Back Reef reduced its coverage in an extensive area towards the north and the east part of the reef. The material moved to areas such as the Lagoon and Reef Crest, with a gain contribution between 5 and 7 m in depth (Table 5 and Fig. 7). In comparison, a decreasing trend was observed in the post-Iota period in the Rock Terrace at the northeast of the barrier reef. In this case, the material shows a contribution to the Sandy Terrace at depths between 7 and 10 m (Fig. 8). Finally, the Lagoon Unit lost coverture in all periods (pre- and post-Iota), but this was more localized in the deeper area of the Rock Terrace (Fig. 8).

Table 5 Losses and gains (km^2) pre-Iota 2019–2020 and post-Iota 2020–2021 in the geomorphic units (GU) at Providencia

Cover class	2019–2020		2020–2021	
	Loss/gain (km^2)	Net	Loss/gain (km^2)	Net
Reef Crest	↓ −3.55/3.02	−0.53	↑ −3.75/3.11	−0.64
Lagoon	↓ −10.40/5.85	−4.55	↓ −11.17/8.90	−2.27
Back Reef	↑ −1.04/4.69	3.65	↓ −6.71/1.27	−5.44
Front Reef	↓ −2.37/0.57	−1.8	↑ −0.47/4.24	3.77
Rock terrace	↑ −5.73/5.74	0.01	↓ −14.43/8.51	−5.92
Sand Terrace	↑ −1.37/3.25	1.88	↑ −1.34/12.75	11.41

4 Discussion

Our geomorphological characterization of recent changes in Providencia offers a synoptic view of the magnitude and power of a category 5 hurricane event and its impact on the tropical shallow coral reef.

4.1 Changes in Shallow Units (Reef Crest, Lagoon, and Back Reef)

The GU with the greatest geomorphological stability was the Reef Crest, where insignificant structure variations were observed (0.64 km^2). The structural stability of the Back Reef could indicate that this unit performed well as a NBS to the beating by the hurricane, and acted as a good barrier against wave power. The role of the Reef Crest as a buffer for storm energy has been extensively documented in various studies describing how much (up to 86%) of the incoming wave energy is dissipated by this feature (86%) (Beck et al. 2018 and references therein). Beck et al. (2018) performed flooding model scenarios of storm events with and without a Reef Crest. The reef scenarios attain only a decrease of 1 m in the height and roughness of Reef Crest. The authors concluded that without reefs, annual damages on the coast in the United States would be more than double (118%) and land flooding would increase by 69%, affecting 81% of the population. Furthermore, the use of natural barriers for sea-level rise and storm surges is becoming increasingly popular as a source of nature-based risk reduction options. Accurate seafloor maps are essential for determining the wave degradation of benthic habitats such as coral reefs, with the necessity of current and repeatable observations of sediment stability and geomorphic complexity (Fourqueran et al. 2020) as the ones presented here.

Regarding the Lagoon, this unit showed overall stability in the extension of the central body, both pre- and post-Iota. However, erosion zones were observed at depths beyond 7 m at the edge of the Lagoon area following the passage of the

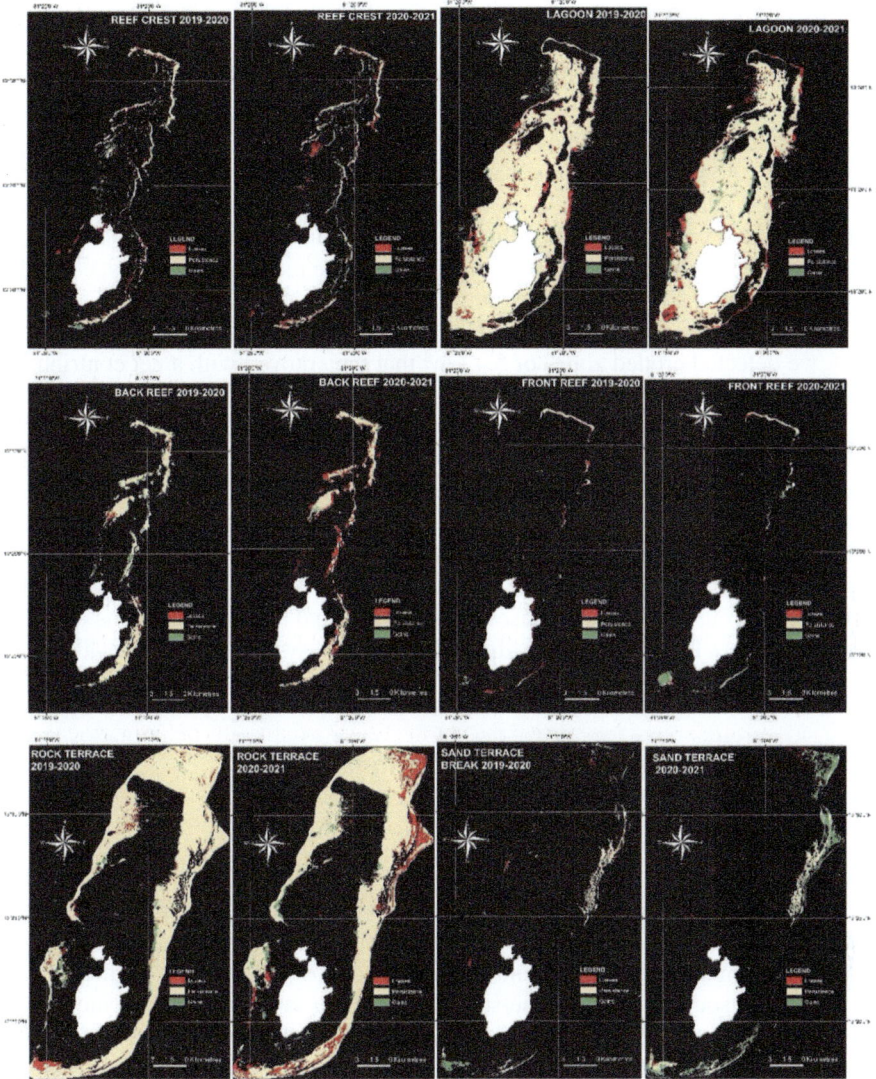

Fig. 8 Geomorphic units' losses, persistence, and gains pre-Iota 2019–2020 and post-Iota 2020–2021 at Providencia

hurricane. This effect was observed all around the island with some focus on greater change to the northeast, southeast, and southwest. In-situ evaluations conducted by INVEMAR (2021) showed a generalized disturbance of 65% in the Lagoon and the Back Reef. As part of the disturbance, large colonies of *Orbicella annularis* and *Orbicella faveolata* were observed overturned in settings shallower than −6 m. Highly impacted sites such as Marcela's place (a monitoring point within the MPA) with overturned colonies, located in transitional zones between the Back Reef and the Lagoon, could be affected by the dynamics of sediments transported from the Reef Crest and the Back Reef to the Lagoon. Similar results were also found by Bries et al. (2004) who reported that larger colonies of a species were more prone to damage than smaller colonies of the same species. They also claimed that larger, more aged colonies are not necessarily more strongly attached to the substratum, because their attachment bases become more likely weakened through bioerosion, rendering them vulnerable to toppling under sufficient wave energy. In some cases, the bioerosion of the colonies can be observed on the reef forming strips, channels, or lobes such as the ones observed on the satellite images in the same GU.

Finally, after the hurricane hit, we also observed the loss of coverture in the Back Reef, we detected eroded zones of 5.4 km^2 that then moved with a net contribution to the Lagoon. These changes contribute to formed lobes or channels detected on the leeward side of the Reef Crest mostly on the northern reef. For instance, in the Lagoon zone, we observed accretion lobes between 50 and 150 m in length and 3 and 7 m in depth. However, under pre-Iota conditions, an increase (3.65 km^2) in the amount of sedimentary material was also observed in the Back Reef, where unconsolidated granular material was transported from the Lagoon. There is evidence that accretion processes on the reef are heterogeneous in space and time and during different intervals (Medina-Valmaseda et al. 2020), and that some sections of a reef, such as the Back Reef, represent a highly depositional system (Kennedy et al. 2021).

4.2 Changes in the Deepest Units (Sand Terrace and the Rock Terrace)

After the impact of Hurricane Iota, a high-sedimentary dynamic was observed in the deepest GU (the Rock Terrace and the Sand Terrace) up to -22 m. Turbulence generated by Iota probably caused that part of the Rock Terrace was covered with sediments from the Sand Terrace and upper slope deposits (11.41 km^2) in the leeward northeast and southeast side of the study area. These sandbanks represented storm ridges and channels up to 7 km long along the Rock Terrace edge (Fig. 7c). The bathymetric model shows the formation of sand channels, 2–3 m deeper than the Rocky Terrace. Compared to windward settings, the northwest leeward side showed very little change in the Rock Terrace. Beyond -22 m, we could not identify changes since the area falls outside the boundary of the detection limit for bathymetry. Similar results were reported by Scoffin (1993) and by Bries et al. (2004), who found that after hurricane impacts, the Reef Crest and Front Reef coral debris accumulate as talus at the foot of the Front Reef slope and on submarine terraces and grooves. These studies also found that carbonate sand and mud move in deep off-reef locations in the Fore Reef. Blanchon and Jones (1995) mentioned the presence of a gently sloping Rock Terrace in the reef complex around Grand Cayman, which is either covered with coral spurs or is a bare rock ground, that has been sculptured by wave scour into low ridges and shallow furrows during the passage of hurricanes. Blanchon (2011) also stated that the Sand Terrace is frequently crossed by a system of widely spaced coral spurs between which thick deposits of skeletal sand and gravel accumulate. He also suggests that this material is only mobilized during storms, and this could explain why at Providencia just a few changes were observed in the Rock Terrace and the Sand Terrace in the pre-Iota period.

5 Conclusions

Hurricanes are short-lived but highly dynamic synoptic events that could intensively impact the sediment dynamics of coral reef ecosystems in the Caribbean Sea. Here, we adapted the factors provided by the BTM such as Bathymetric Position Index (BPI) and slope, incorporating the components of visible bands of the satellite image Sentinel 2, machine learning, and cloud computing in GEE to investigate the rapid response of the reef geomorphic changes on Providencia with the passage of Hurricane Iota in November 2020. To understand these changes, we also incorporated information about the pre-Iota period 2019–2020. Our results indicate that Hurricane Iota enhanced the delivery of the reef re-suspended sediment substantially to the northeast and seaward sediment transport. As a result, high sedimentary dynamics was observed in the deepest GU (the Rock Terrace and the Sand Terrace) up to 22 m. In contrast, the unit with the greatest geomorphological stability was the Reef Crest, where no significant structure variations were observed (0.64 km^2), showing that this GU performed well against the beating of the hurricane, and acted as a good barrier against wave power, illustrating the importance of this GU as a protection service, and as an approach of NBS. The Lagoon zone also showed general stability in the extension of the central body, both pre- and post-Iota. However, erosion zones were observed beyond 7 m of depth at the edge of the Lagoon area after the passage of the hurricane. Thus, GUs changes due to the hurricane produced several distinct patterns based on reef site, reef depth, and colony size (such as those observed in Marcela's place).

The temporal resolution of sensors such as Sentinel 2, the use of the Land Change Modeler and their easy accessibility, make the integration of these approaches a very interesting alternative for monitoring reef geomorphic units covered in extreme events such as hurricanes and where impact mitigation measures can be provided using quality mapping in a rapid manner. Its integration enables long-term monitoring by observing the evolution of changes through time. Thus, providing valuable information to coastal managers and stakeholders in the decision-making process.

Acknowledgements This research was primarily funded by the Royal Academy of Engineering (IAPP 18-19\210 made through the Industry Academia Partnership Programme scheme) and by the Corporation Center of Excellence in Marine Sciences, CEMarin. We are thankful to Aqua & Terra S.A.S for providing the in-situ high-resolution bathymetry to calibrate our Satellite-Derived Bathymetry (SDB).

References

Areendran G, Raj K, Mazumdar S et al (2013) Modelling REDD+ baselines using mapping technologies: a pilot study from Balpakram-Baghmara landscape (BBL) in Meghalaya, India. Int J Geoinf 9(1):61–67

Beck MW, Losada IJ, Menéndez P et al (2018) The global flood protection savings provided by coral reefs. Nat Commun 9:2186. https://doi.org/10.1038/s41467-018-04568-z

Bino GN, Levin S, Darawshi N et al (2008) Accurate prediction of bird species richness patterns in an urban environment using Landsat-derived NDVI and spectral unmixing. Int J Rem Sens 29(13):3675–4370. https://doi.org/10.1080/01431160701772534

Blanchon P, Jones B (1995) Marine-planation terraces on the shelf around Grand Cayman: a result of stepped Holocene sea-level rise. J Coast Res 11:1–33

Blanchon P (2011) Geomorphic zonation. In: Hopley D (ed) Encyclopedia of modern coral reefs: structure, form and process. Springer-Verlag Earth Science Series, pp 469–486

Bries JM, Debrot AO, Meyer DL (2004) Damage to the leeward reefs of Curaçao and Bonaire, Netherlands Antilles from a rare storm event: Hurricane Lenny, November 1999. Coral Reefs 23:297–307. https://doi.org/10.1007/s00338-004-0379-9

Díaz JM, Barrios LM, Cendales MH et al (2000) Áreas coralinas de Colombia. INVEMAR. Serie Publicaciones Especiales, No 5, Santa Marta

ESA (2019) Coral reef monitoring with Sen2Coral | Space4Water Portal. https://www.space4water.org/capacity-building-and-training-material/coral-reef-monitoring-sen2coral

Eastman JR (2006) IDRISI Andes guide to GIS and image processing. Worcester, Clark Labs

Eastman JR, Toledano J (2018) CAMarkov. In: Olmedo MTC, Paegelow M, Mas JF et al (eds) Geomatic approaches for modeling land change scenarios. Springer, pp 481–484

Fuller DO, Parenti MS, Gad AM et al (2012) Land cover in upper Egypt assessed using regional and global land-cover products derived from MODIS imagery. Remote Sens Lett 3(2):171–180. https://doi.org/10.1080/01431161.2011.551847

Halmy MWA, Gessler PE, Hicke JA et al (2015) Land use/land cover change detection and prediction in the north-western coastal desert of Egypt using Markov-CA. Appl Geogr 63:101–112. https://doi.org/10.1016/j.apgeog.2015.06.015

Geister J (1992) Modern reef development and Cenozoic evolution of an oceanic island/reef complex: Isla de Providencia (Western Caribbean Sea, Colombia). Facies 27:1–69. https://doi.org/10.1007/BF02536804

Geister J, Diaz JM (2007) Reef environments and geology of an oceanic archipelago: San Andrés, Providence and Santa Catalina (Caribbean Sea, Colombia). INGEOMINAS

Goes ER, Brown CJ, Araújo TC (2019) Geomorphological classification of the benthic structures on a tropical continental shelf. Front Mar Sci 6:47. https://doi.org/10.3389/fmars.2019.00047

Gorelick N, Hancher M, Dixon M et al (2017) Google Earth Engine: planetary-scale geospatial analysis for everyone. Remote Sens Environ 202:18–27. https://doi.org/10.1016/j.rse.2017.06.031

Hedley JD, Harborne AR, Mumby PJ (2005) Simple and robust removal of sun glint for mapping shallow-water benthos. Int J Remote Sens 26:2107–2112. https://doi.org/10.1080/01431160500034086

Hill JL, Curran PJ (2003) Area, shape and isolation of tropical forest fragments: effects on tree species diversity and implications for conservation. J Biogeogr 30(9):1391–1403

Holmer B, Postgård U, Eriksson M (2001) Sky view factors in forest canopies calculated with IDRISI. Theor Appl Climatol 68(1–2):33–40

Horn BKP (1981) Hill shading and the reflectance map. Proc IEEE 69:14–47. https://doi.org/10.1109/PROC.1981.11918

INVEMAR (2021). Evaluación del estado de los arrecifes coralinos en el archipiélago de San Andrés, Providencia y Santa Catalina posterior al paso del huracán IOTA en el marco de la operación Cangrejo Negro Fase II, Santa Marta. Unpublished manuscript

Kennedy EV, Roelfsema CM, Lyons MB et al (2021) Reef cover, a coral reef classification for global habitat mapping from remote sensing. Sci Data 8(196):1–20. https://doi.org/10.1038/s41 597-021-00958-z

Kumar P, Debele SE, Sahani J et al (2021) An overview of monitoring methods for assessing the performance of nature-based solutions against natural hazards. Earth Sci Rev 217:103603. https://doi.org/10.1016/j.earscirev.2021.103603

Lyzenga DR (1978) Passive remote sensing techniques for mapping water depth and bottom features. Appl Opt 17(3):379–383

Lecours V, Lucieer V, Dolan M et al (2015) An ocean of possibilities: applications and challenges of marine geomorphometry. In: Jasiewicz J, Zwoliński Z, Mitasova H et al (eds) Geomorphometry for geosciences. Adam Mickiewicz University in Poznań - Institute of Geoecology and Geoinformation, Bogucki Wydawnictwo Naukowe, pp 23–36

Medina-Valmaseda AE, Rodríguez-Martínez RE, Alvarez-Filip L et al (2020) The role of geomorphic zonation in long-term changes in coral-community structure on a Caribbean fringing reef. PeerJ 8:e10103. https://doi.org/10.7717/peerj.10103

Poirazidis K, Papageorgiou A, Kasimiadis D (2006) Mapping the animal biodiversity in the Dadia National Park using multi-criteria evaluation tools and GIS. In: Conference proceedings: international conference on sustainable management and development of mountainous and island areas, vol 2, pp 299–304

Post AL (2008) The application of physical surrogates to predict the distribution of marine benthic organisms. Ocean Coast Manag 51:161–179. https://doi.org/10.1016/j.ocecoaman.2007.04.008

Sanchez JA, Zea S, Diaz JM (1998) Patterns of octocoral and black coral distribution in the oceanic barrier reef-complex of Providencia Island, Southwestern Caribbean. Carib J Sci 34(3–4):250–264

Scoffin TP (1993) The geological effects of hurricanes on coral reefs and the interpretation of storm deposits. Coral Reefs 12:203–221. https://doi.org/10.1007/BF00334480

Stumpf RP, Holderied K, Sinclair M (2003) Determination of water depth with high-resolution satellite imagery over variable bottom types. Limnol Oceanogr 48(1–2):547–556. https://doi.org/10.4319/lo.2003.48.1_part_2.0547

UNISDR (2009) United nations office for disaster risk reduction (UNISDR) terminology on disaster risk reduction. https://www.undrr.org/publication/2009-unisdr-terminology-disaster-risk-reduction

Verfaillie E, Degraer S, Schelfaut K et al (2009) A protocol for classifying ecologically relevant marine zones, a statistical approach. Estuar Coast Shelf Sci 83(2):175–185. https://doi.org/10.1016/j.ecss.2009.03.003

Open Access This chapter is licensed under the terms of the Creative Commons Attribution 4.0 International License (http://creativecommons.org/licenses/by/4.0/), which permits use, sharing, adaptation, distribution and reproduction in any medium or format, as long as you give appropriate credit to the original author(s) and the source, provide a link to the Creative Commons license and indicate if changes were made.

The images or other third party material in this chapter are included in the chapter's Creative Commons license, unless indicated otherwise in a credit line to the material. If material is not included in the chapter's Creative Commons license and your intended use is not permitted by statutory regulation or exceeds the permitted use, you will need to obtain permission directly from the copyright holder.

A Light Pollution Assessment in the Fringing Reefs of San Andrés Island: Towards Reducing Stressful Conditions at Impacted Coral Reefs

Andres Chilma-Arias, Sebastian Giraldo-Vaca, and Juan A. Sánchez

Abstract The degradation of the night sky's quality due to artificial light sources negatively affects marine environments, because many organisms use natural light as cues for reproductive and dispersal behaviors, find favorable habitats, and for the biochemistry of their symbiotic microorganisms. Despite the tremendous effect on marine life, measuring the effects of artificial light pollution is difficult because our understanding of natural light brightness coming from celestial bodies like the Moon is minimal. Here, we fill this gap by quantifying the sky's brightness and Artificial Light Pollution at Night (ALAN). This study assessed light pollution along the reefs around San Andrés Island, which Hurricane Iota significantly impacted. We modified and installed Sky Quality Meters (LU-DL) at both leeward and fringing reefs, down to 11 m depth. The results indicate the highest ALAN values in the area of Johnny Cay (18 msas) compared to Acuario (20 msas) and West View (21 msas). Additionally, National Oceanic and Atmospheric Administration NOAA and Unihedron databases show an increase in artificial light on land, where constant artificial light and coastal vegetation loss due to Hurricane Iota (between 15 and 19th November 2020), are the main factors that may be generating this increase in artificial light.

Keywords ALAN · Chronobiology · Light pollution · Coral reefs · Sky quality meter

1 Introduction

When looking at the night sky, we quickly notice the presence of stars; we begin to count them one by one and notice that more of them appear between each count. In this process, we are also able to identify perhaps some planets, the star clusters that gave rise to the Pleiades, and constellations that inspired the great poets of ancient Greece and mythical Arab navigators, and filled with knowledge the astronomers and physicists of modern times, and even, astrologers! The interpretation of zodiacal lights has captivated the curiosity of many people, permeating many areas of daily life. For example, the moon phases are linked to the agricultural sowing and harvesting seasons. However, it has become increasingly difficult to observe and interpret the dynamics of the night sky due to light pollution, the most influential factor affecting astronomical observation (Hamidi et al. 2011).

This chapter focuses on a scientific perspective about natural and artificial light and how light pollution affects organisms from coral reefs around San Andrés Island. Based on astronomical information provided by NOAA and Unihedron Sky Quality Meter, and data collection of the sky's brightness with in situ sensors, we were able to determine that the quality of natural light has been declining with the recent increase of the artificial light sources, due to constant population growth around the coast and Hurricane Iota's impact decimating coastal vegetation, populations of marine species located in the fringe and leeward reefs will be more vulnerable, mainly when synchronizing their reproductive cycle with the vertical light spectrum.

2 Background

The natural light observed during night hours comes from multiple sources such as the Milky Way, stars, and the Moon, which is fundamental in the diurnal, nocturnal, and seasonal cycles, and that plays a particular role in the behavior patterns of marine and terrestrial animals (Gaston et al. 2017; Luarte et al. 2016). However, in the last century, significant anthropogenic light sources gained relevance, disrupting the quality of natural light. This phenomenon is associated with the increase in the world population, specifically in the coastal areas where it is occurring faster and where demand for energy has increased with respect to other zones (Gaston et al. 2013, 2015). Consequently, the periods in which ecosystems are exposed to artificial light are longer and with more incidence, this is known as ALAN (Artificial Light Pollution at Night), and it has widespread negative effects on diurnal and nocturnal organisms (Sanders et al. 2021), including birds (McLaren et al. 2018), Fish (Pulgar et al. 2019), marine turtles (Dimitriadis et al. 2018) and insects communities (Grubisic and Van Grusven 2021). Many coastal marine ecosystems are currently exposed to artificial light at night, affecting the biological clocks of both marine and terrestrial organisms (Tamir et al. 2017). Many reef dweller species naturally synchronize their reproductive events with signals from the night sky (vertical light), mainly linked to changes in the moon's phases.

ALAN can also intervene in the survival of coral symbionts (specifically green algae) in species such as Pocillopora damicornis and Acropora euristoma, that after 120 days of exposure to light, the density of their microbial communities exhibits significant differences (Ayalon et al. 2019; Levy et al. 2020). Studies indicate that lunar irradiance affects both the maturation of gametes and their release, and it is a precursor of speciation in populations distributed along a bathymetric/light gradient. However, ALAN also affects reproductive behavior in corals due to its dependency on lunar cycles. In the long term, with the intervention of natural light patterns, sexual selection, reproductive isolation, gene flow, and genetic drift can be altered (Hopkins et al. 2018), due to the close relationship between the quality and quantity of natural light with the responsible genes of the reproduction, as is the expression of the cryptochromes (CRY genes) (Poehn et al. 2021). Its expression mechanism is based on the perception of a specific wavelength known as blue pulse (Levy et al. 2007; Sweeney et al. 2011), this environmental signal is the most predominant light spectrum within the photic zone of the sea. For the Cryptochromes (Cry genes) to recognize the spectral signals and, thus, for the organism to be able to synchronize for reproduction, this wavelength or blue pulse must be specific and of "good" quality (Kronfeld-Schor et al. 2013).

Artificial light is a new, silent yet highly visible enemy, altering these natural cycles, in turn affecting the resilience capacity of these marine ecosystems. The increase of this anthropogenic pressure is poorly studied in marine ecosystems, even if they mask and deteriorate the spectral quality of the light penetrated in the column water (Davies et al. 2013). Precise moonlight intensities comprise the proximate reproductive cue for many marine sessile invertebrates (Coelho and Lasker 2014). Particularly in coral reefs, part of the reproductive success of benthic organisms is linked to the planulation and synchronized reproduction generated by circadian stimuli, including particular moon phases (Sorek and Levy 2014). Laboratory studies suggest that night light pollution has negative implications, preventing synchronization in their reproductive timing, including tidal (12.4 h), lunarian (24.8 h), semilunar (14.77 days), or lunar (29.53 days) patterns in marine organisms (Naylor 1999; Kaiser and Heckel 2012).

Despite growing evidence on the impacts of light pollution on different species, the areas exposed to ALAN are expected to increase in intensity and spatial extent in the coming years (Davies and Smyth 2018). Based on the current data, the International Hydrographic Organization has defined areas especially vulnerable to contamination, considering places where at least 10% of the reef area is exposed to light levels more than double the brightness of the natural night sky. These locations include the Gulf of California (21.6%), the Persian Gulf (20.4%), the Gulf of Thailand (24.2%), the Gulf of Aqaba/Eilat (18.1%), the Gulf of Oman (17.6%), the South Atlantic Ocean (14.4%), the Malacca Strait (10.0%), and the Singapore Strait (34.5%). However, it is still necessary to expand the coverage of these data since there is almost no information associated with this phenomenon in Colombia. Hence, starting the exploration at the local level shows a first brushstroke of the state of light pollution in marine areas of reef importance in Colombian territory.

Therefore, to assess the impact of ALAN on marine ecosystems, we focused on quantifying, in situ, the artificial light that penetrates the column water in the leeward and fringing reefs from San Andrés Island, Colombia, because the Seaflower Biosphere Reserve is of paramount importance in worldwide, for its role in the biodiversity of marine species and its potential as a marine corridor for Caribbean species and that recently it was impacted for Iota Huracan. The finding of this work suggests that there is a significant incidence of artificial light in (1) the high scattering from the island to the reefs and (2) the depth at which it was assessed, and (3) annually there is evidence of a constant increase in horizontal light. This furthers our knowledge of the natural light alterations in both terrestrial and marine ecosystems. Additionally, as a result of the growing interest in tourism in the Archipelago of San Andrés, Providencia, and Santa Catalina the demand for multiple urban services on the island has increased and, consequently, artificial light sources that support the nightly business routine. As an initiative, the information obtained becomes a fundamental resource when proposing new management and conservation plans that can be effectively maintained in the face of the unstoppable arrival of artificial light in the future of the human population.

3 Methods

To measure artificial and natural skylight levels, we took nocturnal data for twelve days, starting at 19:00 h until 5:00 h, every 5 min, from 9 to 20th December 2020. The sensors were installed with scuba diving equipment in 3 locations around San Andrés Island (Fig. 1), where tourism is the main diurnal and nocturnal activity. One sensor was located on the fringe reef near Johnny Cay (12°35′47″N–81°41′52″E) (Green), where the incidence of artificial light comes from the island's downtown commercial center. The second sensor was installed near the area known as Acuario, or the Aquarium (12°32′47″N–81°41′23″E) (Red), where the incidence of horizontal light comes from the San Luis and Rocky Cay sectors. The last sensor was installed in the West View zone (12°30′49″N–81°43′54″W) (Blue), classified as a place with less artificial light with respect to the other sampled sites.

Measurements of moonlight intensity were conducted using a Sky Quality Meter (SQM-LU-DL, Unihedron) (Fig. 2), which is an electromagnetic device sensitive to the light coming from a night sky (denominated vertical light). The SQM-LU-DL can discriminate against light pollution (denominated horizontal light) because its autonomous data-logging allows it to register and analyze the intensity of the night sky for several days in real time.

According to the supplier's specifications, the SQM-LU-DL works with an HWHM of angular sensitivity at around 10° and uncertainty of ±0.10 (msas) magnitudes per square arcseconds (mag/arcsec^2), indicating that it is a logarithmic measurement. Therefore, significant changes in the brightness of the sky will be interpreted

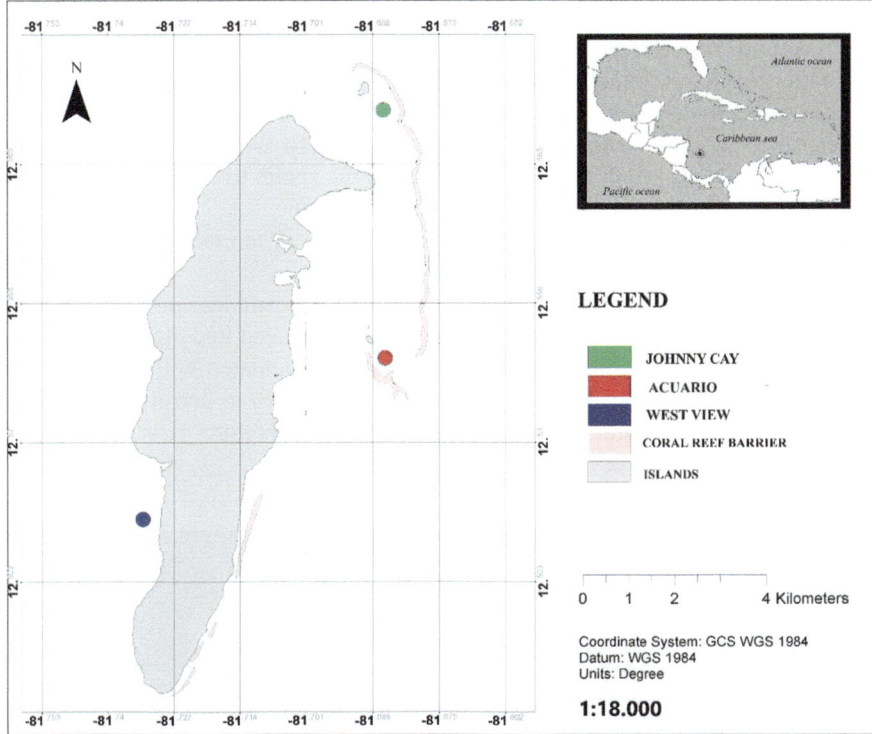

Fig. 1 Sites selected for the installation of the sensors. Johnny Cay (green) and Acuario (red) and West View (blue). Image was taken and modified by Guerra-Vargas et al. (2020)

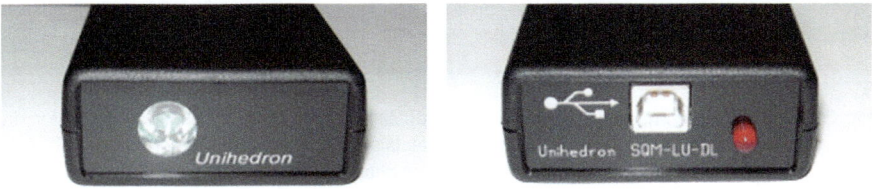

Fig. 2 Sky quality meters with lenses (SQM-LU-DL). Sensors to quantify the lunar irradiance and light pollution

as small numerical values, i.e. a difference of 1 is defined to be a factor of (100) (1/5) in received photons (Unihedron). In terms of the brightness quality, a higher magnitude is indicated by a darker color (24 msas; Fig. 3), and the magnitude decreases as the color becomes brighter. Therefore, it is also possible to measure light levels in the transition between day and night (twilight), but data registered by SQM-LU-DL will be saturated with daylight. Data obtained were analyzed in OriginPro v.9.8 (OriginLab Corporation, USA). In R studio, a Shapiro test was carried out to determine the normality of the data and a Kruskal–Wallis test to see if there were significant differences between each sampled area.

The sensors are designed to be used on land. Therefore, we generated a polycarbonate box for underwater measurements using O-rings to avoid water. Each sensor was installed at 11 m depth and fixed with two leads to avoid water motion destabilizing the system, which was oriented towards the zenith to carry out the vertical measurement (Fig. 4). To contrast our in-situ measurements at 11 m depth, we also obtained maps and graphs (Fig. 5) of light pollution at the terrestrial level from NOAA and Unihedron databases.

On the other hand, with the natural disaster of Hurricane Iota in 2020, much of the coastal vegetation was affected with the most significant loss along the island's west coast. The West View data (in this study) was compared with the light data previously collected by Chilma-Arias et al. (2021) in 2019 to corroborate whether the vegetation is a natural barrier to artificial light. Therefore, it is necessary to consider the following:

Fig. 3 Artificial light and natural light. A value of 24 msas indicates an excellent dark sky site, where the Milky Way, zodiacal light, stars, and planets can be visible. Values less than 15–16 msas are considered an inner-city sky, where the Moon, the planets, and a few of the brightest star clusters can be found. Image elaborated in Adobe Illustrator 2022 v 26.0.3

Fig. 4 Installation of the lunar light sensor to 11 m depth in **a** Johnny Cay, **b** Acuario, and **c** West view. Photos: Andrés Felipe Chilma

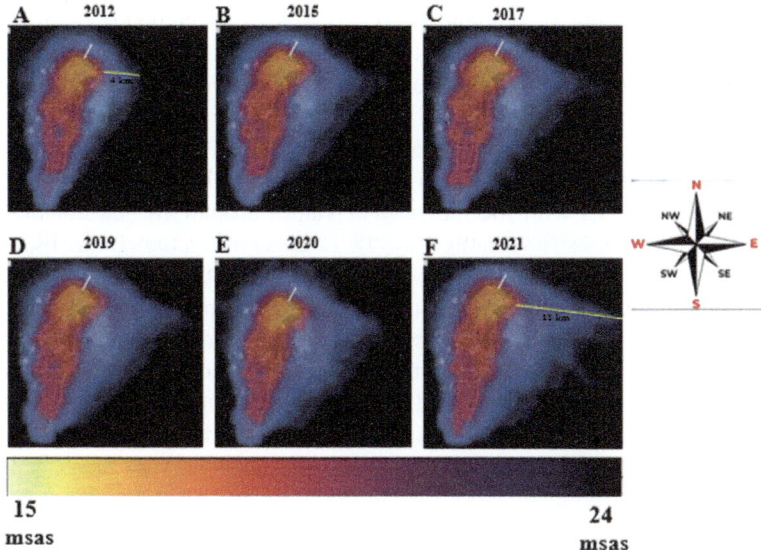

Fig. 5 Light pollution spread between 2012 and 2021. From www.lightpollutionmap.info

- The sensor was installed at a depth of 14 m in 2019 in the locality of West View.
- The data compared between the two studies were taken on specific days (December 9th to 20th).
- Between December 9th and 20th 2019, data were collected in the presence of full and waning lunar phases. Between December 9th and 20th 2020, the information was collected in the new moon and crescent moon phases. This observation is important due to the differential sky brightness.

4 Results

Records collected from the Unihedron and NOAA websites (Fig. 5) indicate that, visually, the amount of artificial light has increased on San Andrés Island in the last nine years (2012–2021). The heat map (Fig. 5) shows that the horizontal light coverage in the northeast area of the island for the year 2012 reached a distance of 4 km (yellow line). In this same place and by the year 2021 (Fig. 5), the artificial light has spread to a distance of 11 km. It is important to mention that the island of Johnny Cay is approximately 2 km from our sensor at the Spratt Bight beach and the commercial center of the island, and the barrier reef in the fringing reef is 3 km away from this area of the island. This same pattern of artificial light was evidenced in the area of Acuario, indicating that, by 2021, the scattering of anthropogenic light sources altered the quality of natural light received by both the fringing reef and that entering 11 km offshore. The horizontal light dispersion was lower than at the previous sites regarding West View. However, visually, there was a slight increase in it between the years 2012 and 2021. It was not possible to find numerical data to carry out the statistical analyses and thus observe if there were significant differences in the measurements made.

Plots of the data collected by the SQM sensors (Fig. 6) during the 12 days of monitoring show that, in West View (blue dots), the sensor recorded a maximum magnitude of 21 msas with the majority of values around 18 and 20 msas. The average was 14.97 msas for Johnny Cay, 18.22 msas for Acuario, and 18.70 msas for West View. In contrast, the data obtained from the Acuario sector (red dots), revealed that the highest concentration of readings was between 16 and 20 msas, which indicates that, visually, there are slight differences between West View and Acuario. The sensor installed in Johnny Cay (green points), the highest incidence of horizontal light coming from the island occurs in this site compared to the other sampling sites because the concentration of the data was between 18 and 14 msas. These results indicate that the leeward reef located in the West View area is where the least damage from artificial light occurs. The opposite occurs in Johnny Cay, since the presence of LED lights from commerce and tourism negatively impacts the quality of natural vertical light. The above is supported by a significant Kruskal–Wallis test ($X2 = 642.56$, $df = 2$ and a p-value $< 2.2e-16$).

When comparing the data recorded between December 2019 and December 2020 (Fig. 7), at a depth of 14 m, there was a maximum magnitude of 22 msas compared to the magnitude recorded at 11 m, which was 21 msas. A Kruskal–Wallis test shows significant differences ($X2 = 66.47$, $df = 1$ and p-value $= 3.54e-16$), which could indicate that during the year and indeed with the loss of vegetation, the quality of

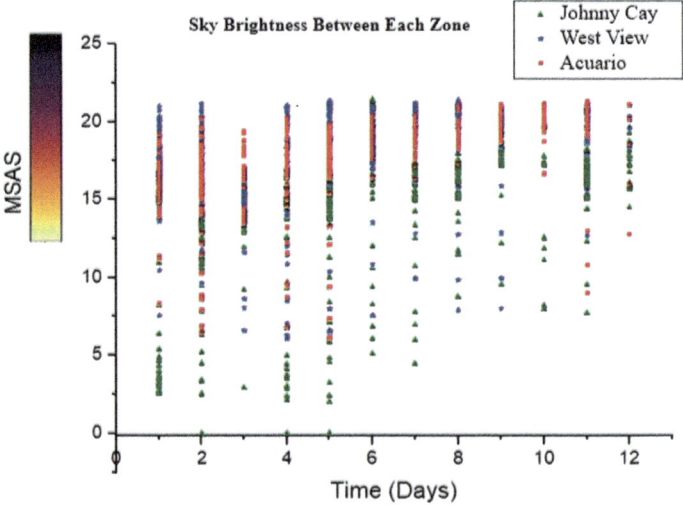

Fig. 6 Sky brightness. A value of zero is the presence of a lot of light, while a value of 25 represents a dark sky. The data collected in Johnny Cay are graphed in green, Acuario in red and west view in blue

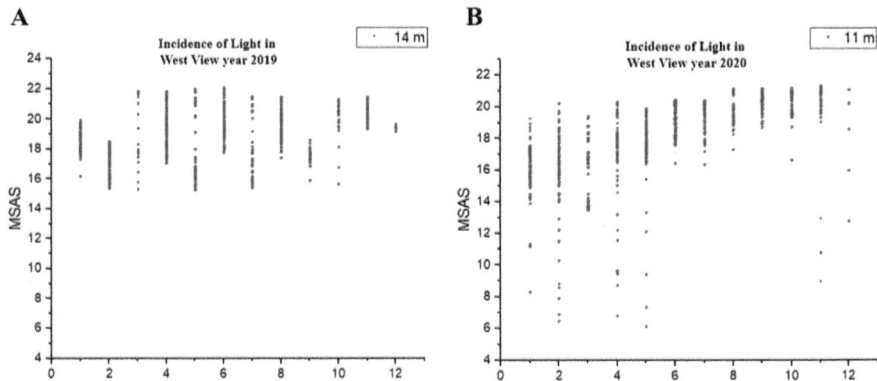

Fig. 7 Lighting comparison in December 2019 and December 2020. **a** The graph shows data taken in December 2019 to 14 m depth by Chilma et al. (in preparation). **b** Data collected in December 2020 to 11 m depth

vertical light on the island decreased by a factor of 1 (from 22 to 21 msas), noting that magnitudes per square arcsecond are a logarithmic measurement. Therefore, the reduction of 1 unit between 2019 and 2020 corresponds to a 20-fold increase in light intensity (see http://www.stjarnhimlen.se/comp/radfaq.html). It is necessary to highlight that the difference between magnitudes is possibly higher, because in December 2019, the data was recorded in the presence of a full moon, while in 2020, the data were recorded in a new moon, indicating that in 2020 the light quality can be less than 22 msas.

5 Discussion of Results

Since the origin of life on the planet, living beings have responded to natural changes and extreme conditions in the history of the earth, which have directed their particular evolutionary trajectories and have allowed them to accumulate adaptations that, over millions of years, modeled organisms with incredible capacities to survive and exploit the resources of each particular niche. However, it is undeniable that the anthropogenic era has generated changes that are so rapid that they have not allowed many species to respond at a rate of adaptation as fast as the innovations that occur throughout the world (Allgeier et al. 2020). The influence of human activities on different natural fields and the understanding of the multiple ecological synergies have become more important day by day. The consequences of our actions on the environment is something largely unexplored, and today we observe new collateral effects resulting from changes in the landscape.

Reef organisms such as corals, fish, echinoderms, mollusks, and crustaceans have evolved to synchronize their reproductive with the lunar periodicity, which is by a set of genes called Cryptochromes (Cry) that is expressed by perceiving a particular specific wavelength and thus carrying out the release of gametes (Levy et al. 2007; Sweeney et al. 2011). However, the spectral dynamics change due to the increase and excessive use of artificial light in coastal and island cities. In this study, it was impossible to quantify the current state of the blue pulse present on the island of San Andrés since the SQM sensors do not quantify each wavelength separately. Nevertheless, we can analyze that there is a significant disruption in the natural light cycles, due to constant incidence of the anthropogenic light sources inside the sea. Therefore, it is possible that the resilience of the coral reef organisms can be highly affected by the unprecedented nature of this pressure (Swaddle et al. 2015).

The levels of understanding of how human actions affect the planet provide tools to implement effective responses. Our findings helped unveil the state of the intervention of artificial light in the marine ecosystems of the Seaflower Biosphere Reserve in points close to urban centers, offering new data on an unexplored threat to marine ecosystems in Colombia. Figure 5 illustrates the patterns of light increase on the island of San Andrés from 2012. We can see a growth in the coverage area of artificial light in the northeastern zone, where the range of extension of the light passed from 4 km in 2012 to 11 km in 2021. In proportion, these phenomena are similar to what occurs in areas such as the Gulf of Eliat in the Red Sea, where the coverage of the artificial light that affects it is 47% brighter than a natural night sky and rises to a maximum of 60 times brighter than starlight on the north shore (reef mean 470%) (Ayalon et al. 2021).

According to the Bortle scale (Sky and Telescope 2020), a class 1 in sky brightness is an excellent dark-sky site, and a class 9 in sky brightness is an inner-city sky. Therefore, the Johnny Cay sector can be categorized in class 5, defined as a suburban sky (magnitude value between 14.5 and 15 msas). At the same time, Acuario and West View probably belong to class 2, equivalent to the typical genuinely dark site. However, it is essential to highlight that this classification system was developed for terrestrial data. Therefore, the underwater data taken here should be adjusted to terrestrial data to observe the change in magnitude from the surface to 11 m depth. There was not an appropriate number of sensors to collect data at the two depths for the three sampled sites in this investigation. Additionally, here we make a possible approximation to the Bortle classification system. Although the Unihedron maps (Fig. 5) are updated, certain liminal information is part of the atlas presented by Falchi et al. (2016).

When exploring the local level, the significant changes in the island's light conditions are in the area where trade is centered (North Zone). The less light-saturated part is in the South Zone, as shown in Fig. 6. The highest light records correspond to Johnny Cay, followed by Acuario, where there is also a tendency for the presence of two peaks of light during a 24-h period (morning and night). The Acuario area presents considerable disturbances but shows maximum darkness values similar to the places less affected by light pollution, which corresponds to West View. Despite this, differences higher than 1 arcsecond are present in the data from the three sites (Shapiro–Wilk > 0.05), with the West View being the place with ideal natural light conditions. These observations generate new hypotheses that must be explored to evaluate the problem at the local level. For example, it would be expected that the processes of coral reproduction would be somehow affected by lunar synchrony, so following the moments in which the reproductive event occurs in the areas where light records were taken, the connection between the problem of artificial light and its influence on corals could be quantified.

The temporary nature of the records made it possible to analyze the consequences of the Iota climate phenomenon, which still has many questions about how it affected marine ecosystems. The data during 2019 showed significant differences compared to those taken in 2020 in the West View area. Figure 7 focuses on the site where less intervention of artificial light had been perceived. We can see how, after the hurricane, a group of data with values less than 10 arcseconds begins to appear, leading to the conclusion that after the hurricane, the penetration of light was greater. One of the explanations for these differences lies in the substantial loss of vegetation found between the sea and the closest sources of light. This strip of trees, no greater than 25 m, served as a natural barrier to artificial light, which was reflected in higher darkness values at the beginning of the study. Given the rapid growth of many developing world economies, future increases are expected to be greater in these regions as compared with the developed world over the coming decades, with unknown consequences for some of the planet's most biodiverse marine ecosystems (Aubrecht et al. 2008), as is the case of the Seaflower Biosphere Reserve.

An additional point to consider is how artificial light affects different members of coral communities. Previous research shows how many reef diseases are generated by the instability of their microbial communities, allowing opportunistic microorganisms to colonize the tissue and cause negative consequences (Boilard et al. 2020; MacKnight et al. 2021). However, the inner relationship between corals and their symbionts is fragile and increasingly threatened by anthropogenic stressors (Hoegh-Guldberg 2014). Artificial light induces photoinhibition of the symbionts, overproduction of reactive oxygen species (ROS), and increased oxidative damage to lipids in coral species (Levy et al. 2020). Similarly, the modification of the symbiont community by artificial light intervenes at the time of reproductive processes (Tamir et al. 2020). This is causing a potential loss of ecological barriers in coral reefs that spawn in similar time frames.

Under continuous light conditions, oxidative stress occurs accompanied by a reduction in the ability to perform photosynthesis, both in the coral and its community of microorganisms. This result indicates the potential danger of artificial light for corals' adaptation to human pressure because of initial characteristics in coral bleaching processes (Suggett and Smith 2020). These observations show that there is indeed an interruption in the coral-symbiont symbiosis, mainly because the excess of light does not allow a complete cycle of the dark part of photosynthesis, where many endosymbionts complete cell recovery processes throughout the night (Hill et al. 2011).

6 Conclusions and Final Reflections

Despite the small number of instruments deployed to measure light and the short sampling window, the results are conclusive in revealing the threat of artificial light in different locations to reef organisms in different locations on the island. As such, the contribution is of value and even provides valuable artificial light management recommendations that should be relatively easy to implement. This study is just an approximation regarding the new techniques that can be useful at the local level, taking data with a higher level of precision is recommended to access more robust conclusions about how we should respond to the new collateral challenges of population growth. Among the recommendations to address the problem of imminent population growth is an appropriate use of lighting resources that allow it to be in harmony with the marine environments, minimizing the effects of artificial light and given that reproductive cycles occur on specific days and times. An alternative could be based on reducing light levels by a certain percentage on spawning days. In this way, the island's tourist activities will not be affected, and marine organisms' dynamics will be less altered. The use of light sources based on high-pressure sodium or fluorescent lights with a wavelength that does not have the same effects as LEDs is one possible solution that can also be explored. Additionally, lower-intensity LED lights could specifically reduce the emission of blue light peaks. Finally, although the island's vegetation has been recovering naturally (a slow succession process), it is pertinent to carry out reforestation programs in the coastal areas and thus recover this natural barrier, which reduces the interaction between horizontal light and vertical light.

Acknowledgements We want to thank the BIOMMAR research group for providing the SQM sensors, diving cases, the Corporation CEMarin for the call for this publication, and the esteemed professors who participated as evaluators of this research. Thanks to DIMAR for providing us with environmental information and the boats to install the sensors, and thanks to the Sea Pride dive center for the logistics and loan of diving equipment.

References

Allgeier JE, Andskog MA, Hensel E et al (2020) Rewiring coral: anthropogenic nutrients shift diverse coral–symbiont nutrient and carbon interactions toward symbiotic algal dominance. Glob Change Biol 26(10):5588–5601. https://doi.org/10.1111/gcb.15230

Aubrecht C, Elvidge CD, Longcore T et al (2008) A global inventory of coral reef stressors based on satellite observed nighttime lights. Geocarto Int 23(6):467–479. https://doi.org/10.1080/10106040802185940

Ayalon I, Benichou JI, Avisar D, Levy O (2021) The endosymbiotic coral algae symbiodiniaceae are sensitive to a sensory pollutant: artificial light at night, ALAN. Front physiol 12:695083

Ayalon I, Rosenberg Y, Benichou JI et al (2019) Coral gametogenesis collapse under artificial light pollution. Curr Biol 31(2):413–419. https://doi.org/10.1016/j.cub.2020.10.039

Boilard A, Dubé CE, Gruet C et al (2020) Defining coral bleaching as a microbial dysbiosis within the coral holobiont. Microorganisms 8(11):1682. https://doi.org/10.3390/microorganisms8111682

Chilma-Arias A, Calixto-Botia I, Ortiz M et al (2021) Population genomics and lunar irradiance: a case of incipient ecological speciation in a Caribbean Octocoral. Universidad de Los Andes. https://repositorio.uniandes.edu.co/handle/1992/53062

Coelho MA, Lasker HR (2014) Reproductive biology of the Caribbean brooding octocoral Antillogorgia hystrix. Invertebr Biol 133(4):299–313. http://www.jstor.org/stable/24697565

Davies TW, Bennie J, Inger R et al (2013) Artificial light alters natural regimes of night-time sky brightness. Sci Rep 3(1):1–6. https://doi.org/10.1038/srep01722

Davies TW, Smyth T (2018) Why artificial light at night should be a focus for global change research in the 21st century. Glob Change Biol 24(3):872–882. https://doi.org/10.1111/gcb.13927

Dimitriadis C, Fournari-Konstantinidou I, Sourbès L et al (2018) Reduction of sea turtle population recruitment caused by nightlight: evidence from the Mediterranean region. Ocean Coast Manag 153:108–115. https://doi.org/10.1016/j.ocecoaman.2017.12.013

Falchi F, Cinzano P, Duriscoe D et al (2016) The new world atlas of artificial night sky brightness. Sci Adv 2(6):e1600377. https://www.science.org/doi/10.1126/sciadv.1600377

Gaston KJ, Bennie J, Davies TW et al (2013) The ecological impacts of nighttime light pollution: a mechanistic appraisal. Biol Rev Camb Phil Soc 88:912–927. https://onlinelibrary.wiley.com/doi/10.1111/brv.12036

Gaston KJ, Visser ME, Hölker F (2015) The biological impacts of artificial light at night: the research challenge. Phil Trans R Soc B 370:20140133. https://doi.org/10.1098/rstb.2014.0133

Gaston KJ, Davies TW, Nedelec SL et al (2017) Impacts of artificial light at night on biological timings. Annu Rev Ecol Evol Syst 48:49–68. https://doi.org/10.1146/annurev-ecolsys-110316-022745

Grubisic M, van Grunsven RH (2021) Artificial light at night disrupts species interactions and changes insect communities. Curr Opi Insect Sci 47:136–141. https://doi.org/10.1016/j.cois.2021.06.007

Guerra-Vargas LA, Gillis LG, & Mancera-Pineda JE (2020) Stronger together: do coral reefs enhance seagrass meadows "blue carbon" potential?. Front Mar Sci 7:628

Hamidi ZS, Abidin ZZ, Ibrahim ZA et al (2011) Effect of light pollution on night sky limiting magnitude and sky quality in selected areas in Malaysia. In: 2011 3rd international symposium and exhibition in sustainable energy and environment (ISESEE), Malacca, Malaysia, pp 233–235. https://doi.org/10.1109/ISESEE.2011.5977095

Hill R, Brown CM, DeZeeuw K et al (2011) Increased rate of D1 repair in coral symbionts during bleaching is insufficient to counter accelerated photo-inactivation. Limnol Oceanogr 56(1):139–146. https://doi.org/10.4319/lo.2011.56.1.0139

Hoegh-Guldberg O (2014) Coral reef sustainability through adaptation: glimmer of hope or persistent mirage? Curr Opin Environ Sustain 7:127–133. https://doi.org/10.1016/j.cosust.2014.01.005

Hopkins GR, Gaston KJ, Visser ME et al (2018) Artificial light at night as a driver of evolution across urban–rural landscapes. Front Ecol Environ 16(8):472–479. https://doi.org/10.1002/fee.1828

Kaiser TS, Heckel DG (2012) Genetic architecture of local adaptation in lunar and diurnal emergence times of the marine midge Clunio marinus (Chironomidae, Diptera). PLoS ONE 7(2):e32092. https://doi.org/10.1371/journal.pone.0032092

Kronfeld-Schor N, Dominoni D, de la Iglesia H et al (2013) Chronobiology by moonlight. Proc R Soc B 280:20123088. https://doi.org/10.1098/rspb.2012.3088

Levy O, Appelbaum L, Leggat W et al (2007) Light-responsive cryptochromes from a simple multicellular animal, the coral Acropora millepora. Science 318(5849):467–470. https://www.science.org/doi/10.1126/science.1145432

Levy O, de Barros LF, Benichou JI et al (2020) Artificial light at night (ALAN) alters the physiology and biochemistry of symbiotic reef building corals. Environ Poll 266:114987. https://doi.org/10.1016/j.envpol.2020.114987

Luarte T, Bonta CC, Silva-Rodriguez EA et al (2016) Light pollution reduces activity, food consumption and growth rates in a sandy beach invertebrate. Environ Poll 218:1147–1153. https://doi.org/10.1016/j.envpol.2016.08.068

MacKnight NJ, Cobleigh K, Lasseigne D et al (2021) Microbial dysbiosis reflects disease resistance in diverse coral species. Commun Biol 4(1):1–11. https://doi.org/10.1038/s42003-021-02163-5

McLaren JD, Buler JJ, Schreckengost T et al (2018) Artificial light at night confounds broad-scale habitat use by migrating birds. Ecol Lett 3(21):356–364. https://doi.org/10.1111/ele.12902

Naylor E (1999) Marine animal behavior in relation to lunar phase. Earth Moon Planets 85:291–302. https://doi.org/10.1023/A:1017088504226

Poehn B, Krishnan S, Zurl M, et al (2021) A Cryptochrome adopts distinct moon- and sunlight states and functions as moonlight interpreter in monthly oscillator entrainment. https://doi.org/10.1101/2021.04.16.439809

Pulgar J, Zeballos D, Vargas J et al (2019) Endogenous cycles, activity patterns and energy expenditure of an intertidal fish is modified by artificial light pollution at night (ALAN). Environ Poll 244:361–366. https://doi.org/10.1016/j.envpol.2018.10.063

Sanders D, Frago E, Kehoe R et al (2021) A meta-analysis of biological impacts of artificial light at night. Nat Ecol Evol 5:74–81. https://doi.org/10.1038/s41559-020-01322-x

Sky and Telescope (2020) The essential guide to astronomy. Obtained 25th January 2022 from: https://skyandtelescope.org/astronomy-resources/light-pollution-and-astronomy-the-bortle-dark-sky-scale/

Sorek M, Levy O (2014) Coral spawning behavior and timing. In: Numata H, Helm B (eds) Annual, lunar, and tidal clocks. Springer, Tokyo, pp 81–97. https://doi.org/10.1007/978-4-431-55261-1_5

Suggett DJ, Smith DJ (2020) Coral bleaching patterns are the outcome of complex biological and environmental networking. Glob Change Biol 26(1):68–79. https://doi.org/10.1111/gcb.14871

Sweeney AM, Boch CA, Johnsen S et al (2011) Twilight spectral dynamics and the coral reef invertebrate spawning response. J Exp Biol 214(5):770–777. https://doi.org/10.1242/jeb.043406

Swaddle JP, Francis CD, Barber JR et al (2015) A framework to assess evolutionary responses to anthropogenic light and sound. Trends Ecol Evol 30(9):550–560. https://doi.org/10.1016/j.tree.2015.06.009

Tamir R, Lerner A, Haspel C et al (2017) The spectral and spatial distribution of light pollution in the waters of the northern Gulf of Aqaba (Eilat). Sci Rep 7(1):1–10. https://doi.org/10.1038/srep42329

Tamir R, Eyal G, Cohen I et al (2020) Effects of light pollution on the early life stages of the most abundant northern red sea coral. Microorganisms 8(2):193. https://doi.org/10.3390/microorganisms8020193

Open Access This chapter is licensed under the terms of the Creative Commons Attribution 4.0 International License (http://creativecommons.org/licenses/by/4.0/), which permits use, sharing, adaptation, distribution and reproduction in any medium or format, as long as you give appropriate credit to the original author(s) and the source, provide a link to the Creative Commons license and indicate if changes were made.

The images or other third party material in this chapter are included in the chapter's Creative Commons license, unless indicated otherwise in a credit line to the material. If material is not included in the chapter's Creative Commons license and your intended use is not permitted by statutory regulation or exceeds the permitted use, you will need to obtain permission directly from the copyright holder.

Ciguatera in the Seaflower Biosphere Reserve: Projecting the Approach on HABs to Assess and Mitigate Their Impacts on Public Health, Fisheries and Tourism

José Ernesto Mancera Pineda ⓘ, Brigitte Gavio, Adriana Santos-Martínez, Gustavo Arencibia Carballo, and Julián Prato

Abstract Microalgae constitute the basis of marine food webs. However, the massive growth of some species and the toxicity of others may represent a serious threat to human health, fisheries, mariculture, and tourism. Evidence shows that global warming, climate change, nutrients, and sewage discharge favor microalgal blooms, which are becoming more frequent, intense, and lasting. In the Caribbean Sea, ciguatera poisoning, one of the syndromes caused by toxic dinoflagellates, has increased its incidence in the past three decades. Despite the potential risks, there is no management plan for this and other harmful algal blooms (HABs) in San Andres island, Colombia. We analyze the presence of toxic dinoflagellates along with the incidence of ciguatera in the Seaflower Biosphere Reserve (SBR). Considering that effective climate change adaptation and mitigation decisions are based on relationships between science and society, involving a wide variety of analytical methods to evaluate associated risks and benefits, we propose to evaluate the potential effects of HABs, focusing on the economic value of their impacts on fishing and

---

J. E. Mancera Pineda (✉) · B. Gavio
Department of Biology, Faculty of Sciences, Universidad Nacional de Colombia, Bogotá, Colombia
e-mail: jemancerap@unal.edu.co

B. Gavio
e-mail: bgavio@unal.edu.co

J. E. Mancera Pineda · A. Santos-Martínez · J. Prato
Corporation Center of Excellence in Marine Sciences, CEMarin, Bogotá, Colombia
e-mail: asantosma@unal.edu.co

J. Prato
e-mail: jprato@unal.edu.co

A. Santos-Martínez · J. Prato
Universidad Nacional de Colombia, Caribbean Campus, San Andrés, Colombia

G. Arencibia Carballo
Havana Fisheries Research Center, Havana, Cuba

© The Author(s) 2025
J. E. Mancera Pineda et al. (eds.), *Climate Change Adaptation and Mitigation in the Seaflower Biosphere Reserve*, Disaster Risk Reduction, https://doi.org/10.1007/978-981-97-6663-5_6

tourism. We propose an early warning system conceptual model, based on a monitoring program, as a strategy to contribute to the governance and the management effectiveness of the different institutions of the SBR.

Keywords Climate change · Marine microalgae · Ciguatera · HABs early warning systems

1 Introduction

Humanity's expectations of development based on oceanic sources have been increasing because of the decrease in terrestrial sources, and although the exploration and use of the ocean for well-being and prosperity is nothing new, the scope, intensity, and diversity of current aspirations are unprecedented (Jouffray et al. 2019). In this way, societies rely more and more on marine ecosystems for food, materials, novel bioactive compounds, space, and recreational resources (GlobalHAB 2021).

The drivers of climate change, along with the intensive use of marine resources, have already altered the dynamics among biotic and abiotic components, rapidly transforming the structure and functions of some marine ecosystems in diverse regions worldwide. In particular, the dynamics of some photosynthetic organisms seem to be particularly affected by climate change (GlobalHAB 2021).

Microalgae, together with seaweeds and phanerogams constitute the basis of marine food webs and are essential for both marine and terrestrial ecosystems, since they produce around 50% of all the planet's oxygen. However, they can also become a serious threat to fishery resources, aquaculture, tourism, and human health in certain circumstances. Of the approximately 5,000 known species of algae globally, more than 135 can produce harmful events. Harmful algal blooms (HABs) may be harmful in two ways: some species produce toxins that affect the food chain, including human health, while other species are non-toxic, but may produce high biomass (GlobalHAB 2021). These latter microalgae produce great amounts of organic matter that, when algae die, sink to the ocean floor. Decomposition of this matter by bacteria can result in oxygen depletion, with dead-zone formation and mass death of marine organisms. Although considered natural events that have been historically documented in marine ecosystems, several studies in recent decades have linked HAB events to local mesoscale oceanographic and atmospheric phenomena (Sunesen et al. 2021), generating concern that global climate-driven changes will exacerbate HABs (GlobalHAB 2021). Likewise, other global problems such as nutrient discharge and the introduction of alien species have been associated with HABs' increase in frequency, intensity, and geographic distribution (Cuellar-Martinez et al. 2018; Heisler et al. 2008). Trends analyzed in Latin America and the Caribbean up to 2019 are related to the increasing awareness of the presence of toxic species, the geographical expansion of already known species, the detection of new toxins for the region, and HAB events' duration and/or impacts (Sunesen et al. 2021).

The GlobalHAB Program of the Intergovernmental Oceanographic Commission (IOC-UNESCO) and the Scientific Committee on Oceanic Research (SCOR) has been strengthening a conceptual framework for the understanding and management of HABs' impacts based on multidisciplinary international coordination and ongoing training (www.globalhab.info). However, this very program recognizes that one of the unknown and main challenges to face is how we can prevent or mitigate future HAB impacts (GlobalHAB 2021).

As has been stated in Chap. 1 of this book, Biosphere Reserves (BR) are "living, dynamic laboratories", and therefore represent ideal places to study and replicate interdisciplinary adaptation strategies. The Archipelago of San Andrés, Providencia, and Santa Catalina (hereafter, the archipelago), located in the southwest Caribbean, typifies the definition of small oceanic islands. It has an important wealth of marine and terrestrial natural capital, the ecological characteristics of its islands favor biodiversity but are limited by freshwater supply and are showing signs of great vulnerability to extreme weather events. In 2000, Seaflower, an area of 180,000 km^2 located in the archipelago, was declared an International Biosphere Reserve by UNESCO, and in 2005, part of the BR (65,000 km^2) was declared a Marine Protected Area by the Colombian government.

Considering that effective climate change adaptation and mitigation decisions are based on the relationship between science and society, involving a wide variety of analytical methods to evaluate the associated risks and benefits, the purpose of this chapter is to evaluate HABs in the Seaflower Biosphere Reserve (SBR), based on the economic cost of its impacts on fishing, tourism, and the cost of monitoring and training programs. We hope that this analysis can serve as input for both the design of the early warning system and for an ambitious social training program.

2 HABs in the Seaflower Biosphere Reserve

In the Colombian Caribbean, 25 reported species appear in the IOC-UNESCO Taxonomic Reference List of Harmful Micro Algae (https://www.marinespecies.org/hab/) (Table 1). Most of them are benthic dinoflagellates, like those of the genus *Gambierdiscus*, known to cause toxic problems such as ciguatera in tropical and subtropical regions (Arencibia-Carballo et al. 2009; Chinain et al. 2021). Ciguatera is an intoxication caused by the ingestion of marine organisms with lipid-soluble ciguatoxins in their tissues; these toxins are originally produced by *Gambierdiscus* and *Fukuyoa*, two genera of dinoflagellates, and are accumulated along the chain web. Ciguatera affects between 10,000 and 50,000 people in the world annually (Friedman et al. 2008; Chinain et al. 2021). However, it is considered underdiagnosed, estimating that less than 10% of cases are reported (Friedman et al. 2008).

The dinoflagellate *Gambierdiscus toxicus* (Adachi and Fukuyo 1979), which lives as an epiphyte of seagrasses and macroalgae colonizing coral reefs (Lehane and Lewis 2000), has been considered for years the main cause of ciguatera. However, a detailed taxonomic and toxin characterization of the species in the genus confirms the

Table 1 Influence of climate change stress factors on different HAB species found in the Colombian Caribbean (Arbeláez et al. 2020; Arteaga-Sogamoso et al. 2021; Sunesen et al. 2021) The symbols suggest the confidence level: + (reasonably likely), ++ (most likely) according to Wells et al. (2015). (1) Temperature increase; (2) Nutrient increase; (3) Stratification increase; (4) pH decrease

Potentially toxic marine microalgae reported in Colombia	HAB type	Environmental factor			
		1	2	3	4
Coolia cf. *malayensis* Leaw, P.-T. Lim and Usup, 2001	Benthic	↕	↑	↑	?
Ostreopsis lenticularis Y. Fukuyo, 1981					
Ostreopsis ovata Y. Fukuyo, 1981					
Gambierdiscus caribaeus Vandersea, Litaker, M. A. Faust, Kibler, W. C. Holland and P. A. Tester, 2009		++		++	
Gambierdiscus spp. Adachi and Y. Fukuyo, 1979					
Prorocentrum cordatum (Ostenfeld) J. D. Dodge, 1976					
Prorocentrum cf. *concavum* Y. Fukuyo, 1981					
Prorocentrum emarginatum Y. Fukuyo, 1981					
Prorocentrum hoffmannianum M. A. Faust, 1990					
Prorocentrum lima (Ehrenberg) F. Stein, 1878					
Prorocentrum rhathymum Loeblich III, Sherley and Schmidt, 1979					
Dinophysis acuminata Claparède and Lachmann, 1859	Fish killing	↑	↑	↑	?
Dinophysis caudata Saville-Kent, 1881					
Gonyaulax spinifera (Claparède and Lachmann) Diesing, 1866			+	++	
Protoceratium reticulatum (Claparède and Lachmann) Bütschli, 1885					
Alexandrium catenella/tamarense complex	Toxic flagellates	↑	↑	↑	↕
Alexandrium minutum Halim, 1960					
Alexandrium monilatum (J. F. Howell) Balech, 1995				++	
Gymnodinium catenatum H. W. Graham, 1943					
Didinium polykrikoides (Margalef) F. Gómez, Richlen & D. M. Anderson, 2017					
Pyrodinium bahamense Plate, 1906					

(continued)

Table 1 (continued)

Potentially toxic marine microalgae reported in Colombia	HAB type	Environmental factor			
		1	2	3	4
Anabaenopsis sp. V. V. Miller, 1923	Cyanobacteria	↑	↑	↑	↕
Dolichospermum sigmoideum (Nygaard) Wacklin, L. Hoffmann and Komárek, 2009					
Microcystis aeruginosa Kützing, 1846		+	++	++	

involvement of other *Gambierdiscus* as well as *Fukuyoa* species as toxin producers to a greater or lesser degree (Litaker et al. 2009). It cannot be discarded that other benthic dinoflagellate genera such as *Amphidinium, Coolia, Ostreopsis*, and *Prorocentrum* (Besada et al. 1982), and some cyanobacterial taxa (Laurent et al. 2008), which are often found in association with *G. toxicus*, may also be involved in ciguatera poisonings. Among them, the genus *Prorocentrum* has great relevance, due to the number of species identified as toxic or potentially toxic, and due to its abundance in natural environments (Delgado et al. 2002; Arbeláez et al. 2020).

The oldest known historical record of ciguatera in the world dates back to 1525 in the Eastern Atlantic, when the captains of seven Spanish ships that anchored in the Gulf of Guinea consumed barracuda. All those who ate the barracuda became ill with diarrhea and fell unconscious (Urdaneta 1580, in Fraga et al. 2011) and died months later due to unknown causes (de Miguel 2009, in Fraga et al. 2011). Since then, ciguatera has caused serious problems, such as the death in 1748 of 1,500 people in the Indo-Pacific islands (Halstead and Cox 1973). It has even been suggested that ancient Polynesian migrations were driven by ciguatera events (Rongo et al. 2009). Cases of ciguatera have been reported in the Caribbean since 1862 when, in the Gulf of Mexico, the crew of a French ship became poisoned by eating parrotfish (Halstead 1967). Patterns of resource use by the Arawak and Caribe groups inhabiting the Eastern Caribbean may indicate that they too faced problems with the intoxication (Price 1966). Like other types of marine poisonings, ciguatera is underestimated in much of the Caribbean, making its study highly pertinent in the region, even more if one considers that less than 0.1% of those intoxicated receive medical attention (Tosteson 1995). Reported symptomatology for ciguatera is variable, including gastrointestinal, neurological, cardiovascular, and neuropsychological disorders, ranging from mild and short-term to severe and long-term, in the worst cases leading to death (Arencibia et al. 2009; Faust 2009).

A significant increase in the incidence of ciguatera has been reported in different regions of the world. Skinner et al. (2011) found that during the past three decades, the incidence of ciguatera in the South Pacific increased by 60%, while Tester et al. (2020) and Celis and Mancera-Pineda (2015) detected an increase of 32% among member countries of the Caribbean Epidemiology Center (CAREC). These increases could be even greater in the future due to changes in global weather patterns, overfishing, and the degradation of marine ecosystems (Tester et al. 2020).

Numerous species of dinoflagellates associated with *Thalassia testudinum* beds, macroalgae, and debris from coral reefs and mangrove forests have been found in the Caribbean Sea (Faust 1993a, b, 2000, 2009; Faust et al. 1999; Valerio González and Díaz 2008; Rodríguez et al. 2010). Taking into account the increase in ciguatera in the Caribbean (Mancera-Pineda et al. 2014; Celis and Mancera-Pineda 2015), the presence of potentially toxic dinoflagellates associated with seagrasses and macroalgae (Rodríguez et al. 2010) and the wide diversity of macroalgae and other substrates that make up the drift on the island of San Andrés (Ortiz and Gavio 2012), it is necessary to expand the evaluation of potentially toxic microalgae with a view to generate a comprehensive risk management plan.

Marine biotoxins threaten both human health and food and nutritional security (FAO 2005). In addition to ciguatera, other forms of seafood-borne poisonings caused by microalgae have been identified, whose toxins can enter food webs and affect human health through the ingestion of fishery products. The best-known toxins are paralytic (PST), diarrheal (DST), amnesic (AST), and neurotoxin (NST) (Lagos 2002). The current concern about the impact generated by potentially toxic microalgae in society is great, given that in recent years poisoning events seem to have increased in frequency, intensity, and geographic distribution (Hallegraeff et al. 2021).

In a recent scoping review aimed at mapping the evidence for associations between marine HABs and observed acute and chronic human health effects, it was found that 58% of the 220 publications made between 1985 and 2019 were related to ciguatera poisoning (Young et al. 2020). But while the public health implications of ciguatera have been established, its true regional and global incidence has been difficult to determine due to underreporting of cases. This underreporting is due both to the difficulty in differentiating its symptoms from other syndromes, and to deficiencies in the epidemiological data recording systems of the affected countries (Friedman et al. 2008, 2017; Skinner et al. 2011). Despite the difficulties of diagnosis and notification, ciguatera is now recognized as a major health problem around the world, in addition to its strong socioeconomic consequences (Chinain et al. 2021).

In a study on the historical incidence of ciguatera in San Andrés and the Caribbean island states (Celis and Mancera-Pineda 2015), the results show that, in the period 1980–2010, there were 10,710 registered cases from 18 CAREC countries, with an average annual incidence of 42/100,000 inhabitants. Likewise, there was an increase between the periods 1980–1990 and 2000–2010, with an annual average calculated from the reported cases of 34.2 and 45.2/100,000, respectively. The island of Montserrat had the highest incidence in the region, 350/100,000, while San Andrés had an incidence of 25/100,000 inhabitants, ranking eighth among the islands in the study. The rate ratio for CAREC countries (average annual incidence from 2000 to 2010/average annual incidence from 1980 to 1990) was 1.36, so there was a 32% increase in average annual incidence across countries, and an increase of nearly 300% between the two time periods. The level of reported incidence of ciguatera in the Caribbean has increased in the last 31 years, mainly in the Eastern Caribbean, since island states such as the Bahamas, Antigua, and Barbuda contribute greatly to the total reported increase. Considering that the development model of much of the region is based on the tourism industry and that fish is an important source of protein for Caribbean communities, we may affirm that ciguatera is a problem expected to increase in parallel with environmental changes.

To generate a baseline to quantify the relationship between climate, its variation, and HAB-related diseases, an interdisciplinary approach is required to determine the true burden of the intoxication, after acute and chronic exposures, including impacts on the environment and human well-being (Young et al. 2020).

Bearing in mind that the problems associated with marine toxins are on the rise and that these not only cause problems in ecosystems, but also to public health and productive activities, it is essential to design a management plan that allows the reduction of vulnerability to, and therefore the risk of, this threat. In the SBR and Cartagena, 166 cases of ciguatera have been confirmed in the period 2010–2020 (INS 2021; Celis and Mancera-Pineda 2015).

Although there is no clear trend in the distribution of cases over time (Fig. 1), this incidence of ciguatera—14.4 cases per year—could have negative implications for tourism, even more so considering the possible high levels of underreporting. Tourism is the basis of the development model of the archipelago, as well as that of a good part of the insular Caribbean, and given that ciguatera is a growing phenomenon worldwide, this syndrome constitutes a risk and therefore must be taken into account in development plans. The design and implementation of a monitoring program that becomes an early warning system should be a priority, as well as the training of health and tourism personnel. In this sense, the island of San Andrés could become a model of risk management in the Caribbean.

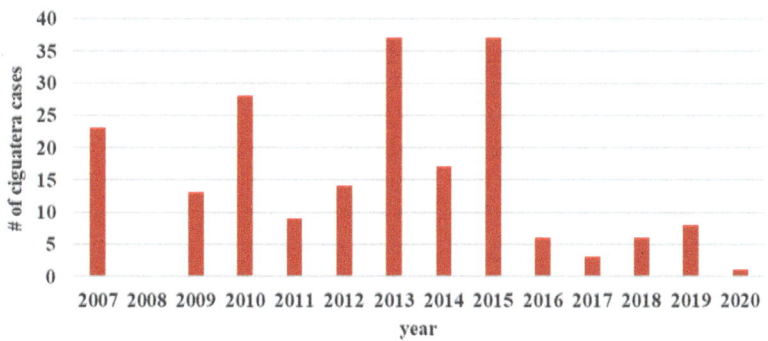

Fig. 1 Ciguatera cases registered in the Seaflower biosphere reserve (INS 2021; Celis and Mancera Pineda 2015)

3 HABs and Climate Change: What to Expect in the Near Future

The effects of climate change will continue to grow for many years, but we cannot ameliorate or manage negative impacts on humans or ecosystems without a better knowledge of those impacts. Understanding the processes thriving HABs has been a very challenging task and one that scientists have not yet fully accomplished. The main reason is that HABs are very complex events: they include a huge variety of species, belonging to evolutionary distinct groups; their life histories are diverse, and may include resistance stages, and sexual as well as vegetative reproduction; the ecosystems involved span from freshwater to brackish and marine, both nearshore and offshore, from tropical to cold temperate and polar latitudes; and their impacts may vary in space and time (Anderson et al. 2015). Furthermore, there are many abiotic and biotic factors triggering HABs (Table 1), including, but not limited to, temperature, salinity, nutrients, biogeochemical cycles, grazing, and anthropic activities such as ballast water discharge and climate change (Anderson et al. 2015; Wells et al. 2015).

Global climate change affects different abiotic factors in marine systems, which, in turn, influence the growth and distribution of bloom-forming algae. With climate change, variations in temperature, salinity, pH, oxygen content, and stratification are expected (Tester et al. 2020; Wells et al. 2015).

3.1 Temperature and HABs

There is no doubt that temperature is increasing at a global scale, although warming is not uniform (Roemmich et al. 2012; Stocker et al. 2013). Temperature is one of the main factors affecting physiological processes in algae, acting at different stages of growth and bloom development (Wells et al. 2015). This increase may impact HABs differently according to their latitude. A growth increase is expected in polar and temperate regions (Moore et al. 2009), while at tropical and subtropical latitudes, temperature rise may not favor algal growth if temperature optima are exceeded (Wells et al. 2015). Therefore, an increase of HABs at higher latitudes, and a decrease in tropical and subtropical regions are expected (Table 1).

3.2 Salinity and HABs

Changes in ocean salinity are expected, due mainly to ice-melting in the polar regions, and changes in precipitation patterns at regional and local scales (Pachauri and Meyer 2014). These changes are predicted to be highly variable across regions, with an expected decrease in salinity in the central Pacific Ocean, eastern and central Indian Ocean, and the Baltic Sea, whereas a salinity increase is predicted in the Gulf of

Mexico, Caribbean Sea, and most of the Atlantic Ocean (Tester et al. 2020). Most toxic dinoflagellates involved in HABs (*Gambierdiscus* spp., *Ostreopsis* spp.) grow better at relatively high salinity, while they fail to thrive well in waters influenced by freshwater runoff. Therefore, the predicted increase in salinity in the Caribbean Sea should not affect negatively these algae (Tester et al. 2020).

3.3 Water Stratification and HABs

With global warming, an increase in surface ocean stratification is expected (Stocker et al. 2013). The changes are anticipated to be more pronounced at mid to high latitudes, while at tropical latitudes stratification changes should be less obvious. Water stratification will change patterns of nutrient availability, a key factor for HAB growth (Marinov et al. 2010), and there is already some evidence linking stratification to low nutrient concentration at low latitudes (Wells et al. 2015). Stratification should favor small species of plankton, which have a higher rate of nutrient uptake (Hein et al. 1995) as well as swimmer taxa (Peacock and Kudela 2014). Several HAB species may prosper in stratified oceanic waters, and at mid-latitudes, some blooms have been associated with stratification variations (Berdalet et al. 2014; Ryan et al. 2014).

3.4 Ocean Acidification and HABs

Increasing atmospheric carbon dioxide (CO_2) leads to ocean acidification, through the dissolution of part of this CO_2 into surface oceanic water. Dissolution of CO_2 in ocean water increases CO_2 availability for photosynthesizing organisms. The effects of a higher concentration of CO_2 on HAB species are not well understood. Some taxa may show an increase in photosynthetic growth rate (Fu et al. 2008), while others show no increase (Cho et al. 2001) or even a decrease (Lundholm et al. 2004). A lower pH may have physiological effects on cell metabolism (Beardall and Raven 2004; Giordano et al. 2005). To date, there is still too little evidence to predict the consequences of ocean acidification on HABs (Table 1).

3.5 Expected Shift in the Caribbean

In general terms, rising temperatures in the Caribbean basin should reach the upper thermal tolerance limit for several species of HABs, resulting in a poleward shift of these species (e.g. *Gambierdiscus carolinianus*) (Tester et al. 2020). However, other species, such as *G. caribaeus*, *G. belizeanus*, and *Fukuyoa ruetzleri*, may become dominant in the Caribbean (Kibler et al. 2015; Tester et al. 2020). For benthic species forming HABs, an increase in coral bleaching events, as well as hurricane damage, may favor the establishment of the benthic population of HABs (Tester et al. 2020).

Therefore, for the archipelago, we can expect no dramatic change in planktonic HABs, since the current predictions assume neutral to unfavorable future scenarios for planktonic HABs in tropical regions. However, with an increase in tropical storm intensity and frequency, along with an intensification in bleaching events due to warming, coral reefs will likely degrade at a faster rate in the upcoming years. With coral loss, the substrate will become available for benthic HAB species, which may increase their population.

On the other hand, there is growing concern about the quality of the SBR's waters. As mentioned, tourism is an important driver of socioeconomic development, while also representing a source of environmental challenges (von Glasow et al. 2013; Abdul Azis et al. 2018). Industrial and domestic wastewater constitute the main threats to water quality in much of the Caribbean (Constanza et al. 1997; Gavio et al. 2010). This contamination reduces the potential of ecosystem services that benefit society while generating public health problems by increasing the load of pathogenic organisms responsible for acute infectious outbreaks, as well as microalgae capable of inducing harmful blooms (WHO 1998; Shuval 2003; Young et al. 2020; Chinain et al. 2021). The costs of dealing with this type of problem are usually very high, not only due to medical requirements but also due to the temporary reduction of workdays in the affected population (Xie et al. 2017).

4 Economic Consequences

The severity of ciguatera impacts in the SBR is poorly understood. However, considering that this syndrome could increase with climate change, it may become a serious development threat. Fishing represents an important source of protein for visitors and local communities, and is therefore essential for food security (Jaramillo-Campuzano et al. 2009; Santos-Martínez et al. 2013), while the economy of the BR is based on tourism, as occurs in other Caribbean locations (Kingsbury 2005; Pantojas 2006; Prato and Newball 2016). The tourism industry in the Caribbean reports more than USD $25,000 billion per year (Burke and Maidens 2005), which represents 20% of GDP. In the SBR, tourism represents around USD $266 million per year (reported in USD 2014) (Prato and Newball 2016). Tourism's associated economic revenues and incomes could be expanded to the trading, fisheries, restaurants, bars, and labor sectors, which makes tourism a very important component of the SBR's market economy. These economic benefits could be threatened by HABs, while their economic impacts could be reduced, avoided, or mitigated through investments in effective monitoring and management programs.

Marine environmental management faces great challenges in maintaining the social benefits provided by ecosystem services (ES) (de Jonge et al. 2003; Elliott 2011; Turner and Schaasfsma 2015). This management must consider the high complexity of the ecological interactions of these systems, which are modulated by the interaction of the atmosphere, the land, and the ocean (von Glasow et al. 2013). HABs could also impacts ecosystems and their ES, modifying food webs and affecting other organisms and ES (Anderson et al. 2000; van Tussenbroek et al. 2017). Non-market values and ES provided by the SBR's marine ecosystems such as seagrasses, mangroves, and coral reefs, represent considerably higher benefits for wellbeing than tourism. An economic valuation of SBR ES estimated that they represent around USD 267 billion per year (Prato and Newball 2016). These benefits could be at risk due to HABs in the SBR and the Caribbean.

The estimation of the economic impacts caused by HAB events has included a wide range of factors such as the loss of gross income in fishery products, public health costs, tourism and recreation impacts, environmental monitoring, and management expenses, or other costs that would not exist in the absence of HABs (Anderson et al. 2000). According to Bernard et al. (2014) significant expenses are sustained annually due to HAB events, which vary by region. Japan may lose more than USD 1 billion dollars, while Europe USD 850 million, and the USA around USD 95 million. Impacts on aquaculture industries could also be very high. For example, in Norway, just one HAB incident impacted tons of Atlantic salmon, generating losses of around USD 300 million (Trainer 2020). In Florida, recurrent *Karenia brevis* blooms (commonly known as "Florida red tides") have been estimated to cause over USD 20 million in tourism-related losses every year (Anderson et al. 2000).

The cost of ciguatera and other HABs varies depending on the conditions of each country, city, or particular territory. Additionally, those costs may have a higher magnitude depending on factors such as the size of the population at risk, the size of each economic activity (aquaculture, tourism, fisheries), and the dependence of people and the economy on each activity.

Economic impacts on public health due to ciguatera cases in US territories vary depending on where those cases occur and the average healthcare cost in each place. For example, the estimated costs for ciguatera related illness treatment and healthcare per reported case in the USA could be around USD 1,000 per reported case and USD 700 per unreported case, while in the specific case of Puerto Rico, the average cost per case might be lower (around USD 530) (Hoagland et al. 2002). Anderson et al. (2000), estimated the economic impacts of ciguatera on public health, and found costs between USD 18 million to USD 24 million per year, averaging USD 21 million per year. Hoagland et al. (2002), also estimated similar economic impacts of ciguatera for public health averaging USD 19 million dollars per year for US tropical territories. Morin et al. (2016) estimated the public health costs of ciguatera cases in the Moorea Island society of approximately USD 50,000 per year.

On the other hand, ciguatera could also affect fisheries due to a reduction in fish sales and consumption as well as other factors that affect artisanal fisheries, sellers, and several actors in the fish market chain. It has been estimated that the economic impacts of ciguatera in Hawaii increased by USD 3 million per year based on the dollars per pound of fish that are unmarketable due to ciguatera (Hoagland et al. 2002). Since fish and shellfish are the main local protein source in the archipelago, ciguatera could be a threat not only to fisheries and the economy, but also to food security in these insular territories.

Ciguatera's economic impacts on public health are likely to be underestimated on different levels for many reasons. For example, (1) unreported cases of sick people that do not go to healthcare providers, (2) misdiagnosed cases considered "standard" poisonings, and (3) unregistered cases due to the insufficient statistical and reporting systems of local healthcare providers. The underreporting of cases may vary from one place to another. Hoagland et al. (2002), presented different "reported" to "unreported" illness ratios, for example, they found rations of 1:4 for Florida, 1:10 for the Northern Mariana Islands and American Samoa, and 1:100 for Hawaii, Guam, Puerto Rico, and the U.S. Virgin Islands. This can have effects on calculations of the economic impacts of HABs. These underreported cases could influence the results of an economic approach, so improved registration, diagnosis, and statistics mechanisms are encouraged for better management.

Economic impacts (economic losses and costs) of HABs such as ciguatera, must be considered from a multi-perspective approach, to better consider their economic impacts and risks, as well as to visualize the benefits of investing in effective management in terms of avoided costs. A cost–benefit analysis could be performed to present to decision-makers, presenting the gains and advantages that effective management and monitoring programs could have. Since there are data limitations and underreporting of ciguatera cases and their impacts, periodically updating calculations of their economic impacts is an appropriate strategy to improve accuracy and to provide more awareness about the importance of avoiding related costs by investing in management plans. These updates must be done as the available data on ciguatera improve in the archipelago and surrounding Caribbean insular territories. This recalls the need for Caribbean territories to invest in better registration and statistics mechanisms for HAB management.

Based on experiences of the assessment and reports of ciguatera and other HABs' economic impacts (Anderson et al. 2000; Sanseverino et al. 2016; Trick et al. 2020), here we provide tools to estimate the economic impacts on the SBR. It is important to consider that the accuracy of economic impact estimates depends on the quality and amount of available data and statistics related to ciguatera cases and related costs. Despite all the limitations due to lack of data, these tools may prove useful to better understand and inform decision-makers about the economic relevance of: (1) the economic impacts of ciguatera; (2) the importance of investing in effective management and monitoring strategies; (3) the importance of investing in improving ciguatera diagnosis, registration, cost, effects, and other related data availability and HABs statistics in the SBR. Here we present some equations as tools to estimate the economic impacts ($E_{impacts}$) of ciguatera and other HABs in the SBR and other Caribbean insular territories:

$$E_{impacts} = (Ti + LFi + PHi + Fi + Si) \qquad (1)$$

in which: Ti: Impacts on Tourism (Eq. 2), LFi: Impacts on Labor Force (Eq. 3), Phi: Impacts on Public Health (Eq. 4), Fi: Impacts on Fisheries (Eq. 5) and Si: Societal impacts (Eq. 6). All the costs and data are suggested to be calculated by year (total or averages per year) to facilitate calculations.

$$Ti = (Tr * Tanu) * Tainc \qquad (2)$$

in which, Tr: Tourism reduction percentage (given in decimal units –0 to 1), Tanu: Average annual number of tourists visiting the island or destination. Tainc: Total average revenues generated by a tourist (including local and national taxes such as the "Tourism card" (*tarjeta de turismo*), VAT and airport taxes, flight tickets, hotel, restaurants, food expenses, drinks, and leisure expenses).

$$LFi = ((Naw * Da) * (Awag)) + (ARw * Naw * Da) \qquad (3)$$

in which, Naw: Number of affected workers (average per year), Da: Average of days of affectation (disease) per worker (days off work), Awag: Average daily wage per worker. ARw: Average revenue per worker (production or revenues for the firm or employer), Naw: Average number of affected workers.

$$PHi = \sum (Cpd * Da) + Cm \qquad (4)$$

in which, Cpd: Cost per day of medical care (including hospitalization if needed and care and medical attention cost, medical exams, and others), Da: Days of affectation per worker, Cm: Costs of medicine. This must be included per person, and for all the registered people affected by ciguatera.

For Eqs. 3 and 4, it is important to consider that the underreporting, under-registration, and underdiagnosis of ciguatera cases may affect the amounts calculated for LFi and PHi, and that the number of workers or patients affected by ciguatera must be higher. It is also important to remember that the under-diagnosed factor has been estimated to be around just 10% of ciguatera cases reported (Friedman et al. 2008) and that less than 0.1% of those intoxicated go to health services (Tosteson 1995).

To consider and include the under-registration and underdiagnosis of ciguatera in the equation to calculate its economic impacts, we can modify Eq. 1, to include and correct the expected real number of cases and their consequent economic impacts in Eq. 1a. These corrections could be performed if well-based "reported-not reported rates" (R) are available:

$$E_{impacts} = \left(Ti + \left(\frac{LFi}{R}\right) + \left(\frac{PHi}{R}\right) + Fi + Si\right) \qquad (1a)$$

in which R: Reported/not reported rate. For example, if the reported ciguatera cases are 10% (1:10), then R = 0.1. Also, if the reported-not reported rate is 3:10 (30%), then R = 0.3.

Since detailed data about healthcare costs could be limited, Eq. 4 could be simplified to Eq. 4a based on averages from the available information:

$$PHi = ((Cpda * Daa) + Cma) * Npa \qquad (4a)$$

in which, Cpda: Average cost per day of medical care per patient, Daa: Average days of affectation, Cma: Average cost of medicine, Npa: Number of people affected.

$$Fi = \sum (Ras * Mas * Nma) + (Caa) \qquad (5)$$

in which Fi is the sum of impacts for all the affected producers (fisherman or fisheries firms). Ras: Reduction of average monthly sales, Mas: Average monthly sales per firm or artisanal fisherman, Nma: Number of affected months, Caa: Cost of additional analyses and controls on fish or shellfish products per firm (tissue toxin detection, special laboratory analyses, monitoring, insurances, and other related costs).

$$Si = \sum_{i=1}^{n} Sim_i \qquad (6)$$

in which, Sim_i: Available data about each indirect economic impact on society, such as indirect wages or job losses related to tourism or fisheries impacts, lawsuits, and extra costs of food substitutes, among others. Indirect economic impacts such as impacts on fuel sales, boat maintenance, jobs, and other related indirect impacts in the fisheries market chain, for example, on wages or revenues, could be difficult to include or consider in these analyses (Anderson et al. 2000). Nevertheless, we included this Si factor in the equation to allow the inclusion and consideration of these inputs in the economic impact estimations.

Other important costs of HABs, usually included in economic assessments of its impacts, are the "monitoring and management" costs (Anderson et al. 2000; Sanseverino et al. 2016). Those must be also considered in the total accounting for HABs' economic impacts. Despite this, we didn't include these costs in Eq. 1, due to our perspective that "monitoring and management" costs should be considered an investment that could mitigate the negative economic impacts of ciguatera and other HAB events. Monitoring and management costs could include water quality testing, the operation of shellfish and fish tissue toxin monitoring programs, plankton monitoring, and other activities.

Regional implementation of management strategies is needed to contribute to reducing the risk of HABs in the Caribbean Sea, especially within the framework of climate change adaptation. Effective actions to improve water quality, reduce nutrients supplied anthropogenically by agriculture, domestic and industrial sewage, as well as those generated by fires that destroy forests and that are later washed into the sea by the rains, which has been identified as one of the mechanisms of the blooms of *Sargassum* (not included in this chapter), which are also affecting tourism, fishing and coastal ecosystems in the Caribbean (Méndez-Tejada and Rosaldo-Jiménez 2019), but to a lesser extent in the SBR. Likewise, the investment in ecosystem-based adaptation of the whole Caribbean Basin will be vital for a better present and future for island territories.

5 HABs Risk Management

During the third UN World Conference on Disaster Risk Reduction held in 2015 in Sendai, Japan, the Sendai Framework for Disaster Risk Reduction 2015–2030 was adopted. This framework proposes that to effectively protect livelihoods, health, cultural heritage, socioeconomic assets, and ecosystems, and to build resilience, it is essential to anticipate and plan for risk. It also emphasizes the need to improve the understanding of risk in its different dimensions, characteristics of exposure, vulnerability, and hazard. The framework also raises the importance of strengthening governance, the resilience of health infrastructure, international cooperation, and accountability processes (UNISDR 2015).

Risk conditions are determined by a relationship between threats and vulnerabilities. Threats are external factors represented by the potential occurrence of a dangerous natural or anthropogenic phenomenon. The analysis of the threat entails knowing its dynamics, characteristics, historical behavior, potential, and area of influence. Vulnerability, on the other hand, is an internal, intrinsic factor, determined by its own characteristics, represented by the limitation or inability to withstand, avoid, mitigate, adapt, and/or resist adverse events and recover from them. Risk management requires a broad approach, focused on people, with multi-risk and multi-sector practices that are inclusive and accessible. It also requires effective collaboration between the public and private sectors, civil society organizations, academia, and research institutions (UNISDR 2015).

The set of material or information resources existing in each place constitutes its Natural Capital (Constanza et al. 1997). The flows of this capital or its interaction with human beings constitute ES, that is, they are the elements of an ecosystem that are used actively or passively to produce the well-being of human populations (Fisher et al. 2009). ES are exposed to damage or deterioration, which can be evaluated through risk theory (Schäfer 2012). In addition to the threats caused by climatic phenomena, inappropriate use can compromise the sustainability of marine resources whenever they deteriorate ecosystem functions (Lozoya et al. 2011).

Due to the high environmental and economic dependence on coastal areas, small islands such as San Andrés, Providencia, and Santa Catalina correspond to the regions with the highest risk from the impacts of climate change and extreme weather events. Marine-coastal ecosystems such as coral reefs, seagrass and macroalgal meadows, beaches, and mangrove forests, provide coastal protection as an ES and, due to their high biodiversity, represent natural farms and great tourist attractions, which allows development and the sustainability of economic activities. Therefore, success in marine-coastal environmental management is a permanent challenge that makes it essential to improve scientific knowledge of the relationships between natural and social systems to reduce vulnerability and increase resilience, that is, the ability to face disturbances such as those produced by HABs and continue generating ES. Considering the effects of climate change, greater adaptive responses will be required to deal with the impacts of natural phenomena in coastal areas.

Table 2 List of HAB monitoring variables, recommended by GlobalHAB (2021)

Abiotic variables	Biotic variables
Temperature	Chlorophylla
Salinity	Composition and abundance of phytoplanktonic and phytobenthonic communities
Precipitation	Microzooplankton and microzooplankton biomass and community composition
Winds and wave heights	Toxins
Light attenuation	Abundance and composition of cysts in the sediment
Nutrients: N, P, Si	Markers for quantifying cell numbers
Dissolved oxygen	
Carbonate system	

As a result of the efforts of the ANCA-IOCARIBE network, part of the HAB Program of IOC-UNESCO, several Colombian institutions have begun a collaborative initiative aimed at managing the risk of HABs on the country's Pacific and Caribbean coasts. Since it is essential to collect information on the appearance of toxic microalgae and to describe their temporal variability, a monitoring program is being carried out, the main results of which were included as part of the Global HAB Status Report (Sunesen et al. 2021).

Considering the vulnerability of the archipelago to HABs, it is urgent to design and apply an early warning system (EWS) for risk reduction. This EWS must include an effective monitoring plan with strategic actions to face and mitigate the challenges of intoxications transmitted by organisms. Considering that early warnings are critical and fundamental elements in risk reduction and/or mitigation, an important factor for EWS effectiveness lies in the level of citizen participation. The communities at risk must be an active part of the system, receiving timely information, training, and exchanging knowledge with other stakeholders.

EWSs must integrate monitoring data, which, in the case of HAB events, correspond to the presence of species, oceanographic and atmospheric variables, and toxicity levels (Table 2). Comprehensive analysis of these data should lead to the generation of information on risk forecasting and prediction. The specific risk assessment is the basic input for the authorities to make decisions and to communicate them to the community and other potentially affected organizations in a timely manner. EWSs, then, must prevent, prepare for, and address the negative impacts generated by HABs (Fig. 2).

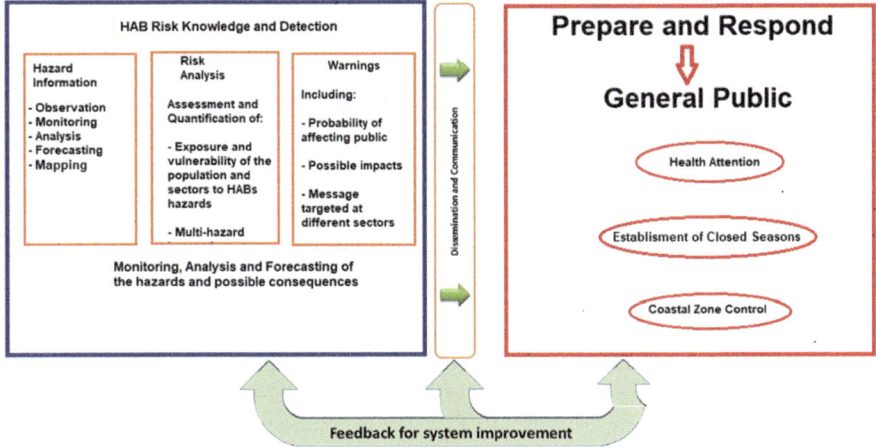

Fig. 2 A model for a harmful algal blooms (HABs) early warning system. Elaborated by the authors based on the general WMO (2018) multi-hazard early warning system

References

Abdul Azis P, Mancera Pineda JE, Gavio B (2018) Rapid assessment of coastal water quality for recreational purposes: methodological proposal. Ocean Coast Manag 151:118–126. https://doi.org/10.1016/j.ocecoaman.2017.10.014

Adachi R, Fukuyo Y (1979) The thecal structure of a marine toxic dinoflagellate *Gambierdiscus toxicus* gen. et sp. nov. collected in a ciguatera-endemic area. Bull Jpn Soc Sci Fish 45:67–71. https://doi.org/10.2331/suisan.45.67

Anderson CR, Moore SK, Tomlinson MC et al (2015) Living with harmful algal blooms in a changing world: strategies for modeling and mitigating their effects in coastal marine ecosystems. In: Shroder JF, Ellis JT, Sherman DJ (eds) Coastal and marine hazards, risks, and disasters. Elsevier, pp 495–561. https://doi.org/10.1016/B978-0-12-396483-0.00017-0

Anderson DM, Hoagland P, Kaoru Y et al (2000) Estimated annual economic impacts from harmful algal blooms (HABs) in the United States. WHOI-2000-11. Department of Biology, Woods Hole Oceanographic Institution, Woods Hole, Massachusetts. https://www.whoi.edu/cms/files/Economics_report_18564_23050.pdf

Arbeláez N, Mancera Pineda JE, Reguera B (2020) Structural variation of potentially toxic epiphytic dinoflagellates on *Thalassia testudinum* from two coastal systems of Colombian Caribbean. Harmful Algae 92:101738. https://doi.org/10.1016/j.hal.2019.101738

Arencibia-Carballo G, Mancera Pineda JE, Delgado Miranda G (2009) Ciguatera: potential risk for humans: frequent questions. Published in Spanish and English, Universidad Nacional de Colombia

Arteaga-Sogamoso E, Rodríguez F, Mancera-Pineda JE (2021) Morphological and molecular characterization of *Gambierdiscus* caribaeus (Dinophyceae), with a confirmation of its occurrence in the Colombian Caribbean Tayrona National Natural Park. Bot Mar 64(2):149–159 https://doi.org/10.1515/bot-2020-0070

Beardall J, Raven JA (2004) The potential effects of global climate change on microbial photosynthesis, growth and ecology. Phycologia 43:26–40. https://doi.org/10.2216/i0031-8884-43-1-26.1

Berdalet E, McManus MA, Ross ON et al (2014) Understanding harmful algae in stratified systems: review of progress and future directions. Deep Sea Res Part II Top Stud Oceanogr 101:4–20. https://doi.org/10.1016/j.dsr2.2013.09.042

Bernard S, Kudela R, Velo-Suarez L (2014) Developing global capabilities for the observation and predication of harmful algal blooms. In: Djavidnia S, Cheung V, Ott M et al (eds) Oceans and society: blue planet. Cambridge Scholars Publishing, UK, pp 46–52

Besada EG, Loeblich LA, Loeblich AR (1982) Observations on tropical, benthic dinoflagellates from ciguatera-endemic areas: *Coolia, Gambierdiscus* and *Ostreopsis*. Bull Mar Sci 32(3):723–735

Burke L, Maidens J (2005) Arrecifes en Peligro. World Resources Institute, Washington

Celis JS, Mancera-Pineda JE (2015) Análisis histórico de la incidencia de ciguatera en las Islas del Caribe durante 31 años: 1980–2010. Bol Invest Mar Cost 44(1):7–32

Chinain M, Gatti CMI, Darius HT et al (2021) Ciguatera poisonings: a global review of occurrences and trends. Harmful Algae 102:101873. https://doi.org/10.1016/j.hal.2020.101873

Cho ES, Kotaki Y, Park JG (2001) The comparison between toxic *Pseudo-nitzschia multiseries* (Hasle) Hasle and non-toxic *P. pungens* (Grunow) Hasle isolated from Jinhae Bay. Korea. Algae 16:275–285

Constanza R, Arge R, De Groot R et al (1997) The value of the world's ecosystem services and natural capital. Nature 387:253–260

Cuellar-Martinez T, Ruiz-Fernández AC, Alonso-Hernández C et al (2018) Addressing the problem of harmful algal blooms in Latin America and the Caribbean—a regional network for early warning and response. Front Mar·Sci 5:409. https://doi.org/10.3389/fmars.2018.00409

de Jonge VN, Kolkman MJ, Ruijgrok ECM et al (2003) The need for new paradigms in integrated socio-economic and ecological coastal policy making. In: Proceedings of 10th international Wadden sea symposium, pp 247–270. Ministry of Agriculture, Nature Management and Fisheries, Department North, Groningen

Delgado G, Popowski G, Pombo MC (2002) Nuevos registros de dinoflagelados tóxicos epibénticos en Cuba. Rev Invest Mar. 23(3):229–232

Elliott M (2011) Marine science and management means tackling exogenic unmanaged pressures and endogenic managed pressures—a numbered guide. Mar Pollut Bull 62:651–655. https://doi.org/10.1016/j.marpolbul.2010.11.033

FAO (2005) Biotoxinas marinas. United Nations Food and Agriculture Organization

Faust MA (2009) Ciguatera-causing dinoflagellates in a coral-reef-mangrove ecosystem, Belize. Atoll Res Bull 569:1–32. https://doi.org/10.5479/si.00775630.569.1

Faust MA (2000) Dinoflagellate associations in a coral reef-mangrove ecosystem: Pelican and associated cays, Belize. Atoll Res Bull 473:137–150. https://doi.org/10.5479/si.00775630.473.137

Faust MA, Larsen J, Moestrup Ø (1999) Potentially toxic phytoplankton. 3. Genus *Prorocentrum* (dinophyceae). In: Lindley JA (ed), ICES identification leaflets for plankton. Natural Environment Research Council. Copenhagen, Denmark

Faust MA (1993a) Three new benthic species of *Prorocentrum* (Dinophyceae) from twins Cays, Belize: *P. maculosum* sp. nov., *P. foraminosum* sp. nov and *P. formosum* sp.nov. Phycologia 32(6):410–418. https://doi.org/10.2216/i0031-8884-32-6-410.1

Faust MA (1993b) Surface morphology of the marine dinoflagellate *Sinophysis microcephala* (Dinophyceae) from a mangrove island, Twin Cays, Belize. J Phycologia 29:355–363. https://doi.org/10.1111/j.0022-3646.1993.00355.x

Fisher B, Turner K, Morling P (2009) Defining and classifying ecosystem services for decision making. Ecol Econ 68(3):643–653. https://doi.org/10.1016/j.ecolecon.2008.09.014

Fraga S, Rodríguez F, Caillaud A et al (2011) Gambierdiscus excentricus sp. nov. (Dinophyceae), a benthic toxic dinoflagellate from the Canary Islands (NE Atlantic Ocean). Harmful Algae 11:10–22. https://doi.org/10.1016/j.hal.2011.06.013

Friedman MA, Fernandez M, Backer LC et al (2017) An updated review of ciguatera fish poisoning: clinical, epidemiological, environmental, and public health management. Mar Drugs 15(3):72. https://doi.org/10.3390/md15030072

Friedman MA, Fleming LE, Fernandez M et al (2008) Ciguatera fish poisoning: treatment, prevention and management. Mar Drugs 6(3):456–479. https://doi.org/10.3390/md6030456

Fu FX, Zhang YH, Warner ME et al (2008) A comparison of future increased CO_2 and temperature effects on sympatric *Heterosigma akashiwo* and *Prorocentrum minimum*. Harmful Algae 7(1):76–90. https://doi.org/10.1016/j.hal.2007.05.006

Gavio B, Palmer-Cantillo S, Mancera Pineda JE (2010) Historical analysis (2000–2005) of the coastal water quality in San Andrés Island, Seaflower biosphere reserve, Caribbean Colombia. Mar Pollut Bull 60:1018–1030. https://doi.org/10.1016/j.marpolbul.2010.01.025

Giordano M, Beardall J, Raven JA (2005) CO_2 concentrating mechanisms in algae: mechanisms, environmental modulation, and evolution. Annu Rev Plant Biol 56:99–131. https://doi.org/10.1146/annurev.arplant.56.032604.144052

GlobalHAB (2021) Guidelines for the study of climate change effects on HABs. In: Wells M et al (eds) IOC manuals and guides no 88. Paris, UNESCO-IOC/SCOR

Hallegraeff GM, Anderson DM, Belin C et al (2021) Perceived global increase in algal blooms is attributable to intensified monitoring and emerging bloom impacts. Commun Earth Environ 2:117. https://doi.org/10.1038/s43247-021-00178-8

Halstead BW (1967) Poisonous and venomous marine animals of the world, US Government Printing Office, Washington, DC, USA

Halstead BW, Cox KW (1973) An investigation on fish poisoning in Mauritius. Imprimerie Commerciale

Hein M, Pedersen MF, Sandjensen K (1995) Size-dependent nitrogen uptake in micro- and macroalgae. Mar Ecol Prog Ser 118(1–3):247–253

Heisler J, Glibert PM, Burkholder JM et al (2008) Eutrophication and harmful algal blooms: a scientific consensus. Harmful Algae 8:3–13. https://doi.org/10.1016/j.hal.2008.08.006

Hoagland P, Anderson DM, Kaoru Y et al (2002) The economic effects of harmful algal blooms in the United States: estimates, assessment issues, and information needs. Estuaries 25:819–837. https://doi.org/10.1007/BF02804908

INS (2021). Boletín Epidemiològico Semanal—BES. Semana Epidemiològica 16. 18 al 24 de abril de 2021. Instituto Nacional de Salud. Dirección de Vigilancia y Análisis del Riesgo en Salud Pùblica. https://doi.org/10.33610/23576189.2021.16.

Jaramillo Campuzano L, Polania Vorenberg J, Hayes Mathias L (2009) Canasta básica de alimentos de la población en el año 2005, del departamento archipiélago de San Andrés. Universidad Nacional de Colombia—Sede Caribe, Providencia y Santa Catalina

Jouffray J-B, Crona B, Wassénius E et al (2019) Leverage points in the financial sector for seafood sustainability. Sci Adv 5(10):2aax3324. https://doi.org/10.1126/sciadv.aax3324

Kibler SR, Tester PA, Kunkel KE et al (2015) Effects of ocean warming on growth and distribution of dinoflagellates associated with ciguatera fish poisoning in the Caribbean. Ecol Model 136:194–210. https://doi.org/10.1016/j.ecolmodel.2015.08.020

Kingsbury P (2005) Jamaican tourism and the politics of enjoyment. Geoforum 36:113–132. https://doi.org/10.1016/j.geoforum.2004.03.012

Lagos N (2002) Principales toxinas de origen fitoplanctónico: identificación y cuantificación mediante cromatografía líquida de alta resolución (HPLC). In: Sar EA, Ferrario ME, Reguera B (eds) Floraciones algales nocivas en el cono Sur Americano. Instituto Español de Oceanografía, pp 21–52

Laurent D, Kerbrat A, Darius T et al (2008) A new ecotoxicological phenomenon related to marine benthic Oscillatoriales (cyanobacteria) blooms. Ciguatera and Related Biotoxins Workshop, Noumea, New Caledonia, 27–31 October 2008. Secretariat for the Pacific Community, Noumea, New Caledonia, pp 17–22

Lehane L, Lewis RJ (2000) Ciguatera: recent advances but the risk remains. Int J Food Microbiol 61(2–3):91–125. https://doi.org/10.1016/S0168-1605(00)00382-2

Litaker RW, Vandersea MW, Faust MA et al (2009) Taxonomy of *Gambierdiscus* including four new species, *Gambierdiscus caribaeus, Gambierdiscus carolinianus, Gambierdiscus carpenteri* and *Gambierdiscus ruetzleri* (Gonyaulacales, Dinophyceae). Phycologia 48(5):344–390. https://doi.org/10.2216/07-15.1

Lozoya JP, Sarda R, Jiménez JA (2011) A methodological framework for multi-hazard risk assessment in beaches. Environ Sci Pol 14:685–696. https://doi.org/10.1016/j.envsci.2011.05.002

Lundholm N, Hansen PJ, Kotaki Y (2004) Effect of pH on growth and domoic acid production by potentially toxic diatoms of the genera *Pseudo-nitzschia* and *Nitzschia*. Mar Ecol Prog Ser 273:1–15. https://doi.org/10.3354/meps273001

Mancera-Pineda JE, Montalvo-Tagua M, Gavio B (2014) Dinoflagelados potencialmente tóxicos asociados a material orgánico flotante (drift) en San Andrés isla, reserva internacional de la biosfera—Seaflower. Caldasia 36(1):139–156. https://doi.org/10.15446/caldasia.v36n1.43896

Marinov I, Doney SC, Lima ID (2010) Response of ocean phytoplankton community structure to climate change over the 21st century: partitioning the effects of nutrients, temperature and light. Biogeosciences 7(12):3941–3959. https://doi.org/10.5194/bg-7-3941-2010

Mendez-Tejeda R, Rosado Jiménez GA (2019) Influence of climatic factors on Sargassum arrivals to the coasts of the Dominican Republic. J Oceanogr Mar Sci 10(2):22–32. https://doi.org/10.5897/JOMS2019.0156

Morin E, Gatti C, Bambridge T et al (2016) Ciguatera fish poisoning: incidence, health costs and risk perception on Moorea Island (society archipelago, French Polynesia). Harmful Algae 60:1–10. https://doi.org/10.1016/j.hal.2016.10.003

Moore SK, Mantua NJ, Hickey BM et al (2009) Recent trends in paralytic shellfish toxins in Puget sound, relationships to climate, and capacity for prediction of toxic events. Harmful Algae 8(3):463–477. https://doi.org/10.1016/j.hal.2008.10.003

Ortiz JF, Gavio B (2012) Notes on the marine algae of the international biosphere reserve Seaflower, Caribbean Colombia II: diversity of drift algae in San Andrés island, Caribbean Colombia. Caribb J Sci 46(2–3):313–321. https://doi.org/10.18475/cjos.v46i2.a19

Pachauri RK, Meyer LA (2014) Climate change 2014: synthesis report, contribution of working Groups I, II, and III to the fifth assessment report of the intergovernmental panel on climate change. IPCC, Geneva, Switzerland

Pantojas GE (2006) De la plantación al resort: el Caribe en la era de la globalización. Rev Cienc Soc 15:82–99. https://revistas.upr.edu/index.php/rcs/article/view/5502

Peacock MB, Kudela RM (2014) Evidence for active vertical migration by two dinoflagellates experiencing iron, nitrogen, and phosphorus limitation. Limnol Oceanogr 59(3):660–673. https://doi.org/10.4319/lo.2014.59.3.0660

Prato J, Newball R (2016) Aproximación a la valoración económica ambiental del departamento Archipiélago de San Andrés, Providencia y Santa Catalina—Reserva de la Biósfera Seaflower. Secretaría Ejecutiva de la Comisión Colombiana del Océano—SECCO, Corporación para el desarrollo sostenible del Archipiélago de San Andrés, Providencia y Santa Catalina—CORALINA. Bogotá

Price R (1966) Caribbean fishing and fishermen: an historical sketch. Am Anthropol 68(6):1363–1383. https://doi.org/10.1525/aa.1966.68.6.02a00020

Rodríguez A, Mancera Pineda JE, Gavio B (2010) Survey of benthic dinoflagellates associated to beds of *Thalassia testudinum* in San Andrés island, Seaflower biosphere reserve, Caribbean Colombia. Acta Biolo Colomb 15(2):231–248

Roemmich D, Gould WJ, Gilson J (2012) 135 years of global ocean warming between the challenger expedition and the Argo programme. Nat Clim Change 2(6):425–428. https://doi.org/10.1038/nclimate1461

Rongo T, Bush M, van Woesik R (2009) Did ciguatera prompt the late Holocene Polynesian voyages of discovery? J Biogeogr 36:1–10. https://doi.org/10.1111/j.1365-2699.2009.02139.x

Ryan JP, McManus MA, Kudela RM et al (2014) Boundary influences on HAB phytoplankton ecology in a stratification-enhanced upwelling shadow. Deep Sea Res II—Top Stud Oceanogr 101:63–79. https://doi.org/10.1016/j.dsr2.2013.01.017

Sanseverino I, Conduto D, Pozzoli L et al (2016) Algal bloom and its economic impact. JRC Technical Report. EUR 27905 EN. https://doi.org/10.2788/660478

Santos-Martínez A, Mancera Pineda JE, Castro E et al (2013) Propuesta para el Plan de Manejo pesquero de la zona sur del área marina protegida en la reserva de biósfera Seaflower. Universidad Nacional de Colombia, Sede Caribe, Archipiélago de San Andrés, Providencia y Santa Catalina, Caribe colombiano

Schäfer RB (2012) Biodiversity, ecosystem functions and services in environmental risk assessment: introduction to the special issue. Sci Total Environ 415:1–2. https://doi.org/10.1016/j.scitotenv.2011.08.012

Shuval H (2003) Estimating the global burden of thalassogenic disease-human infectious disease caused by wastewater pollution of the marine environment. J Water Health 1(2):53–64. https://doi.org/10.2166/wh.2003.0007

Skinner M, Brewer T, Johnstone R et al (2011) Ciguatera fish poisoning in the Pacific Islands (1998 to 2008). Plos Negl Trop Dis 5(12):1–7. https://doi.org/10.1371/journal.pntd.0001416

Stocker TF, Qin D, Plattner G-K et al (2013) Climate change 2013: the physical science basis. In: Contribution of working group I to the fifth assessment report of the intergovernmental panel on climate change. Cambridge University Press, UK and New York, USA

Sunesen I, Méndez SM, Mancera-Pineda JE et al (2021) The Latin America and Caribbean HAB status report based on OBIS and HAEDAT maps and databases. Harmful Algae 102:101920. https://doi.org/10.1016/j.hal.2020.101920

Tester PA, Litaker RW, Berdalet E (2020) Climate change and harmful benthic microalgae. Harmful Algae 91:101655. https://doi.org/10.1016/j.hal.2019.101655

Tosteson TR (1995) The diversity and origins of toxins in ciguatera fish poisoning. P R Health Sci J 14:117–129

Trainer VL (Ed) 2020 GlobalHAB. Evaluating, reducing and mitigating the cost of harmful algal blooms: a compendium of case studies. PICES Sci Rep No. 59

Trick CG, Anderson L, Beausoleil D et al (2020) An economic assessment of ciguatera outbreaks—an island model. In: Trainer VL (ed) GlobalHAB. Evaluating, reducing and mitigating the cost of harmful algal blooms: a compendium of case studies. PICES Sci Rep No. 59, pp 55–65

Turner RK, Schaafsma M (2015) Coastal zones ecosystem services: from science to values and decision making. In: Studies in ecological economics, vol 9. Springer, Switzerland. https://doi.org/10.1007/978-3-319-17214-9

UNISDR (2015) Sendai framework for disaster risk reduction 2015–2030

Valerio González L, Díaz J (2008) Distribución de dinoflagelados epífitos potencialmente tóxicos asociados a praderas de *Thalassia testudinum* en la isla La Tortuga, la Bahía de Mochima y el Golfo de Cariaco, Venezuela. Boletín Instituto Oceanográfico De Venezuela 47(1):47–58

van Tussenbroek BI, Arana HAH, Rodríguez-Martínez RE et al (2017) Severe impacts of brown tides caused by *Sargassum* spp. on near-shore Caribbean seagrass communities. Mar Pollut Bull 122(1–2):272–281. https://doi.org/10.1016/j.marpolbul.2017.06.057

von Glasow R, Jickells TD, Baklanov A et al (2013) Megacities and large urban agglomerations in the coastal zone: interactions between atmosphere, land, and marine ecosystems. Ambio 42:13–28. https://doi.org/10.1007/s13280-012-0343-9

Wells ML, Trainer VL, Smayda TJ et al (2015) Harmful algal blooms and climate change: learning from the past and present to forecast the future. Harmful Algae 49:68–93. https://doi.org/10.1016/j.hal.2015.07.009

WHO (1998) Draft guidelines for safe recreational-water environments: coastal and fresh waters. Draft for Consultation, Geneva, October 1998. World Health Organization, Geneva (EOS/DRAFT/98.14)

WMO (2018) Multi-hazard early warning systems: a checklist. https://library.wmo.int/records/item/55893-multi-hazard-early-warning-systems-a-checklist

Xie Y, Qiu N, Wang G (2017) Toward a better guard of coastal water safety-Microbial distribution in coastal water and their facile detection. Mar Pollut Bull 118:5–16. https://doi.org/10.1016/j.marpolbul.2017.02.029

Young N, Sharpe RA, Barciela R et al (2020) Marine harmful algal blooms and human health: a systematic scoping review. Harmful Algae 98:101901. https://doi.org/10.1016/j.hal.2020.101901

Open Access This chapter is licensed under the terms of the Creative Commons Attribution 4.0 International License (http://creativecommons.org/licenses/by/4.0/), which permits use, sharing, adaptation, distribution and reproduction in any medium or format, as long as you give appropriate credit to the original author(s) and the source, provide a link to the Creative Commons license and indicate if changes were made.

The images or other third party material in this chapter are included in the chapter's Creative Commons license, unless indicated otherwise in a credit line to the material. If material is not included in the chapter's Creative Commons license and your intended use is not permitted by statutory regulation or exceeds the permitted use, you will need to obtain permission directly from the copyright holder.

Society, Seaflower Marine Ecosystem Services, and Climate Change Adaptation

The Biosphere Reserve Concept, Seaflower, and Climate Change

Germán Márquez

Abstract Based on UNESCO's biosphere reserve concept and on the paper originally proposing an archipelago biosphere reserve, this chapter supports going deeper into implementing the Seaflower Biosphere Reserve as a social, economic, and environmental sustainability model. To this, it proposes some actions, from reconsidering its regulatory status to its integration with national development plans, including payment schemes for ecosystem services (PES), as Seaflower ecosystems provide society with many goods and services, estimated to be huge, but not reflected in their management and financing. Seaflower's meaning has not been properly understood and is not taking advantage of this status. The current situation is worrying and unsustainable; it threatens the natural, historical, social, and cultural heritage of the Archipelago of San Andrés, Providencia, and Santa Catalina, stressed by a questionable mass tourism development model and worsened, mainly in Providencia, by hurricanes Eta and Iota and because of climate change whose impact, mainly in coral reefs, could be extreme. Some of the ideas developed in this chapter were proposed by the author with the name Seaflower Initiative; now, could be integrated with Gran Seaflower Initiative, a recent proposal for the creation of a transboundary biosphere reserve in the western Caribbean.

Keywords Sustainability model · Ecosystem services · Hurricane Iota · Seaflower initiative · Coral reefs

1 Introduction

The biosphere reserve concept was proposed by UNESCO in the early 1970s as a model for both conservation and development; it has been the subject of various interpretations and some misrepresentations (Ishwaran et al. 2008). Originally a

---

G. Márquez (✉)
Sea, Land and Culture Old Providence Initiative PROSEALAND Foundation, Providence island, Colombia
e-mail: germanrquez@gmail.com

proposal for relevant examples of the world's natural areas management and research in harmonious coexistence with society, it was elevated to a sustainable development model or limited to an honorary category, among other cases. This chapter discusses this conceptual landscape and proposes a return to the original concept as a basis for reconsidering and resizing Seaflower in the future to attain sustainability goals. This is convenient because, in Colombia, an interpretation of biosphere reserves as simple international honorary categories of conservation (Colombia 2015) has prevailed, to the detriment of their use as a model for the study and search for sustainability in accordance with the original proposal for a biosphere reserve of the Archipelago of San Andrés, Providencia, and Santa Catalina (hereafter, the archipelago). This was understood by the local community, which saw in it an alternative model at a time of crisis due to economic openness.

Even so, the archipelago entered the labyrinth of tourism monoculture (Márquez and Márquez 2016), collapsed with the pandemic, and is now hardly recovering with the same social, economic, and environmental risks implicit in that kind of tourism. In all cases, the Seaflower idea had demonstrated a significant capacity to influence the islands, and remains strong, not even with the weak support by the Colombian state, which only now seems to be trying to understand its relevance, as Nicaragua recently did, in the context of the territorial conflict between these countries. In these conditions, Seaflower is nowadays playing a more political role. This reinforces the possibility of making it the desirable model for the islands, based on the role of islands, their people, and their ecosystems as providers of ecological goods and services: fishing, biodiversity, and tourism attraction, among many others (Fig. 2; relating to Seaflower see Prato and Newball 2015). In this sense, the role of coral reefs in climate change processes is relevant, both in their role as accumulators of enormous amounts of calcium carbonate (Frankignoulle and Gattuso 1993, in Kault et al. 2022), contributing to reducing the greenhouse effect, and for the risk they face given their extreme fragility to temperature rises (Kault et al. 2022).

Proper management of the archipelago's ecological complex, mainly reefs, after the destructive impact of Hurricane Iota, would make a significant contribution to climate change mitigation, providing the world with a significant environmental service that should be compensated. Payments for Environmental Services (PES) offer a possibility to be considered to finance the proper management of the Seaflower and as an economic option for the archipelago. PES, as well as the creation of a fund, are reviewed as opportunities for Seaflower and its people, maintaining and recovering a very important socioecological heritage for local and global well-being. The educational process that must accompany Seaflower management will also help recover and re-evaluate ancestral cultural practices significant to maintaining the environmental balance that can still be found on the islands. In conclusion, it is proposed to adopt a more rigorous conceptual perspective and to try to implement it and relaunch Seaflower, so that it can fulfill its purpose of contributing to the sustainability and well-being of the archipelago, its people, and its culture. Rebuilding after Hurricane Iota should not mean a return to the previous unsustainable situation and represents an opportunity to make Seaflower what it could be.

2 The Biosphere Reserve Concept

Biosphere Reserves (BR) are areas of natural importance inhabited by significant human populations, which UNESCO (United Nations Educational, Scientific and Cultural Organization) designates as such to ensure their conservation in harmony with the cultural, social, and economic development of those populations, as well as investigating how to achieve it (Márquez 1992). Biosphere reserves are "learning places for sustainable development" (UNESCO 2022). Batisse (1986) proposed focusing the BR concept on three complementary aspects and their confluent functions, according to the following scheme (Fig. 1).

Thus, according to UNESCO (2022), BRs must fulfill the functions of:

- Conservation of biodiversity and cultural diversity,
- Economic development that is socio-culturally and environmentally sustainable,
- Logistical support, underpinning development through research, monitoring, education, and training.

Different BRs emphasize, according to their characteristics, some of these aspects. Developed countries have placed more emphasis on ecosystem conservation and logistics functions; but from the perspective of developing countries, where economic circumstances push towards excessive use of resources, the emphasis has been placed on the need to harmonize culture, conservation, and development (Halffter 1984). BR designations must be the result of voluntary agreements between countries and the residents of the area, who undertake giving it special management; UNESCO contributes with its experience and with the support of an international research and monitoring network.

The Biosphere Reserves program was launched by UNESCO in the early 1970s, with the initial purpose of protecting representative samples of the main types of ecosystems of the Planet's natural regions (Batisse 1986). It is based on Vernadsky's biosphere concept, perhaps the first integral vision of the Earth as a structural and functional unit where humanity is an integral part. Therefore, one BR characteristic was to include society, and their productive activities, as integral parts, contrasting with the idea of natural national parks that, at least at that time, were almost by

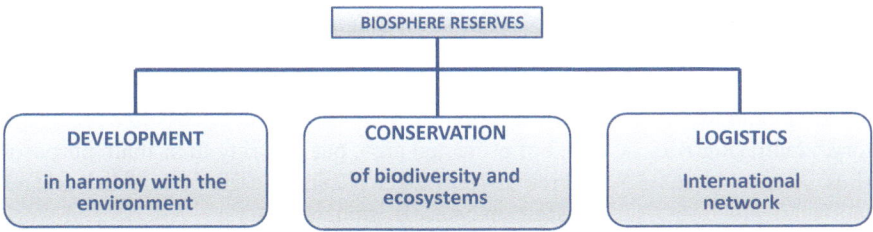

Fig. 1 Outline of the concept of biosphere reserves. Elaborated by the author based on Batisse (1986), in Márquez (1992)

definition areas dedicated only to conservation, whereas as far as possible there should be no human inhabitants.

This implied another important conceptual difference: while parks are for strict conservation, BRs combine the needs for conservation with those for productive activities in harmony with the environment. Therefore, it was sought that BRs include not only significant ecosystems but, especially, real samples of the harmonious coexistence of society with its natural environments, of cultural patterns that could serve as models and examples of the coexistence of society and nature (Batisse 1986).

Another distinctive aspect of the BR concept is that it includes research and education as a main component of its objectives and management strategies. This implies recognition of the fact that it is not yet well known how it is possible to achieve sustainability. So, BRs are practical laboratories for sustainability studies. Finally, it is important to note that BRs are also symbols and models of cooperative resource use for collective well-being.

The BR concept prefigures and anticipates the concept of sustainable development, understood as one capable of turning limited resources into a permanent basis with which to achieve human well-being, based on the harmonious relations of society with nature. This anticipation of BR to the sustainable development concept, which arrived some ten years later, was inspired by the discussion around eco-development, a concept proposed in France, which would evolve into those of sustainability and sustainable development. For more on the BR concept and its evolution, see also Ishwaran et al. (2008).

2.1 Other Perspectives on Biosphere Reserves

The Biosphere Reserves program was very well received worldwide and many countries, including Colombia, rushed to achieve UNESCO BR designations for areas of interest. However, the prestige of BRs as an international honorary category prevailed at least in Colombia, and involvement with the BR model and its purposes was scarce. Indeed, being declared a BR by UNESCO and becoming part of the Global BR network gives an area a prominent connotation, as a recognition of its importance: it contributes to the visibility of the area and its natural and human heritage at the national and international levels, but it is not appropriate that, in many BRs, the commitments for special management and the harmonization of nature-society relationships are not fulfilled.

Colombian legislation brings this honorary characteristic of BRs to a legal category, identifying BRs as a kind of protected area, but ignoring their main purposes. Thus, Article 2.2.2.1.3.7. Decree 1076 of 2015 (Colombia 2015) establishes international distinctions "such as Ramsar Sites, Biosphere Reserves, AICAS and World Heritage Sites", and indicates that "they are not categories of management of protected areas, but complementary strategies for the conservation of biological diversity". It points out that "the authorities in charge of the designation of protected areas must prioritize these sites according to the international importance recognized with the distinction" (Colombia 2015). Despite the limited nature of this perspective, it is a fact that the international importance of BRs plays a very important role in the case of Seaflower, as discussed below.

3 Prehistory of the Seaflower Biosphere Reserve

The Seaflower BR proposal was based both on the basic concept of harmonization of nature and culture, and on the idea of making the archipelago more visible, recognizing its natural and cultural importance as well as an alternative development model for the islands, at a time of crisis (Márquez 1992). The significant work that CORALINA, the local environmental authority, carried out with UNESCO, under the leadership of its director June Marie Mow, to achieve recognition of the Seaflower BR, according to Law 99 of 1993, is well known. It was a task that took several years and significant efforts to meet all the requirements of UNESCO and to overcome the Colombian state's poor understanding of the issue. But the history of where the idea of a BR in the archipelago came from, and how it was included in Law 99—one could say its prehistory—is less known.

The creation of an archipelago BR was originally proposed in a paper (Márquez 1992) forming part of a book (Márquez and Pérez 1992) prepared in the development of the Multinational Project on Environment and Natural Resources, sponsored by the Organization of American States (OAS), through Colciencias, and advanced in Colombia by the National and Javeriana universities. The paper, based on the BR concept, proposed a reorganization of the BR system in Colombia. This system had already declared several important sites in Colombia as BRs, such as the Sierra Nevada de Santa Marta and the Tuparro National Natural Park, but never really applied the BR concept. Colombian BRs were honorary names for sites that otherwise deserved the title but were not managed as such. The case of the Sierra Nevada is especially interesting, as it brings together, in an exemplary way, its enormous natural importance as the largest coastal mountain massif in the world, with the representation of all the main types of terrestrial, freshwater, and marine tropical ecosystems, and the wisdom of indigenous cultures coexisting in harmony with their natural environment.

The paper proposed the creation of a new BR in the archipelago, on the islands of Providencia and Santa Catalina, on the basis that these islands met very well the BR requirements, by combining an important natural heritage and a population with traditional cultural patterns of harmonious coexistence with nature. Moreover, it proposed, as reiterated in other parts of the book, that the BR could incorporate other parts of the archipelago, as was finally done. The book (Márquez and Pérez 1992) said in its introduction:

> The constitution of the islands of Providencia and Santa Catalina and their adjacent coral reef platform into a world-class natural monument is fully justified by the great natural importance of the islands and the reef complexes that surround them. Such a constitution would attract worldwide attention and interest in knowing and preserving them. This document supports

this possibility as a tool for Providencia's development reorientation to sustainability, while increasing its economic potential, since the BR seeks harmonizing development and environment, but above all improving living conditions of their inhabitants. The application of the proposed scheme in broader sectors of the Archipelago is an open possibility (p. 11).

The Multinational Project allowed several more activities to promote the idea of the archipelago BR, including a workshop held in Providencia in June 1993, where both the idea and the book were presented to the community (Pérez-García and Márquez 1993); the idea was also explained to the representative of the archipelago in the House of Representatives of the National Congress, Mr. Julio Gallardo and his assessor, Mr. Arne Britton, that incorporated the BR, including all the terrestrial and marine areas of the archipelago, as Article 37 of Law 99 of 1993 (Colombia 1993), which created the national environmental system (SINA, using its Spanish initials) and the Ministry of the Environment. Thus, the project of the archipelago BR became a part of Colombian law. Other activities were then carried out to socialize the archipelago BR, including an International Workshop on Biosphere Reserves, also with the support of COLCIENCIAS, OAS-CYTED (OAS Science and Technology Iberoamerican Program), and the National University of Colombia's Environmental Studies Institute (IDEA-UN). This workshop, which took place on the islands of San Andrés and Providencia between 27 and 30th June 1994, was attended by representatives of several Latin American and Caribbean BRs (Argentina, Peru, Suriname, Barbados, Jamaica, Costa Rica, among others), who helped to explain the idea to the island's community.

By then, the process had already been assumed by CORALINA, which got broad community participation until the designation of the BR in the year 2000 by UNESCO (UNESCO 2022). CORALINA remained very involved with the BR concept, even after the Seaflower designation; many papers and booklets socializing the new BR are clear on that concept. Collaborating in this process was Catalina Toro, a researcher who worked with Batisse, the father of the BR concept, who at his advanced age lent vigorous support for the nomination of the archipelago BR. There was a very inspired proposal to call it Seaflower, a beautiful name that also evokes the ship on which the first English settlers arrived in Providencia in 1629.

In 2006, the original article was published again (Márquez et al. 2006), to reinforce Seaflower, but only recently has it become available online (see Márquez 1992).

4 Seaflower Advances

The Seaflower BR has already completed 20 years since its creation, leading one to ask: what could be the balance of its achievements and failures? First, it is significant that the idea and its vision keep moving, and many people see the Reserve as the way to sustainable development in the archipelago. In these years, far from disappearing, Seaflower has been gaining strength, even if it cannot be said that it has achieved its objectives. However, it is also far from failing. This is what makes this analysis pertinent, as it questions what could be done to make Seaflower more consistent with its concept. The main change introduced by Seaflower is in the vision of the archipelago, which until its designation as a BR was only that of a tourist destination whose main attractions were not its beauty or its cultural interest, but the possibility of acquiring goods at a lower price. San Andrés was, par excellence, the "*sanandresito*", the Colombian name for places where products, not always legally imported, are traded. Today, San Andrés, Providencia, and Santa Catalina are recognized as a privileged destination, a tropical paradise, as is often said, especially referring to Providencia and Santa Catalina, where nature and Caribbean culture come together in a harmonious way, without deformations introduced by mass tourism or large hotels. The BR designation has contributed to this because, although many do not even know exactly what it is, it sounds good. Somewhat paradoxically, much of Seaflower's progress so far is due to this change in mentality with respect to the archipelago, due more to prestige than to sustainability achievements. That is, it is the international distinction by UNESCO, rather than the concept of BR itself, that has played the main role.

However, it should be noted that mentality change is of great importance, since it creates the conditions for deeper actions. This is a very significant advance, since it has required years of work, such as that of CORALINA, which has led to events and the publication of numerous pedagogical and informative documents on Seaflower. Likewise, significant advances have been made in the knowledge of BRs, even if they have put some emphasis on its natural aspects, as in the Seaflower Expeditions, rather than on the social ones that should be compensated for in the future. What has not been properly understood is that BRs present a model of sustainable development that should be integrated, even being the very basis of the island's development model. This has led to a divorce between national and local development projects or government plans, and the BR project, managed as a CORALINA project, with no clear connection to the development plans of the Governorate or departmental and municipal authorities. Hence, this chapter draws attention to the need to return to the BR concept, from which the future of Seaflower can be reconsidered.

Nowadays, Seaflower is playing another important role as one of Colombia's arguments in the territorial dispute with Nicaragua, as presumed proof of Colombia's involvement in the protection of the environment. This argument could help Colombia, and create conditions for a real involvement of the country in Seaflower. However, the International Court of Justice 2012 verdict divided Seaflower between Colombia and Nicaragua, so that its future administration will require collaboration, which could be an option for reducing conflictive interactions between the countries.

The Gran Seaflower Initiative, a recent proposal to create a large, multinational BR centered in Seaflower, could play a significant role in this scenario (see Sect. 5.9).

5 The Future of the Seaflower Biosphere Reserve

5.1 Rethinking Seaflower

The central idea of this section is the possible resizing of Seaflower, in line with the BR concept and its initial purposes, and some actions believed necessary to do this. Some of these actions are enunciated and briefly described here, most of which were outlined in articles developing the idea of a Seaflower Initiative (Márquez 2014a, b, 2016). It is important to state that the Seaflower Initiative is prior and different from the Gran Seaflower Initiative, an important idea that will be considered later.

The Seaflower Initiative proposal was based on the conviction that it is convenient, possible, and necessary to preserve the exceptional heritage of the archipelago, and that this is also the best and most enduring business we can do in it. Thus, it is an initiative for the protection of the BR, its beauty, and its possible future. In this sense, an initial step is a deep reflection on what is possible and desirable, and hence, to begin work on a model based on the valuation of the natural and human patrimony of the archipelago, which is, at the same time, the means and the end to achieving sustainability and common well-being. By 2030, the islands, depending on the management of environmental and social problems, maybe an increasingly precarious tourism destination, or, on the contrary, models of social, economic, and natural sustainability, and providers of scarce ecosystem goods and services which, by then, will be even more scarce and necessary. This requires many actions, some of which are outlined below.

5.2 Revision of the Seaflower BR's Legal Status

An important step is the revision of the legal status of Colombian BRs, in order to change current legislation that reduces BRs only to international distinctions that do not imply much state commitment and lack mechanisms for effective action. The new status should make involvement with UNESCO legally binding, and link the BRs to national, regional, and local development plans. Likewise, it must incorporate the role that Seaflower is playing in the framework of Colombia's international relations, mainly in the conflict with Nicaragua, as Seaflower is now divided between Colombia and Nicaragua and both countries should decide how to manage the situation created by the International Court of Justice verdict. Colombia also must decide, relating to this international context, what to do about the Gran Seaflower Initiative.

5.3 Sustainable Development Pilot Project

Developing the BR concept, the current Seaflower BR Management Plan must be restructured as a Sustainable Development Plan for the archipelago, to turn it into a prototype that can be replicated in other parts of the country and the world. It should be a pilot project where different sustainability strategies that have been applied around the world can be applied and tested. Examples of these sustainability strategies include:

- Conservation, recovery, restoration, and research of ecosystems and biodiversity,
- Cultural and nature tourism,
- Sustainable artisanal fishing,
- Alternative energies (solar, wind, marine),
- Payment schemes for environmental and ecosystem services (PES),
- Circular economy,
- Zero emissions,
- Carbon capture,
- Good diving practices.

Such a project looks not only for the needs of the harmonious development of the BR but for that of the archipelago department, integrating and managing both as the real unit they are. In this sense, the logistical role of BRs is highly important, being laboratories in which to implement scientific, technical, social, economic, and political strategies for sustainability, which can in turn be replicated. Of course, this also deals with the conservation functions of BRs.

5.4 Socialization of Benefits

To address the need to improve livelihood conditions of local populations, the sustainability pilot project must include systematic efforts to socialize its benefits, so that it benefits everyone, not only a privileged few. This is an important and complex step that involves reorganizing the islands' economy. Nowadays, some receive great benefits from the islands, even if they hardly repay or contribute to the protection of the natural and social bases that support their activities; that is the case with large hotel chains and airlines. Some steps are being taken towards a more equitable distribution of tourism benefits through native inns, for example, although this model does not compensate for the use and abuse of its natural and cultural bases either. It is therefore necessary to design measures to correct this imbalance, such as specific taxes on tourist activities, as considered later.

5.5 Ethnicity and Culture

The ethnic and cultural issue deserves special attention, because the original population of the archipelago constitutes an ethnic group, the Raizal People, with ethnic, cultural, and territorial rights, and a fundamental role in the sustainable development of the archipelago; this is recognized by the Political Constitution of Colombia in its Article 310 (Colombia 1991). In this regard, it is important to bear in mind what Judgment C-053 of 1999 states:

> The Court admitted that the territory of the native community of the archipelago is constituted by the islands, cays and islets included within that territorial entity. The eventual withdrawal of the Raizal population in certain areas of the islands is nothing more than the symptom of the need to provide real protection to the cultural rights of the Raizales (Constitutional Court of Colombia 1999).

To protect traditions, it is also necessary to promote forms of land use and production in accordance with local customs.

5.6 Territorial Planning

In this sense, territorial planning processes are very important. The BRs have a scheme based on three categories of use: conservation, mitigation (buffer), and sustainable use, which must be integrated into the Territorial Planning Scheme (EOT using its Spanish initials), in the Basins Management Plans (POMCAs) and especially in the Management Plan for the Marine Coastal Environmental Unit of the archipelago (POMIUAC), which includes the management of the adjacent sea. Marine conservation and proper management are an integral part of a sustainable development project, since the islands depend on the sea, and mainly on the reefs, for many of their economic activities. In this sense, Seaflower's Marine Protected Areas System and its zoning of the sea, play a fundamental role that must be fully fulfilled to achieve its purposes.

5.7 Education and Participation in Sustainability

Education is a fundamental component of a process of rethinking development, because without the active, conscious, and prepared participation of the population, there is little chance for success.

5.8 Other Possible Actions

- Constitute a Land Bank for the purchase of land within the Seaflower BR, to avoid alienation and the loss of local control over territory.
- Give preferential access to the Raizal population over island resources, especially fishing, which must be entirely artisanal with industrial fishing prohibited. Diplomatic and legal actions to prevent illegal fishing must be undertaken.
- Regulate investment in the archipelago so that investors reinvest in the islands part of the profits obtained in or through them.
- Protect and promote associative forms of real estate ownership for companies present in the archipelago (Article 58 of the Political Constitution) and prohibit the privatization of goods and services such as aqueducts, docks, and the airport.
- Give administrative autonomy to the Raizal people for the management of resources, plans, programs, and projects, without dependence on the national government and under the supervision of international entities.
- Two other important aspects concern payment schemes for environmental services (PES) and the creation of a Seaflower Fund for the financing of the Reserve, which are discussed below, in the context of climate change analyses.

5.9 The Gran Seaflower Initiative

As mentioned, what I called the Seaflower Initiative is prior and different from the Gran Seaflower Initiative, a more recent and widely diffused proposal. According to its promoters:

> Gran Seaflower is an environmental and cultural region within the Southwest Caribbean inhabited by a diversity of people and cross-border ethnic identities. In addition, it is the most biodiverse marine-coastal place in the Western Hemisphere. Its heart is the Seaflower Biosphere Reserve recognized by UNESCO in the San Andrés archipelago… Harmony with nature in this crucial marine-coastal area can only be achieved through a regional consensus – this is the Gran Seaflower initiative. It's about creating high-level partnerships between six Caribbean countries – Colombia, Costa Rica, Honduras, Jamaica, Nicaragua and Panama (Gran Seaflower Initiative 2020).

The Gran Seaflower Initiative could assume, support and realize some of the ideas and actions previously explained as the Seaflower Initiative but, until now, the two initiatives, even if coherent, remain independent.

6 Climate Change

Climate change, that is the alteration in the behavior of the planetary climate system, is a fact. As the IPCC report 5 (AR5) indicates, the human influence on the climate system is clear, especially through the emission of greenhouse gasses, leading to an "unequivocal warming of the climate system the atmosphere and oceans have warmed… and sea level has risen" (IPCC 2014, p. 2) and a situation where "Climate changes have caused an impact on natural and human systems on all continents and across the oceans" (p. 6). Growing evidence of extreme weather events such as floods, droughts, and heat waves, and of course, hurricanes such as Iota which devastated Providencia and Santa Catalina in November 2020, point to the need to act on climate change. Given this, adopting complementary adaptation measures is proposed, such as better preparation for climate extremes, as well as mitigation measures, such as emissions control, that help reduce and manage the risks associated with climate change. Projects to protect and recover the still well-preserved coral reef complexes of the archipelago are an important possibility to improve conditions for carbon sequestering and for enhancing reef health to resist climate change.

6.1 Seaflower, Coral Complexes, and Climate Change

The Seaflower BR includes the largest coral reef areas in Colombia and some of the largest in the Caribbean (Díaz et al. 2000). These reefs play a dual role in the context of climate change because they are among the ecosystems most threatened by global warming and climate change, but can also contribute significantly to its mitigation. Corals are extremely sensitive to global warming because they have evolved under stable tropical temperature conditions and their thermal tolerance ranges are narrow. Under thermal stress, they expel their symbiotic algae and lose their colors, in a phenomenon called bleaching; when bleaching is very intense and prolonged, corals can die. Several episodes of bleaching have been reported in the archipelago (Díaz et al. 2000), however, they have not, so far, been extreme and recovery has been good (Navas-Camacho et al. 2019), suggesting that Seaflower could be a climate refugia.

But in the context of climate change, increasingly extreme events and the risk of mass die-offs that could be very serious and affect reefs in different ways are expected (NOAA 2021). Indeed, if many corals die, dangerous processes would be triggered, since most reef biodiversity depends directly or indirectly on corals (Díaz et al. 2000). Mass deaths of corals and other organisms, and their subsequent decomposition, increase the demand for oxygen and thus tend to deplete it, causing more deaths, in a snowball process of deterioration. Decomposition would accelerate the processes of water acidification which are already occurring in the oceans, leading to the dissolution of the calcareous skeletons of corals and the huge calcareous structures that make up reefs (Kault et al. 2022; NOAA 2021). This would release enormous amounts of CO_2 into the atmosphere, which could generate a vicious spiral in which higher temperatures kill corals, generating more CO_2 release, and thus causing even higher temperatures and more deaths.

From a very pessimistic perspective, it can be thought that an imbalance of this nature would lead to a total imbalance of the Earth's climate, as coral reefs around the world die and decompose. This process can be even more serious than deforestation, as it is also more difficult to control once it is triggered. Thus, care for the extensive

Seaflower reef area would play a very important role, without even considering the many other implications for local society through impacts on fishing or tourism, so important in the Caribbean and other reef areas.

Considering the other role of reefs in climate change—as carbon accumulators—in contrast with the apocalyptic perspective, if reefs are preserved they will continue accumulating carbon, contributing to reducing its concentration in the atmosphere. With a plus: terrestrial vegetation, especially tropical rainforests, once they accumulate a large amount of biomass and carbon, enter a state of equilibrium in which almost all the new carbon they remove from the atmosphere through photosynthesis, is returned to it through respiration; mature reefs behave in a similar way (Kault et al. 2022), but, on the other hand, reefs accumulate as much carbon as calcium carbonate. This can accumulate indefinitely, not only in the form of reefs but as calcareous sands and sediments that, in fact, make the ocean the largest carbon reservoir on the planet (Biologydictionary.net 2017). This reservoir is an almost tight reservoir unless climate change and ocean acidification destroy it.

6.2 Adaptation and Mitigation Measures in Seaflower

As explained, coral reef conservation is a very important part of the efforts needed to mitigate global climate change, and Seaflower has an important role to play in this. The questions are: what can be done? And above all, how can it be done?

To know what to do we also have to know what is happening and what is affecting the reefs. To do this, it can be useful to differentiate local from global agents of change. Among these is the general phenomenon of climate change, which does not depend on, or is not very influenceable directly by the inhabitants of a given area, in this case Seaflower. Instead, we can try to manage local factors, for example, local sources of pollution and sedimentation, direct damage to reefs by tourism (boats, divers), and other human activities (dredging, docks, maritime transport), including industrial and illegal fishing and overfishing, or more specific issues, such as parrotfish protection.

For this type of local agents, it is more feasible to propose effective actions which, in turn, raises the question of how. Economics is very important in this regard, as any action will have associated costs of at least two types: the direct costs of proposed interventions, and those generated by action or by omission of actions. The first would be, for example, compensation for actions of conservation, regeneration, and restoration of ecosystems, control, and surveillance to avoid deterioration, or those works that are required (for example, research, restoration of reefs, sewage treatment plants).

The latter, very important, arises from the question: what would be the cost of doing nothing? The likely cost is enormous, to the extent that climate change would end up destroying the reefs and with them many economic possibilities, and even island life itself, including tourism, trade, and fishing.

7 Environmental or Ecosystem Goods and Services

An approximation to this cost was made through a value estimation of Seaflower environmental or ecosystem goods and services, that is, those that Seaflower's ecosystems provide to society. Goods refer to tangible objects that have an economic value, like fishes or beaches, while services are intangible activities and functions (like climate regulation, biodiversity refugia or the attraction of tourism) that meet human or societal needs, and represent, likewise, an economic value. According to Prato and Newball (2015), services refer to the benefits that ecosystems provide to people or "direct and indirect ecosystems contribution to human wellbeing". Figure 2 (Forest Trends and The Katoomba Group 2010) classifies and presents some examples of services. In the case of environmental goods and services, it happens that their economic value has been little recognized and therefore they are not paid for, that is, in a certain way, their status as such is not recognized, in what constitutes a so-called market failure (Forest Trends and The Katoomba Group 2010).

Since the end of the last century, economists have been trying to solve this market failure, beginning with the identification of said goods and services which have been classified into categories, as presented in Fig. 2, and with valuation exercises of their economic value. In this sense, the classic work of Constanza and collaborators (1997) proposes a global valuation with very questioned results, but highlighting the enormous contribution of nature to humanity that, according to their estimates, is several times the gross domestic product (GDP) of the formal world economy: the contribution of nature is equivalent to USD 33 trillion, around 20 of those coming from marine ecosystems (Constanza et al. 1997). It should be noted that many people

Fig. 2 Types of ecosystem services. Elaborated by the author based on Forest Trends and the Katoomba Group (2010)

do not agree with the notion of ecological goods and services, mainly because it means monetizing the contribution of nature, and thus creates the implicit risk of its privatization—for instance, of water—to the detriment of society. However, it is also true that not knowing the value of these contributions has allowed the destruction of nature for the sake of an alleged development measured in currency. It is worth bearing in mind these considerations when raising the issue of PES, as they have risks, although they should have to be used because they are, at least for the moment, one of the main economic alternatives for defending, preserving, and restoring nature.

7.1 Seaflower Goods and Services

In a study published by the Colombian Commission for the Ocean (CCO) (Prato and Newball 2015), it is pointed out that Seaflower ecosystems, including both marine and terrestrial ones, provide ecosystem goods and services that would have an enormous value: USD 267,000 to 353,000 million a year. Many of these goods and services come from coral reefs, such as food production, tourism, and protection against coastal erosion and natural phenomena such as hurricanes. It also includes biodiversity maintenance services, climate regulation, carbon sequestration, and oxygen production, among others. This suggests a significant market failure, since it is not accounted for or entered into the accounts of Colombia or any other state, even though goods and services favor not only the archipelago and its inhabitants, but also the country and the world. This consideration leads one to think that correcting, at least partially, this market failure is part of the how, now being tried worldwide through what is called Payment for Environmental Services (PES), among alternatives that are analyzed below, including direct costs.

7.2 Payment for Environmental Goods and Services

Payment for environmental services (PES) is being made according to what is called the PES scheme, which consists of a voluntary social and economic agreement wherein an actor—the buyer—acquires a given environmental service (which must be very well defined), by paying another or others—the supplier(s)—in exchange for ensuring that the ES continues to be provided (Forest Trends and The Katoomba Group 2010). This has also been subject to criticism because of the mentioned risks of misuse for the privatization of nature, so one must be cautious. Nevertheless, PES is also a way to recognize natural services.

PESs are regulated in Colombia (Colombia 2017), where payments are essentially to compensate for ceasing an environmentally harmful economic activity, paying for what is no longer earned. For example, if a farmer grows potatoes in an area someone (mainly the state) wants to protect, earnings of potato crop production are calculated, and the farmer is paid that amount. In this case, the state is the buyer of the service and the farmer is the supplier, and both accept the arrangement voluntarily and by mutual agreement, which is highly important. The payment can be made with state resources for conservation and for a limited time, up to five years, after which the idea is that the land, now recovered, is purchased and integrated into the corresponding reserve or protected area (Colombia 2017).

PESs are rather new in most of the world, but have antecedents in the so-called *swaps* or debt-for-nature exchanges (Dogsé and von Droste 1990) that have been tried, with limited success, since the 1980s. In swaps, external debt was exchanged for conservation programs, for instance, designations of protected areas. The logic is similar, since the protected area ceases to be used for direct productive purposes (as in PES), but the problem was that payments were very low, since it was a one-off and not based on the value of the permanent goods and services offered, which were generally underestimated. It also implied a certain abuse of the economic fragility of the countries involved, pressuring them, even with the best of intentions, to take on unwanted obligations (see Sevilla 1990; Márquez 1992). Despite all of this, swaps are still in use.

7.3 Applications of PES in Marine Ecosystems

In the case of marine ecosystems, there are already some experiences that shed light on what could be done in Seaflower. In a work already cited (Forest Trends and The Katoomba Group 2010), some examples are identified.

- One of them, an example of what is called public payments, is a fund in Tanzania, East Africa, created with resources from fishing licenses, ecotourism, and taxes on oil and gas exploitations. The Fund pays coastal inhabitants for habitat protection and sustainable use of coastal marine resources and supports conservation activities. As will be seen, something similar has been proposed for Seaflower in the past.
- Another is an example of what they call open trade regulation, consisting of buying, but not using, fishing rights that are granted to fishermen and companies, according to the fishing quota. As a system of quotas is also granted to fishermen in the archipelago, the feasibility of such a mechanism could be studied.
- The third example refers to private schemes or agreements and mentions the leasing of reefs in the Fiji Islands. In Colombia, there is no ownership of reefs or marine areas, and the rights of owners in the coastal zone are limited, so this model would not seem to have much application. However, this does not prevent the possibility that, since these are voluntary and private agreements, their application can be explored in the archipelago, for example, in the case of mangroves that are next to private properties.
- A fourth example refers to PES in marine protected areas (MPAs) and the case of Bonaire, where a marine national park protects the reefs, which in turn supports diving, the basis of tourism to that island. A charge is established for the right

to dive and for the anchoring and entry of boats, and voluntary donations are also received. As an MPA system is established and there are marine parks in the archipelago, conditions are established for PES schemes on this basis, which has also been raised in the past.

7.4 PES and Seaflower

To apply PES to the protection of Seaflower, a study would have to be done and PES schemes designed, both of which go beyond the scope of this article. However, some possible features can be sketched. The objective would be to protect terrestrial or marine ecosystems, mainly reefs, in order to fight climate change. What would have to be resolved is, first, what kind of activities (done or not done) would be paid and to whom (see Fig. 2), and second, where the funds for such payments would come from.

For the archipelago, in an initial exploration of the subject, it has been thought that a possible payment would be to compensate the fishermen of the islands. A hypothetical case for Providencia and Santa Catalina could be to stop fishing in Queena (Quitasueño) or Roncador cays, according to fishermen leader Edgar Jay. Queena is the main lobster fishing area of the archipelago, whose lobster catch value could be estimated, and then paid for through a PES scheme. An agreement to remove fishing pressure and let Queena rest for some time could be applied by areas, for example, Serrana and Roncador, or by sectors of the marine platform of Providence and Santa Catalina, or the no capture (*no take*) zones of the marine protected areas (MPAs).

On its own, such an agreement would have a great effect on reef and reef resource conservation, and could justify a significant payment for fishermen who could thus access resources that, amid the humanitarian crisis generated by Hurricane Iota, are now greatly needed. In the longer term, it would bring environmental, social, and economic benefits as reefs and their resources recover. The scheme would have to include many complementary aspects, such as the control of fishing pressure in other areas, since it makes no sense to leave Queena resting while overfishing Serrana or Providencia, as well as agreements with the authorities to enforce non-fishing agreements because it also makes no sense to stop artisanal fishing if industrial and illegal fishing continues.

There are also more activities that could be identified as susceptible to PES, for example (to name some):

- Recovery and restoration of mangroves affected by Iota (another urgent need that would also have effects on climate).
- Protection of other ecosystems, such as seagrasses.
- Reintroduction and repopulation with sea turtles, a key element for the recovery of the ecological balances of archipelago reef complexes.
- Surveillance and control to ensure PES schemes compliance, and to reinforce authorities in areas with prohibitions on fishing.

- Specific checks and controls on sensitive species such as parrotfish, groupers, and chernas, among others.
- Reduction and negotiation of fishing quotas.

7.5 Seaflower Fund

The issue of how to finance activities is critical. For a successful long-term PES project, it is necessary to seek sufficient and permanent resources. This could perhaps be achieved through the creation of a fund like the Tanzanian fund mentioned previously, complemented with ideas that had been proposed for Seaflower (Márquez 2016), including possible funding sources like:

- The Colombian state through budgets must be invested in the archipelago's development.
- An aggregated value tax (IVA in Spanish) for tourist activities (10% per room per night, for example).
- Specific participation in income per tourist card, which is a tax that people pay to enter the archipelago.
- Voluntary donations, from an extra dollar contributed by a tourist to contributions by companies or philanthropists.
- Contributions by the inhabitants of the archipelago.
- Contributions by companies already established on the islands and which benefit from them, including, for example, airlines, hotel chains, supermarkets, merchants, and fuel distributors.

Another very important aspect is the negotiation of international support for Seaflower conservation, as the environmental services it provides benefit not only the archipelago but also the country and the world; this has been proposed more than once by the present author. A similar idea was very recently proposed by the Colombian government but based on debt-for-nature swaps. As previously explained, this is not the most favorable option, even if it has allowed countries like the Seychelles Islands to reduce their external debt and substantially increase their system of MPAs. However, PES is much fairer. Colombia established BRs and its MPA system without negotiating a specific agreement, and assumed risks and costs to protect goods and services that benefit many: people, countries, international organizations—especially those interested in marine issues, tourists, divers, researchers, and of course those who benefit from business on the islands, among other beneficiaries that should retribute some of the benefits. Only local people pay with the potential detriment of their living conditions, submitted to restrictions in the use of their territory, even if they also benefit from Seaflower.

8 Final Considerations

The Seaflower Biosphere Reserve is a project that is in force and appears to be becoming increasingly important, although it is far from having been fully understood and properly incorporated in both the archipelago's and Colombia's development processes. It is playing a key role in the context of the territorial conflict with Nicaragua. Under these conditions, it seems pertinent and possible to reassume and reinforce Seaflower in accordance with the original Biosphere Reserve concept, so that it reaches its highest goals of contributing to the protection of a very important natural and cultural heritage, for the benefit not only of its inhabitants, but also for those of the whole country and the whole world. In this context, the Gran Seaflower Initiative should be considered. Alternatives analyzed here are only part of the possible actions: the most important thing is to undertake them, taking advantage of the opportunity offered, somewhat paradoxically, by the ongoing crisis after the COVID-19 pandemic and aggravated by Hurricanes Eta and Iota. As said in a recent documentary (Welcome 2021), it's the moment to "Arise Seaflower!"

References

Batisse M (1986) Developing and focusing the biosphere reserve concept. Nat Resour 22(3):2–11

Biologydictionary.net Editors (2017) Carbon cycle reservoirs. Biology Dictionary, 25 September 2017. https://biologydictionary.net/carbon-cycle-reservoirs/

Colombia (1991) Constitución Política. Gaceta Constitucional No. 116, 20 Jul 1991

Colombia (1993) Ley 99 de 1993. Por la cual se crea el Ministerio del Medio Ambiente, se reordena el Sector Público encargado de la gestión y conservación del medio ambiente y los recursos naturales renovables, se organiza el Sistema Nacional Ambiental, SINA y se dictan otras disposiciones. Diario Oficial 41146, 22 Dec 1993

Colombia (2015) Decreto 1076 de 2015 Por medio del cual se expide el Decreto Único Reglamentario del Sector Ambiente y Desarrollo Sostenible. Ministerio de Ambiente y Desarrollo Sostenible, 26 May 2015. http://www.ideam.gov.co/documents/11769/46844622/Dec+1076_2015.pdf/8c28b13e-0937-42bd-b4a2-4b99114f9362

Colombia (2017) Decreto 870 de 2017 (25 de mayo) Por el cual se establece el Pago por Servicios Ambientales y otros incentivos a la conservación, 25 May 2017. https://encolombia.com/medio-ambiente/normas-a/pago-servicios-ambientales/

Constanza R, Arge R, De Groot R et al (1997) The value of the world's ecosystem services and natural capital. Nature 387:253–260

Constitutional Court of Colombia (1999) Sentencia C-053 de 1999, 2 Feb 1999. MP: Eduardo Cifuentes Muñoz

Díaz JM, Barrios LM, Cendales MH et al (2000) Áreas coralinas de Colombia. INVEMAR, Serie Publicaciones Especiales, 5, Santa Marta, Colombia

Dogsé P, Von Droste B (1990) Debt-for-nature exchanges and biosphere reserves: experiences and potentials. MAB Digest Series 6. UNESCO. Paris

Forest Trends and the Katoomba Group (2010) Payments for ecosystem services: getting started in marine and coastal ecosystems - a primer. https://www.forest-trends.org/wp-content/uploads/imported/marine-coastal-pes-getting-started_2010-pdf.pdf

Frankignoulle M, Gattuso J-P (1993) Air-sea CO_2 exchange in coastal ecosystems. In: Wollast R, Mackenzie FT, Chou L (eds) Interactions of C, N, P and S biogeochemical cycles and global change. NATO ASI Series 4. Springer, Berlin, pp 233–248. https://doi.org/10.1007/978-3-642-76064-8_9

Gran Seaflower Initiative (2020) ¿Qué es Gran Seaflower? https://granseaflower.com/about/

Halffter G (1984) Las Reservas de la Biosfera: Conservación de la naturaleza para el hombre. Acta Zool Mex (ns) 5:31–50. Instituto de Ecología: Mexico City

Intergovernmental Panel on Climate Change (IPCC) (2014) Climate change 2014: synthesis report. Contribution of Working Groups I, II and III to the Fifth Assessment Report of the Intergovernmental Panel on Climate Change [Core Writing Team, Pachauri RK, Meyer LA (eds)]. IPCC, Geneva, Switzerland. https://www.ipcc.ch/site/assets/uploads/2018/02/SYR_AR5_FINAL_full.pdf

Ishwaran N, Persic A, Tri NH (2008) Concept and practice: the case of UNESCO biosphere reserves'. Int J Environ Sust Dev 7(2):118–131. https://doi.org/10.1504/IJESD.2008.018358

Kault J, Jacob F, Detournay O (2022) Carbon balance in corals. Coral Guardian. https://www.coralguardian.org/en/carbon-balance-in-corals/

Márquez G (1992) Desarrollo sostenible y conservación: propuesta de Reservas de Biósfera y un caso de estudio para Colombia: las islas de Providencia y Santa Catalina. In: Britton Howard AD, Archbold Ramírez JF, Newball Bryan C et al (eds) El archipiélago posible: ecología, reserva de biosfera y desarrollo sostenible en San Andrés, Providencia y Santa Catalina. Universidad Nacional de Colombia, Bogotá, Colombia, pp 51–78

Márquez G (2014a) Iniciativa Seaflower. El Isleño (21 May 2014). https://www.xn--elisleo-9za.com/index.php?option=com_content&view=article&id=7675:iniciativa-seaflower&catid=47:columnas&Itemid=86

Márquez G (2014b) Iniciativa Providencia, Iniciativa Seaflower o Iniciativa Colombia. Revista Cuadernos del Caribe 17. Universidad Nacional de Colombia. Sede Caribe. San Andrés, Colombia

Márquez G (2016) Ideas para una iniciativa llamada Seaflower. El Isleño (24 January 2016). https://www.xn--elisleo-9za.com/index.php?option=com_content&view=article&id=10804:ideas-para-la-iniciativa-seaflower&catid=47:columnas&Itemid=86

Márquez G, Pérez ME (eds) (1992) Desarrollo sostenible del Archipiélago de San Andrés y Providencia: Perspectivas y acciones posibles. Proyecto Multinacional del Medio Ambiente y los Recursos Naturales. OEA - COLCIENCIAS - IDEA/UN. Bogotá, Colombia

Márquez G, Pérez ME, Britton A et al (2006) El Archipiélago posible: ecología, reserva de biosfera y desarrollo sostenible en el Archipiélago de San Andrés. Providencia y Santa Catalina. Universidad Nacional de Colombia. Sede Caribe, San Andrés, Colombia

Márquez G, Márquez AI (2016) Turismo y medio ambiente en el Caribe: Pros y contras. In: Mantilla S, Velásquez C, Román R et al (comps) Desarrollo y turismo sostenible en el Caribe. Universidad Nacional de Colombia

Navas-Camacho R, Acosta-Chaparro A, González-Corredor JD et al (2019) 20 years (1998–2017) of coral formations monitoring in San Andrés and Providencia. General Publications Series No. 106. INVEMAR-CORALINA, Santa Marta. https://n2t.net/ark:/81239/m9sd44

NOAA (National Ocean Service) (2021) ¿How does climate change affect coral reefs? https://oceanservice.noaa.gov/facts/coralreef-climate.html

Pérez-García ME, Márquez G (1993) Gestión del desarrollo en las Islas de Providencia y Santa Catalina. Memorias del Taller en Providencia Isla, 18 & 19 June 1993. OEA - COLCIENCIAS-IDEA/UN. Bogotá, Colombia

Prato J, Newball R (2015) Aproximación a la valoración económica ambiental del departamento Archipiélago de San Andrés, Providencia y Santa Catalina – Reserva de la Biósfera Seaflower. Secretaría Ejecutiva de la Comisión Colombiana del Océano SECCO, Corporación para el desarrollo sostenible del Archipiélago de San Andrés, Providencia y Santa Catalina -CORALINA. Bogotá, Colombia

Sevilla R (1990) El canje de la deuda por conservación: los casos de Bolivia, Ecuador y Costa Rica. In: Maihold G, Urquidi VL (eds) Diálogo con Nuestro Futuro Común: Perspectivas latinoamericanas del Informe Brundtland. Ediciones Nueva Sociedad: Caracas, Venezuela, pp 138–161

UNESCO (2022) Biosphere Reserves. https://en.unesco.org/node/314143

Welcome (2021) ¡Arise Seaflower! Documentary. Casa Editorial Welcome: San Andrés, Colombia

Open Access This chapter is licensed under the terms of the Creative Commons Attribution 4.0 International License (http://creativecommons.org/licenses/by/4.0/), which permits use, sharing, adaptation, distribution and reproduction in any medium or format, as long as you give appropriate credit to the original author(s) and the source, provide a link to the Creative Commons license and indicate if changes were made.

The images or other third party material in this chapter are included in the chapter's Creative Commons license, unless indicated otherwise in a credit line to the material. If material is not included in the chapter's Creative Commons license and your intended use is not permitted by statutory regulation or exceeds the permitted use, you will need to obtain permission directly from the copyright holder.

Marine Ecosystem Services for Climate Change Adaptation and Mitigation Strategies in the Seaflower Biosphere Reserve: Coastal Protection and Fish Biodiversity Refuge at Caribbean Insular Territories

Julián Prato, Adriana Santos-Martínez, Amílcar Leví Cupul-Magaña, Diana Castaño, José Ernesto Mancera Pineda, Jairo Medina, Arnold Hudson, Juan C. Mejía-Rentería, Carolina Sofia Velásquez-Calderòn, Germán Márquez, Diana Morales-de-Anda, Matthias Wolff, and Peter W. Schuhmann

Abstract Insular and coastal territories like those in the Seaflower Biosphere Reserve are exposed to strong winds, waves, storms, and hurricanes. In November 2020, Hurricanes Eta and Iota provided a costly reminder of the risks facing Seaflower's people and ecosystems. Coral reefs and mangroves are natural shields, reducing wind and wave strength during normal and extreme conditions. These

---

J. Prato (✉) · A. Santos-Martínez · D. Castaño
Universidad Nacional de Colombia, Caribbean Campus, San Andrés, Colombia
e-mail: jprato@unal.edu.co

A. Santos-Martínez
e-mail: asantosma@unal.edu.co

D. Castaño
e-mail: dcastano@unal.edu.co

J. Prato · A. Santos-Martínez · J. E. M. Pineda · J. Medina · J. C. Mejía-Rentería
Corporation Center of Excellence in Marine Sciences, CEMarin, Bogotá, Colombia
e-mail: jemancerap@unal.edu.co

J. Medina
e-mail: jhmedinac@unal.edu.co

A. L. Cupul-Magaña · D. Morales-de-Anda
Universidad de Guadalajara, Centro Universitario de La Costa, Puerto Vallarta, México

J. E. M. Pineda
Universidad Nacional de Colombia, Bogotá Campus, Bogotá, Colombia

J. Medina
Universidad Nacional de Colombia, Caribbean Campus, San Andrés Botanical Garden, San Andrés, Colombia

© The Author(s) 2025
J. E. Mancera Pineda et al. (eds.), *Climate Change Adaptation and Mitigation in the Seaflower Biosphere Reserve*, Disaster Risk Reduction,
https://doi.org/10.1007/978-981-97-6663-5_8

coastal protection ecosystem services (ES) are vital for human safety and well-being, and become more important given the heightened vulnerability of low-lying insular islands to climate change impacts. These ecosystems also provide biodiversity refuge ES for fishes and shellfish, key for food security and resilience to global challenges like hurricanes, sea level rise, and global warming. Despite their importance, these valuable ecosystems are threatened by anthropogenic pressures, jeopardizing the survival and well-being of islanders; their restoration and recovery require improved management and decision-making, and heightened societal awareness of our dependence on marine ecosystems and their potential as climate change adaptation solutions. We identify ES provided by coral reefs and mangroves, interdisciplinary management tools, and recommendations to motivate society and decision-makers to expand efforts for the protection, restoration, and use of these ecosystems as Nature-based Solutions for climate change adaptation and mitigation in Seaflower.

Keywords Coastal management · Climate change · Ecosystem-based adaptation · Marine ecosystem services · Nature-based solutions

1 Introduction

Ecosystem services (ES) encompass a wide variety of direct and indirect contributions from ecosystems to human well-being (Burkhard and Maes 2017). An appreciation of the value of ES allows society to better understand our dependence on nature and biodiversity (Sánchez 2021), and provides a framework for policy and decision-making related to the sustainable management and use of natural resources, including environmental protection, Nature-based Solutions (NbS), climate change

---

A. Hudson
San Andrés Raizal Community, San Andrés, San Andres and Providencia, Colombia

J. C. Mejía-Rentería
Department of Biology, ECOMANGLARES Research Group, Universidad del Valle, Cali, Colombia

C. S. Velásquez-Calderòn
Department of Geography, Florida State University, Tallahassee, FL, US

G. Márquez
PROSEALAND Foundation, Bogotá, Colombia

M. Wolff
ZMT Leibniz Centre for Tropical Marine Research, Bremen, Germany
e-mail: matthias.wolff@leibniz-zmt.de

P. W. Schuhmann (✉)
Department of Economics and Finance, University of North Carolina Wilmington, Wilmington, NC, US
e-mail: schuhmannp@uncw.edu

adaptation, and disaster risk reduction (Waite et al. 2014; Prato and Newball 2016; Burkhard and Maes 2017).

As hazards associated with climate change increase, disaster risk, and disaster-related losses are expected to increase in frequency and magnitude (IPCC 2014, 2021). Investments in the protection and maintenance of coastal and marine ecosystems can help coastal communities mitigate these losses and adapt to climate change. The natural assets contained within the Archipelago of San Andrés, Providencia, and Santa Catalina (hereafter, the archipelago) are a perfect example of the potential for proper management of natural capital to overcome vulnerability to climate change and generate Ecosystem-based Adaptation (EbA) strategies to safeguard economic activity and human well-being. Declared a Biosphere Reserve (BR) by UNESCO in 2000 due to its natural richness, culture, and sustainable management opportunities, the Seaflower BR (SBR) contains around 77% of Colombian coral reefs, with 9 main reef islands partially protected by barrier reefs, multiple cays, and a volcanic basement (Sánchez et al. 2005; Guarderas et al. 2008; Coralina-Invemar 2012; Prato and Newball 2016).

The archipelago's ecosystems include mangroves, seagrass beds, coral reefs, and the open ocean, all of which harbor a wide variety of marine species (Friedlander et al. 2003; Coralina-Invemar 2012) and play a crucial role in providing food, tourist attractions, and protection from waves and winds. The coastal protection ES provided by the archipelago's ecosystems is especially valuable due to the remote, oceanic context and the low-lying nature of the islands and cays (Prato and Newball 2016), which put the islands and their inhabitants at heightened risk from tropical storms and hurricanes (IPCC 2021; Coralina-Invemar 2012). Indeed, the SBR has been recognized as the most vulnerable region of Colombia (Ideam et al. 2017).

In 2005, Category 1 Hurricane Beta affected Providencia and Santa Catalina, with loss assessment in coral reefs showing 20% mortality, mainly in the northern part of the islands, and coral bleaching on the west side of the island (Taylor et al. 2008b). As a result, multiple projects were implemented for the recovery and restoration of beaches, mangroves, and coral reefs. Sixteen years later, in 2020, Category 4 Hurricane Iota hit Providencia and Santa Catalina, destroying nearly 98% of the island's houses. The Institute for Marine and Coastal Research, INVEMAR, estimated damage to 80% of mangroves, 90% of tropical forests, and severe damage to coral reefs located approximately 12 m deep. While highly affected, these ecosystems provided a clearly important protection for human life, by reducing wind and waves during the storm. They also found coral bleaching signs in 41% of the sites evaluated (INVEMAR 2021). Due to the importance of marine ecosystems for islanders' well-being and biodiversity, CORALINA and the Universidad Nacional de Colombia's Caribbean campus, led the elaboration of restoration protocols for marine ecosystems after hurricanes with the participation of other local, national, and international institutions, to provide guidelines to recover the natural capital that is vital for the SBR's resilience and climate change adaptation strategies (Velásquez-Calderón et al. 2022).

In addition to protection from hazardous events, healthy coral reefs, mangroves, seagrasses, and the ocean are vital for food security for the archipelago's people

(Coralina-Invemar 2012; Santos-Martínez et al. 2013). These ecosystems provide provisioning ES by supplying protein-rich food sources like fish, queen conch, lobster, octopus, and other shellfish (Cooper et al. 2009; Rueda et al. 2010; Prato and Newball 2016). Given that more than 90% of food supplies in the archipelago are imported—particularly in the two most-populated islands, San Andrés and Providencia—the need for sustainable food sources to safeguard human well-being during times of crisis is critical.

Despite their obvious importance and economic value, only 1% of the value estimated from marine ES in the Colombian Caribbean and the SBR has been reflected in national statistics and accounts (Prato and Reyna 2015; Prato and Newball 2016) and most non-market values of marine ES are not considered in national and local institutional accounting systems. This gap of around 99% of real values and benefits from coastal and marine ecosystems (Prato and Newball 2016) can lead to misinformed decision-making with negative consequences for ecosystem functioning, economic activities, and human well-being (Ranganathan et al. 2008). Together, these facts highlight the need to estimate the economic value of ES provided by the natural assets of the archipelago so that their true worth can be incorporated into decision-making processes, climate change adaptation strategies, and national accounting systems (Waite et al. 2014; Sánchez 2021).

In this chapter, we present an interdisciplinary approach to better understand the importance of ecosystems and ES for well-being, sustainability, and ecosystem-based climate change adaptation strategies in the SBR since it is time for action. Our review forms the basis for recommendations for future research to guide efficient and sustainable management of the SBR's natural assets.

2 Mangroves for Coastal Protection and Human Well-Being in Seaflower

Mangroves are the predominant wetlands along the world's tropical and subtropical coasts, occupying around 13,586,000 ha (Worthington et al. 2020) and representing 0.7% of the planet's total tropical forests (Spalding et al. 2010). Mangroves are widely recognized for providing valuable ES critical for the social, economic, and cultural development of communities in coastal and insular areas.

The main ES provided by mangroves includes biodiversity, refuge, and trophic subsidy for multiple biological groups, many of which support important fisheries (Lee et al. 2014). Likewise, mangroves sequester and store large amounts of carbon (C)—called blue carbon—for very long periods, thus contributing to climate regulation and sediment stability (Donato et al. 2011). Numerous investigations have quantified the C retention capacity of mangroves, showing that it can be up to five times greater than that stored in tropical forests (McLeod et al. 2011). Globally, C stocks in mangroves vary between 50 and 2,200 Mg C/ha (Bindoff et al. 2019). Worldwide, mangrove forests store the equivalent of 22.86 gigatons of CO_2, the loss of just 1% of remaining mangroves could release 0.23 gigatons of CO_2, equivalent to the annual emissions of 49 million cars in the USA (Leal and Spalding 2022). In 2019, the World Bank noted that 51% of the emissions covered by C pricing initiatives were below US$10 per ton (t) of CO_2 equivalent (1 t/CO_2 = 1 C credit). In 2017, the Colombian government estimated the value of potential CO_2 emissions at US$5.08/t. San Andrés' mangroves are recognized for their high carbon storage capacity, storing around 2,658 Mg C/ha, meaning they can store around 37% of C emissions produced by the roughly 900 annual flights (7,170 Mg C) that operate on the island (Medina 2022).

Mangroves also function as connectors between terrestrial and marine environments, regulating the water quality of adjacent ecosystems (Feller et al. 2010). They contribute to the mitigation of coastal erosion, and protect against sea level rise and extreme weather events (McLeod et al. 2011; Sánchez-Núñez et al. 2019). The complex structures of mangrove vegetation, including aerial roots and associated biota, dissipate wave energy between 5 and 39% per meter of displacement (Morris et al. 2018; Sánchez-Núñez et al. 2020) and facilitate the deposition and retention of sediments (Sánchez-Núñez et al. 2020).

Healthy mangroves have even been shown to reduce economic impacts after extreme weather events (Hochard et al. 2019). Economic analysis of 22 cyclone-impacted countries showed that activity declined less and recovered more quickly in places with more extensive mangrove forests along the coastline (Morris et al. 2018). This has made it possible to estimate the economic value of this ES between US$3,679 and 693 ha/year, and has led to concepts such as "building with nature" or "living coasts" in which coastal erosion is controlled through natural marine ecosystems (Morris et al. 2018).

The mangrove forests of the SBR represent critical natural capital that directly benefits both resident and visiting populations. On the islands of Providencia and Santa Catalina, the largest extension of mangroves is found in the McBean Lagoon National Natural Park. The mangroves of southeast Santa Catalina—Manchineel Bay, South West Bay, and Old Town—also stand out. The mangroves on San Andrés currently occupy an extension of 96.98 ha, distributed across six main forests plus other minor mangrove areas. A multi-temporal analysis over 66 years (1944–2010), based on aerial photographs and satellite images, revealed a general growth of around 100%, with four of the six main mangrove forests expanding their coverage, which could be related to differences in sensors and methods between years; only the Smith Channel mangrove swamp presented a loss of 26.3%. Some of the observed changes could be explained by anthropogenic factors such as the construction of roads, houses, and buildings, sand dredging, construction of spurs, hydraulic fills, and the felling of trees (Mancera-Pineda et al. 2019).

Mangroves are key ecosystems for coastal protection and risk management facing natural disasters. For example, the Asian tsunami in 2004 caused severe impacts on mangrove ecosystems but, at the same time, the tidal wave energy was substantially reduced, protecting the inland population (Barbier 2006). After the tsunami, several governments, including Indonesia, Sri Lanka, and Thailand, announced plans for widespread replanting and rehabilitation in degraded and deforested mangrove areas to bolster coastal protection (Barbier 2006). Hauser et al. (2015) revealed how Hurricane Sandy caused major losses to a large coastal wetland area in New Jersey, USA. They showed that erosion, sediment deposition, and marsh salinization caused severe degradation of 40% of the wetland area and long-term degradation of 50%. Additionally, Hurricane Sandy caused significant losses of flood regulating services, water filtration, and water supply ES (Hauser et al. 2015). In South Florida in September 2017, Hurricane Irma, with winds of more than 52 MPs (116 mph) and storm surge as high as 3 m, triggered one of the largest recorded mangrove dieback events in the region, with 10,760 ha of mangroves showing evidence of complete dieback (Lagomasino et al. 2021).

The coastal protection services provided by mangroves were observed on Providencia on November 16, 2020, when the eye of Hurricane Iota passed just 10 km north of Santa Catalina, generating winds between 213 km/h and 250 km/h, and waves of more than 5 m, with hurricane-force winds persisting for approximately 7 h (Stewart 2021). Inhabitants from the northern coast of Providencia noted that mangroves helped maintain water levels, reduced flooding, and acted as traps for debris and vessels that were dragged by the wind and waves, services that have been recognized in the literature (Zhang et al. 2012; Spalding et al. 2014). It is important to note that observations correspond to mangroves north of Providencia, where mangrove forests (*Rhizophora mangle*) between 10 and 50 m wide provided considerable protection during Iota (Fig. 1). After the hurricane, total defoliation and mass mortality of these mangroves were observed, with no leaf recovery observed even six months after the hurricane, this presents a need for immediate and effective restoration actions in order to recover mangroves and their ES for Providencia and its people (Fig. 1).

Fig. 1 *Rhizophora mangle* trees on the Northern coastline of Providencia, defoliated after Hurricane Iota. Note the gray color of the trees without green leaves. Mangroves provided a natural shield for houses and the people that lived behind them. Inhabitants from Jones Point town noted the importance of these trees in trapping debris and reducing water levels during the hurricane. Photograph: Julián Prato, June 2021

In 2020, Prato et al. (2020) conducted field measurements at three locations in San Andrés, and found that just one single mangrove tree reduced wind speed by 59% in the case of *Rhizophora mangle* roots and up to 88.9% for *R. mangle* and *Conocarpus erectus* canopies, producing an average wind speed reduction of up to 70%. These results suggest that mangroves can reduce wind speeds and maintain non-damaging conditions even under category 2 hurricane winds. Because mangroves are a key natural defense system, protecting coastlines, coastal infrastructure, and human lives from storm damage, their protection and restoration in the archipelago are essential to regain this vital ES in the face of extreme cyclonic events that are expected to increase in frequency and magnitude (IPCC 2021).

Despite progress in recent decades regarding the knowledge and management of the resources provided by mangroves—developing research projects, and strengthening knowledge networks and social appreciation—there is clearly still a long way to go to incorporate mangrove ES into socioeconomic development in the SBR. It is recommended that response and restoration management plans be designed for the archipelago's mangroves. To ensure that the value of the ES provided by mangroves is broadly appreciated, these plans should be developed with input from diverse groups of stakeholders, including members of the community, academia, and local and national governments. The first step is to identify priority areas for intervention and restoration, based on vulnerability to damage and the importance of coastal protection and other ES. Care must also be taken to develop strategies around best practices, considering the successes and failures of restoring mangrove ecosystems in other locations (Ellison et al. 2020). Recognizing that restoring the coastal protection service provided by mangroves will provide other key ES, such as food provision, water quality improvement, and biodiversity refuge, which contribute to the health and well-being of coastal communities (Prato and Newball 2016; Prato et al. 2020), mangrove conservation and restoration can serve as a critical component of Nature-based Solutions (NbS) to climate change (Spalding et al. 2014).

3 Coral Barrier Reefs for Coastal Protection and Human Well-Being in Seaflower

Corals are living colonial animals that secrete calcium carbonate as an exoskeleton that forms tridimensional hard structures with a variety of sizes and shapes. Aggregations of corals form reefs that diminish wave energy and, like mangroves, protect coastlines from erosion and storm damage (Mumby et al. 2014). Specifically, wave attenuation and energy dissipation by coral reefs occur due to different processes such as wave reflection, refraction, and bottom friction, which cause wave breaking by shoaling (Monismith 2007). The steep-sloped structures of reefs provide rapid changes in water depth and reef bathymetry, leading to wave breaking by shoaling from the fore reef to the reef crest (Fig. 2), where most of the wave energy, up to 95%, is dissipated (Lowe et al. 2005; Quataert et al. 2015). This dissipation is increased

Fig. 2 Wave breaking at the barrier reef's crest at San Andrés. Calmer water conditions can easily be observed on the right of the picture after the wave break on the reef, due to wave height and energy attenuation, while higher energy waves are seen on the left (fore reef and open ocean). Photograph: Julián Prato, 2020

by bottom friction, which is positively associated with the tridimensional structural complexity of reefs (Franklin et al. 2013; Monismith et al. 2015; Rogers et al. 2016). Because healthier reefs are more complex, tridimensional, and higher in terms of distance from the bottom, they provide more coastal protection ES than degraded reefs, which tend to be flatter and less complex (Mumby et al. 2014).

The annual economic benefits of coral reefs in terms of avoiding damage from flooding, erosion, or storms have been estimated at between USD 0.8 to USD 1.3 million for Tobago (Burke et al. 2008), USD 120 to USD 180 million for Belize (Cooper et al. 2009), and USD 700 to USD 2,200 million for the Caribbean (Burke and Maidens 2005). Reguero et al. (2021) found that benefits from coral reefs for US coastal protection exceed USD 1.8 billion annually, where many highly developed coastlines in Florida and Hawaii receive annual benefits of over USD 10 million km^{-1}. In addition to coastal protection ES, coral reefs provide many other important contributions to economic activity and human well-being such as biodiversity, food provision, and revenues and jobs associated with tourism (Costanza et al. 1997).

These valuable contributions to economic activity and well-being are at risk due to coral reef degradation (Burke et al. 2011). Globally, coral reefs have been affected by anthropogenic threats including warmer sea temperatures, ocean acidification, pollution, eutrophication, overfishing, and the growth of high-impact mass tourism in coastal zones (Hoegh-Guldberg et al. 2007). Additionally, marine litter, mainly plastic pollution, has become a global problem in recent decades, increasing the risk of disease in corals (Lamb et al. 2018). Marine litter also impacts other marine fauna, including reef fish, through ingestion (Kroon et al. 2018). Massive bleaching events, diseases, and mortalities have severely impacted many Caribbean reefs (Cramer et al. 2020). Coral reef degradation has been widely reported in the region (Burke et al. 2011), with coral cover losses of approximately 80%, resulting in considerable ES losses (Gardner et al. 2003).

The multiple impacts on coral reefs and their associated organisms can cause the loss of coral cover and further translate into the degradation of reef ecosystems and the decline of structural complexity, a phenomenon that has been observed throughout the Caribbean in recent decades (Alvarez-Filip et al. 2009). Coral cover losses that alter reef complexity and roughness jeopardize natural processes, resulting in 'reef flattening', whereby reef-building species are replaced by flat-growing weedy species (Alvarez-Filip et al. 2009; Mumby et al. 2014). As reef health, cover, and structural complexity degrade, associated ES and ecological functions are affected, threatening the well-being of coastal communities, and increasing their exposure, vulnerability, and risk of harm from erosion, inundation, and other natural hazards (Sheppard et al. 2005; Ferrario et al. 2014; Reguero et al. 2018).

Despite the ecological and economic relevance of the reef ecosystems and their associated fish assemblages, coral reefs in the SBR face the same array of threats. Indeed, the decrease in structural complexity found throughout the Caribbean has also been observed for some sites with the most extensive coral cover and colony size in the SBR: the Serrana and Roncador Banks (Sánchez et al. 2019). These losses jeopardize critical ES.

In the SBR, reefs are known to provide significant wave attenuation (Ortiz-Royero et al. 2015). The archipelago and SBR are known for having the third biggest barrier reef complex in the world after Australia and the Mesoamerican barrier reef complex (Coralina-Invemar 2012). With an extension of 142,005 ha, it represents 77% of Colombia's coral reefs. San Andrés has 6,340 ha and Providencia has 18,144 ha of coral reefs; these two islands, along with the other seven islands of the archipelago, have a barrier reef located at the East side of each with longitudes between 55 and 12 km (Prato and Newball 2016). We have recorded wave height attenuation of up to 90% due to the coral barrier reef at San Andrés, with a reduction of waves up to 4.5 m height (Hs) at the fore reef to 0.5 m in the reef lagoon thanks to the coastal protection of the barrier reef to the East of the island (Prato et al. 2020). These results suggest the need to increase investment in restoring and maintaining healthy reefs around the islands, especially on the barrier reef areas to the East of the islands for coastal protection, and to also put efforts into strengthening the health of the West side's coral reef areas to improve as much as possible the bottom friction provided by corals as a protection strategy to reduce waves, since that side of the islands is unprotected by a barrier reef formation. This coastal protection service is vital for human safety and well-being, especially with the expectation of more extreme events such as hurricanes in the SBR in the future (IPCC 2014, 2021).

Ecosystems should be included in the definition of exposure and vulnerability as essential for climate risk management (Walz et al. 2021). It is also necessary to make ES an integral component at the heart of risk management and development strategies. There is an intrinsic link between ecosystems and risk management.

Ecosystems themselves can offer sustainable solutions for reducing the severity of disaster impacts, while adapting to and coping with extreme events. Mitigating the economic losses that will occur as coastal protection ES decrease as a consequence of reef degradation, and protecting or enhancing the economic value created by other ES benefits from healthy coral reefs must be considered by decision-makers when creating nature-based management plans for climate change mitigation and adaptation strategies in the SBR. Because healthy reefs provide significant coastal resilience and numerous other valuable ES, spending on coral reef management and restoration should be considered a wise investment strategy (Ortiz-Royero et al. 2015; Reguero et al. 2018).

4 Food Provision, Refuge for Fish Biodiversity, and Fisheries in Seaflower

The waters and reefs of the SBR provide habitat for 731 fish species (Vides et al. 2016) and its mangroves provide habitat for at least 100 species of resident or migrant birds, 10 reptiles, and more than 60 fish, either adults or juveniles (Riascos 1999; López et al. 2009). Some of these species, and the ecosystems that support them, provide the ES of food provision, which is essential for human health, well-being, and food security. Functioning in close connection with seagrasses and deep and open ocean ecosystems, mangroves and coral reefs provide habitat for numerous species of fishes and shellfish (mainly Queen Conch *Aliger gigas*, spiny lobster *Panulirus argus*, black crab *Gecarcinus ruricola* and weelks *Cittarium pica*) which are the principal local protein source for food security of the Raizal people and islanders in the SBR. In addition to supporting human life and health, marine resources such as fish and invertebrates also provide protein for other species across the trophic web (Swartz et al. 2010). Protecting the ability of SBR ecosystems to sustainably provide these valuable sources of food and trophic support is an essential priority.

4.1 Relationship Between Coral Reef Structural Complexity and Fish Metrics

In combination with other associated organisms (i.e., sponges, calcareous algae), corals, particularly those coral species with traits such as branching morphology, increase the habitat's tri-dimensionality or structural complexity (Graham and Nash 2013). Sites with higher structural complexity are essential for many species, including reef fish which depend on live coral cover as a food source, refuge, or in a stage of their development (Komyakova et al. 2013). This link between structural complexity and reef fishes is observed across reef sites worldwide (González-Rivero et al. 2017) and has also been reported in reefs from the Serranilla island in the SBR.

Here, structural complexity explained fish variability by up to 60%, whereas sites with higher structural complexity and coral cover also had higher biomass and reef fish diversity (Castaño et al. 2021).

The fish biodiversity reported in the SBR represents more than one-third of the species registered for the Greater Caribbean region (Acero et al. 2019; Bolaños Cubillos et al. 2015; Robertson and Van Tassell 2019). Furthermore, recent studies' new registers of fish species in the SBR highlight the relevance of continuing monitoring efforts to characterize all the reef fauna in the SBR (Robertson and Van Tassell 2019). Although the SBR harbors a significant percentage of biodiversity, studies in the area emphasize the need to protect specific fish groups with ecological and economic relevance for the SBR, such as herbivores (i.e., parrotfishes) and mesopredators (i.e., groupers) (Prada et al. 2007; Acero et al. 2019; Castaño et al. 2021).

Herbivorous fish are one of the most studied functional groups in reef ecosystems, particularly parrotfish, as they have a relevant role in mediating space competition between corals and macroalgae (Plass-Johnson et al. 2015). Parrotfish also participate in carbon flux and represent a vital, high-value economic activity in the SBR and many Caribbean sites (Hawkins and Roberts 2004). Likewise, mesopredators (i.e., groupers, snappers, and sharks) are a relevant food resource in the region: they contribute to trophic web control and their presence indicates a healthier condition in reef ecosystems (Frisch et al. 2016; Salas et al. 2011).

Studies have shown that habitat loss through a decline in coral cover and structural complexity is one of the primary drivers of reef fish composition and structure (Graham and Nash 2013). Thus, preserving structural complexity in the SBR is crucial for coral ecosystems and their associated organisms, especially for reef fish assemblages. Combined with habitat loss, fishing pressure is one of the primary threats to reef fish (Hawkins and Roberts 2003).

Reef sites in the SBR are exposed to anthropogenic pressure, predominantly through illegal fishing. These fishing practices have been observed during scientific expeditions into the SBR, making reef fish assemblages vulnerable, even in this remote site (Friedlander et al. 2003). Parrotfish are one of the primary target fisheries in many sites, including the Caribbean, so their biomasses have been drastically reduced in some areas with habitat degradation and unregulated fishing practices (Hawkins and Roberts 2003). Although reported parrotfish biomasses in the SBR are not as low compared to other sites, recent surveys in Serranilla found that the average parrotfish size is below that reported for the Caribbean (Castaño et al. 2021). Since fisheries commonly target this species, the absence of large-bodied fish species generally indicates intense fishing pressure (Wilson et al. 2010). Mesopredator fishes are also a heavily impacted fish group in the SBR, with depleted grouper populations (Prada et al. 2007; Acero et al. 2019). Due to the ecological and economic relevance of these large-bodied functional groups, a loss or decrease in their population can heavily impact both the ecosystem and regional well-being (Prada et al. 2007; Edwards et al. 2014). The SBR represents a unique opportunity to protect and develop management strategies that benefit the inhabitants and the ecosystem. It is crucial to generate strategies to cope with current and ongoing threats to coral reef ecosystems. Integrating research efforts to generate valuable information and community involvement can be critical to the success of management strategies (Matera 2016).

Management strategies must include interdisciplinary actions to protect herbivorous fishes such as parrotfish in order to facilitate reef resilience, health, and functionality (Ferrari et al. 2012), including determination of the non-market value through economic valuation assessments for this species, to increase consciousness of its importance for economies and well-being. This must include monitoring parrotfish diversity, abundance, and biomass as indicators that reveal reef health, to thereby identify geographical areas that require special management attention. Broader education and commitment of fishermen regarding the ecological importance of parrotfish will contribute to protecting these fishes and to improving the reestablishment of healthy coral reef ecosystems, even after extreme events such as hurricanes. In the SBR, there are currently laws and control measures to protect parrotfish and penalize their catching and commercialization; more holistic mechanisms than penalties, such as environmental education, economic stimulus, and awareness, will empower communities to become part of the solution and contribute to the health of these important coastal ecosystems (Mumby et al. 2014).

Castaño et al. (2021) found a positive relationship between reef structural complexity (rugosity) and fish abundance and biomass at Serranilla, with more richness, biomass, and fish abundance at more complex reefs. From a fisheries perspective, structural complexity also can have an impact on fish catches. Rogers et al. (2014) found that complexity losses could cause more than a 3-fold reduction in fisheries productivity. The recovery of structural complexity must be an essential component of coral reef restoration objectives and monitoring programs in the SBR. This must provide more refuge and increase fish biomass and abundance, which could help food provision and food security for human well-being in the SBR, and be more attractive for ecotourism and diving in the sea of seven colors. Maintaining healthy and structurally complex reefs provides refuge for key species such as parrotfishes, which could also increase reef resilience to climate change and extreme events. Regarding the importance of fish and shellfish provided by healthy reefs and the importance of those as the main local protein source for Seaflower, here we suggest that *Food security from local sources based on healthy ecosystems* is a key strategy for climate change EbA.

4.2 Fisheries and Management in Seaflower

Traditionally, community fishing has been a practice used to obtain food. However, both artisanal and industrial fishing activities have negatively impacted ecosystems and the main target species populations are overfished (Jackson et al. 2001; Pauly et al. 1998). Fishing fleets worldwide are currently operating with a boundless "fishing effort" that covers 55% of the marine territory, leaving a spatial and temporal impact that has not yet been fully evaluated (Kroodsma et al. 2018). Globally, 80% of fishery resources are fully exploited, overexploited, or depleted, and only 20% of assessed fish stocks are moderately exploited or recovering (FAO 2018). Marine resources such as snappers, groupers, tuna, horse mackerel (fish), shrimp, lobster (crustaceans), and queen conch (mollusks) represent the majority of catches in the case of the Greater Caribbean region and the SBR, where several of the main fishing populations are overexploited due to increasing fishing efforts followed with temporal later reduction due to resource scarcity (Acosta et al. 2020; Santos-Martínez et al. 2013, 2020a).

In the SBR, artisanal fishing is undertaken mainly by the Raizal community, the islands' native population, but illegal artisanal and industrial boats also undertake fishing operations in the territory (Santos-Martínez et al. 2020a). An average of approximately 1,201 fishermen fish in the SBR per year (72% San Andrés; 28% P&SC), using an average of 217 artisanal boats (66% San Andrés; 34% P&SC). Being a multi-species fishery, most fishing is done with hand lines or by freediving in seagrass areas and near coral reefs, with more than 100 fish species captured, in addition to mollusks like the queen conch (*Aliger gigas*) and crustaceans like the spiny lobster (*Panulirus argus*) (Santos-Martínez et al. 2013).

Research has been carried out to characterize the SBR's fisheries and calculate maximum yields, to provide management measures. Interannual artisanal fishing production in the SBR shows drastic decreases, mainly associated with overfishing. Fisheries landings data from the two main islands shows that in San Andrés (2004–2018), estimated annual catches were between 46.2 and 251 tons/year, with fish contributing 97.7% (104 species) of total catches, and in Providencia (2012–2018) landings were between 5.1 and 59 tons/year, with fish representing 78% (90 species) of total catch landings. The Catch per Unit of Effort (CPUE), as a relative index of abundance, as well as the Maximum Sustainable Yield (MSY), showed a downward trend, suggesting decreases in resources and that fisheries are in a state of full exploitation (Santos-Martínez et al. 2013).

To find and promote sustainable policy solutions, joint efforts are underway between authorities, artisanal fishermen, and the community (Santos-Martínez et al. 2020a). For the archipelago, fulfilling the purposes and characteristics of being a BR has been proposed, in terms of ES conservation and development based on sustainable models for community well-being and biodiversity conservation (Santos-Martínez et al. 2009), to achieve adequate fishery resource management from natural, social, and economic sustainability and with clear policies with a transnational emphasis (Santos-Martínez et al. 2013, 2020a). However, despite the vital lessons learned from recent experiences, climate change and the effects of hurricanes (Santos-Martínez et al. 2020b) have created great challenges for the islands in transitioning to a model of sustainability and resilience (Santos-Martínez and Velásquez 2009; Santos-Martínez et al. 2020a, b; Santos-Martínez and Prato 2020; Velásquez and Santos-Martínez 2020).

4.3 Fisheries Management Tools

While fisheries resources may be well- or badly managed depending on the knowledge and expertise of the fisheries scientists and managers involved, even the best management advice will not suffice to rebuild the resource biomass if the causes of shrinking catches do not lie within the fishery itself, but in the deteriorating habitat of the resources. It is thus of the utmost importance to monitor and maintain ecosystem health and counteract habitat loss and deterioration. Holistic research on fisheries and ecosystem health indicators is imperative, as is the application of the ecosystem-based fisheries management approach (EBFMA).

It is widely known that single-species resource management has often led to unsustainable exploitation of fisheries resources because social, economic, and ecological objectives cannot be met simultaneously, and target species were not placed in the ecosystem context to allow assessment of fisheries' effects on the ecosystem. EBFMA has been developed to ensure the sustainability of fisheries by preserving the habitat in which the target resources and interacting species live. The main goal of EBFMA is to maintain high yields in fisheries while ensuring sustainability, and the structure and function (health) of the ecosystem.

While this appears to be a convincing goal, it requires adequate research and management tools, as well as indicators to qualify the state of the (fished) ecosystem. The trophic modeling software EwE (Ecopath with Ecosim) (Polovina 1984; Christensen and Walters 2004) is the most widely-used tool for holistic system description, system modeling, and fisheries management and ecosystem health assessment. The basic principle is to group the main system biota into functional groups (compartments), link them via a diet matrix (who eats who), and quantify the biomass flows within the food web and to the fishery. The trophic model represents a snapshot of balanced energy flows for the time period modeled but can also be used to simulate ecosystem changes over time if time series of catches, biomass changes of compartments, and/or other variables like fishing effort and primary productivity are available. Many of those time series-forced changes of different ecosystems have allowed the identification of the relative importance of fisheries and environmental drivers for observed changes (Wolff et al. 2012; Alms and Wolff 2020; Taylor et al. 2008a, b, among many others). ECOSPACE, another module of EwE, allows for spatial ecosystem modeling if data on the spatial distribution of the compartments are available to be mapped over a spatial grid (base map) of the model. This model can then be used to simulate biomass changes in time and space forced by fisheries management and/or conservation actions, such as the designation of a Marine Protected Area (Romagnoni et al. 2015).

In recent years, Ecopath models have also been used directly for fisheries management and stakeholder engagement. In the Philippines (Bacalso et al. 2016), for example, scenarios were modeled for the reallocation of fishing efforts from illegal to legal gears to explore the ecological and economic implications of different management options. In another study in Costa Rica (Sanchez-Jimenez et al. 2019), different degrees of fishing effort reductions were simulated to explore the effects on total catches and the biomass rebuilding of the key resources of the artisanal fishery. While this EwE modeling and management tool is freely available, its application requires data availability and a group of well-trained scientists who know how to gather data and construct, use, and interpret the models. While it is pivotal to have at hand and use these mathematical modeling and fisheries management tools, we should be aware that parallel monitoring of the ecosystem and its health status (Halpern et al. 2008) is always needed, since the best modeling tools do not help if the ecosystem is being lost or degraded due to background anthropogenic and/or environmental drivers.

The message here is twofold: we need an ecosystem-based approach to fisheries management and the application of modeling tools as described above, but we should also be aware that coastal systems are social-ecological systems, with ecosystems embedded in a social, political, and economic context. All these different realms may drive changes in the system. For this reason, we may extend the EBFMA to follow the Sustainable Livelihoods Approach in tropical coastal and marine social–ecological systems (Ferrol-Schulte et al. 2013), which provides a framework to understand and guide policymaking, considering system complexity. This framework, as well as the related Social-Ecological-Network analysis (Kluger et al. 2015), has stimulated the development of different modeling approaches such as Social Network Analysis (SNA), Bayesian Network models (BNs), and Loop Analysis (LA) to analyze Social-Ecological Systems (SES).

5 Trends and Needs in Seaflower Ecosystems

According to the Ministry of Environment and Sustainable Development (MADS 2015), the SBR terrestrial area (nearly 5,000 ha) includes around 3,000 ha (about 60%) of natural and semi-natural areas. This includes 491 ha of highly valuable terrestrial dry forests, which may be the most endangered terrestrial ecosystem in Colombia and the Caribbean, still covering 22.5% of Providencia and Santa Catalina (P&SC), and 210 ha of mangroves, 60 ha in P&SC and 150 ha in San Andrés (Prato and Newball 2016). Despite severe damage by Hurricane Iota, these dry forests and other terrestrial cover are mostly recovering, except for P&SC mangroves whose condition is extremely poor and will require major restoration efforts to recover. San Andrés, though deeply transformed by urban expansion, maintains good coverage as most forests have been replaced sustainably with fruit tree ground yards in a traditional land use form. P&SC also has many yards but with significant extensions of pastures for cattle farming. Urban expansion has also transformed coastal lands in P&SC, with some parts recovering since the start of agricultural decline many years ago. On balance, the terrestrial ecosystem status is satisfactory but careful management is needed to preserve valuable forests, recover mangroves, maintain fruit tree yards, and maintain and recover natural landscapes.

Marine ecosystems include the most extensive coral reef areas in Colombia. Reef complexes include barrier reefs, atolls, seagrasses, sandy and muddy bottoms, rocky and sandy shores, mangroves, and pelagic systems, in an extremely diverse and beautiful mix. SBR coral reefs in Serranilla, studied by the Global Reef Expedition, have the second highest mean coral cover ($13 \pm 7\%$) among the sites studied in the Caribbean, and the highest fish densities are around 77 ± 19 individuals/ 100 m^2 (Carlton et al. 2021). A study of the state of Providencia and San Andrés coral formations (Navas-Camacho et al. 2019) did not find a statistically significant decrease in coral cover nor an increase in algal coverage since 1998, meaning that the current situation appears stable, and it also reported that bleaching affected no more than 3% of coral cover. Nevertheless, Zea et al. (1998) reported losses of up to 90% of coral cover for San Andrés from the 1970s to 1995. Invading lionfish populations are around 0.8 ± 1.3 ind/250 m^2 (32 ind/ha) on average, with maximum density values reaching the maximum values for the Colombian and western Caribbean areas (Chasqui et al. 2020).

These results could be considered favorable but reveal something that must be considered in relation to the SBR: the need for a historical perspective and the effect of the shifting baseline syndrome that makes people believe that current lower thresholds for environmental conditions have always been the same. In general, this applies to mangroves, coral reefs, seagrasses, and particular fish, shellfish, and turtle species among others. Most monitoring studies in the archipelago began in the late 1990s, after major changes of up to a 50% decrease in living coral, and increased algal coverage due to different drivers of degradation, mainly of wide distribution, like coral and sea urchin diseases, Caribbean Sea pollution, coastal development and sedimentation, destructive overfishing and even global warming causing bleaching (Díaz et al. 1995 in Díaz et al. 2000; Garzón-Ferreira and Kielman 1995; Gil-Agudelo et al. 2009; Rodríguez-Ramírez et al. 2010; Navas-Camacho et al. 2019). Larger changes to Caribbean coral reefs probably began with extensive turtle hunting since the eighteenth century (Jackson et al. 2001) but only became evident in the 1970s with increased reef research. Hence, there is a lack of historical information upon which to interpret present conditions in the SBR.

It must be said that, in Colombia, there has been an early involvement with deterioration. The first reports are from Prahl (1983) for Pacific reefs. Díaz et al. (1995) reported recent mortality of about 50% in San Andrés, and Garzón-Ferreira and Kielman (1995) mentioned that live coral cover in the Colombian Caribbean declined to around 20–30% of the hard substrate, mainly in the 10 years before their report, that is from around 1984, concluding: "Evidence suggests that coral mortality has had its origin principally from agents of wide distribution (i.e., bleaching events and pathogenic diseases like BBD Black band disease and WBD White band disease) as a part of a generalized reef deterioration process in the wider Caribbean". Diseases have also been studied (Gil-Agudelo et al. 2009) and monitoring continues (Rodríguez-Ramírez et al. 2010; Navas-Camacho et al. 2019). Current researchers must keep in mind this historical perspective, as changes since 1998, when monitoring began, have been rather light, meaning their baseline mainly began with high algal and low living coral coverages, the latter at least half that in 1980, masking the real dimension of deterioration, similar to other Caribbean areas (Gardner et al. 2003; Alvarez-Philip et al. 2009).

Another significant issue relates to seagrasses and turtles. The sharp decline in turtle populations, due to hunting since the eighteenth century, affects seagrass beds due to lack of grazing. Reduced primary production from seagrasses affects their natural capacity for run-off retention and transformation of excess nutrients into biomass (Nellemann et al. 2009). Moreover, sea diseases affecting reef complexes seem to be related as excess organic materials favor microbial outbreaks (Jackson et al. 2001). This suggests that actions to recover turtles and seagrasses will have positive spillover effects on reef health (Guzmán-Hernández et al. 2022). In line with the benefits of an EBFMA, this means that an integrative approach to coral reef recovery is needed. As Jackson et al. (2001) note: "The central point for successful restoration is that loss of economically important fisheries, degradation of habitat attractive to landowners and tourists, and emergence of noxious, toxic, and life-threatening microbial diseases are all part of the same standard sequence of ecosystem deterioration that has deep historical roots. Responding only to current events on a case-by-case basis cannot solve these problems". So, to face coral reef deterioration, the restoration of some coral species, even if very important and useful, does not guarantee coral reef restoration. As with fisheries, it is necessary to restore the health of the entire system by restoring key species and key ecological functions as a strategy to increase resistance to diseases and pollution, and resilience to climate change.

In the SBR, actions must involve local people in the management process by fostering and taking advantage of local community knowledge and awareness of natural resources. The creation of a special fund dedicated to providing financial resources to implement solutions and support actions such as coral restoration, banning industrial fishing, controlling illegal fishing and overfishing, reducing pollution and run-off from land, restoring dry forests, agricultural lands and mangroves, and recovering turtle populations to restore seagrass beds and reduce diseases of corals is also important (Jackson et al. 2001). Seaflower's BR status should be leveraged to secure the resources required to sustainably manage its ecosystems, allowing the SBR to become a model of sustainability and recognizing nature as a basis for well-being, biodiversity conservation, and social, cultural, and economic prosperity.

6 Mapping, Remote Sensing, and Ecosystem Services for Climate Change Adaptation

Coral reefs and mangrove forests represent typical coastal ecosystems in tropical areas of the planet, as is the case of the archipelago. Part of the study of these ecosystems consists of spatial analyses related to their location, extension, composition, and structure, not only to establish spatiotemporal changes but also to quantify these resources and their connections to ES like coastal protection, biodiversity, and fisheries production (Nagelkerken et al. 2015). Spatial analyses, ecosystem mapping, and ES can also be an important communication tool to raise awareness of human dependence on marine ecosystems, to identify priority areas for conservation, and to identify risks, opportunities, and strengths for planning, designing, and implementing climate change EbA and NbS (Burkhard and Maes 2017) (see Fig. 3).

Because the waters in which many of the coral reefs are found are clear and shallow, they can currently be mapped using remote sensing or recent technologies for coastal and marine mapping that require indirect sampling (i.e., satellite data, lidar, multibeam sound, etc.) (Li et al. 2011; Pittman et al. 2009). Free public-use mapping applications such as Google Earth Engine, SPOT, Copernicus Sentinel, and Landsat imagery can be useful for large scales, but their spatial resolution limits their use on small reefs (Mumby et al. 1997). These satellite images have been used since

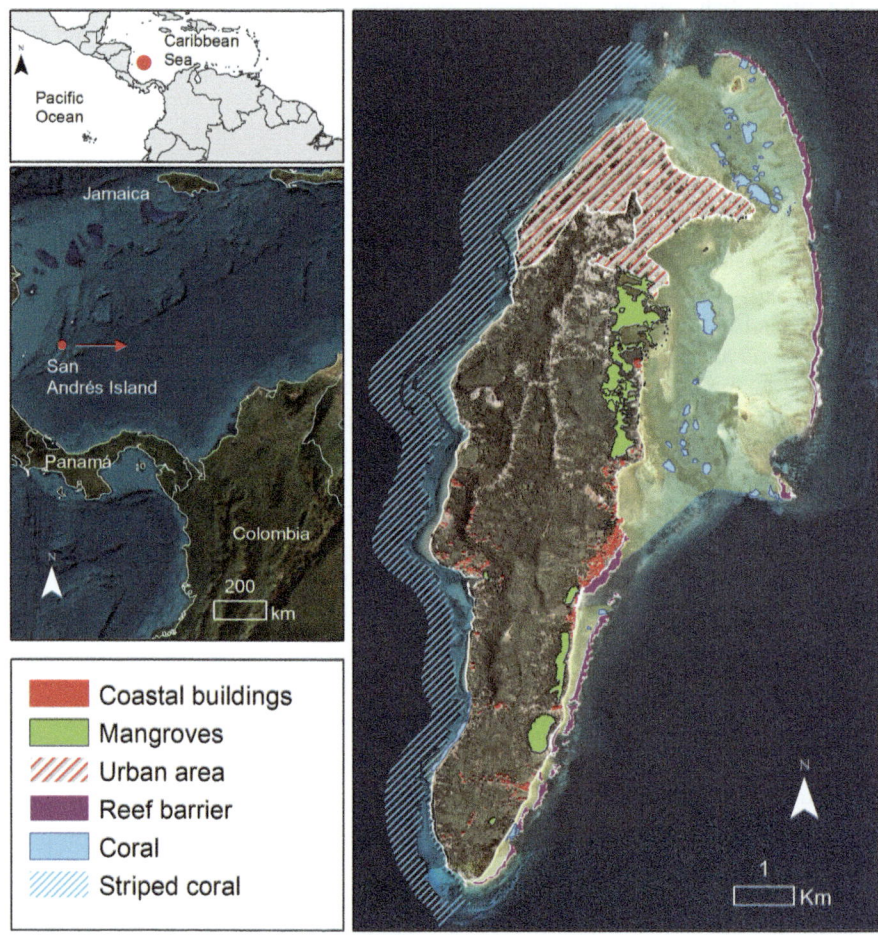

Fig. 3 Map of coral reefs and mangroves at San Andrés Island, highlighting the coastal protection ES provided by the barrier reefs complex located to the East of the island (light yellow). San Andrés' barrier reef complex offers coastal protection to urban (red stripes) and rural (red) areas, including Raizal people's houses, hotels, tourism infrastructure, and beaches

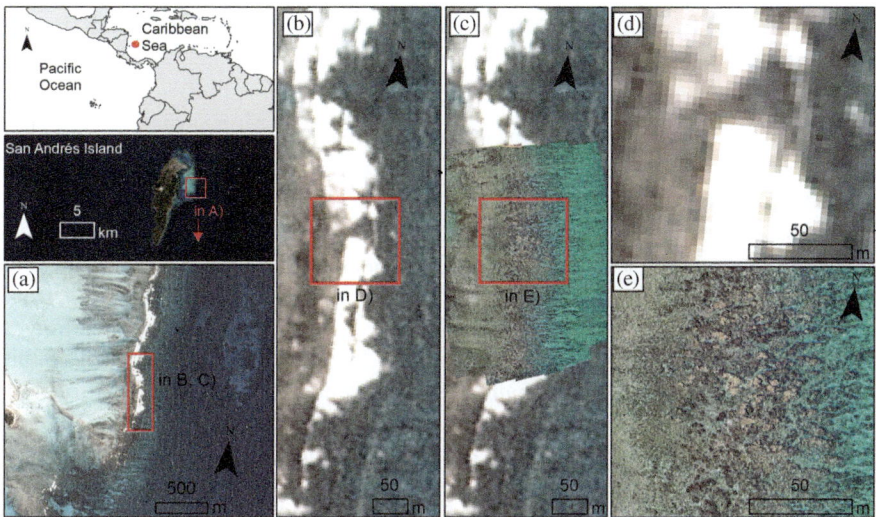

Fig. 4 Differences between satellite imagery and orthomosaics based on drone technology. **a, b** San Andrés barrier reef section, planet 3 m resolution satellite image; **c** same barrier reef section with drone's orthophotomosaic overlapped on satellite image planet; **d** zoom to satellite image in (**b**), **e** zoom to orthomosaic in **c**. Mosaics by Juan Carlos Mejía-Rentería and Julián Prato, 2021

the early 1970s, with resolutions of 10–30 m, but have limitations because infrared radiation does not penetrate the waters and cloud cover is very high in regions like the tropics (Nagendra and Rocchini 2008) (see Fig. 4). With improved spatial resolution in satellite sensors, the utility of these images has greatly improved, and there are already several initiatives to map the world's coral reefs at finer spatial scales, like the Allen Coral Atlas (Allen Coral Atlas 2022).

The use of Unoccupied Aerial Vehicles (UAVs) or drones to map coastal ecosystems provides a novel opportunity that overcomes many other remote sensing limitations, including higher resolution that enables the differentiation of mangrove tree species, identify coral zones, and register even small-scale temporary changes in these ecosystems. Here we present a practical example of drone mapping exercises carried out in 2020 to accomplish research on ES from barrier reefs at San Andrés (Prato 2018). A comparison is made with a 3 m resolution planetscope satellite image provided by Planet (https://www.planet.com/products/planet-imagery/) for research purposes. Despite being a very high-resolution satellite image, there are marked differences in the detail and information that can be generated with respect to drone results. Through the use of this orthophotomosaic (Fig. 4e) it is possible to identify, quantify, and geo-localize coral species and other benthic cover categories following the methodologies described in Castellanos-Galindo et al. (2019) and Komarkova and Jech (2020). These mapping tools offer a novel, reliable, and fast resource to characterize and monitor these ecosystems, inform management decision-making, and design climate change EbA and NbS.

7 Climate Change Adaptation and Ecosystem-Based Risk Management

Disasters affect all three dimensions of sustainable development: society, economy, and the environment (Walz et al. 2021). In the environmental dimension, extreme events are causing massive impacts on marine ecosystems, which means ES losses and increased climate risk. Because coastal and marine ecosystems and their services are needed for society's functioning and human survival, they play a central role in disaster risk reduction (DRR). The Sendai framework urges preserving ecosystems and reducing environmental losses and calls for transboundary cooperation on EbA to reduce disaster risk. The International Union for Conservation of Nature (IUCN) (2022) highlighted five main reasons why ecosystems are central to DRR: (1) Human well-being depends on ecosystems that provide multiple livelihood benefits; (2) Healthy ecosystems provide cost-effective natural buffers against natural events and climate change impacts; (3) Ecosystem degradation reduces the ability of natural ecosystems to sequester carbon, and unhealthy ecosystems are less resilient to extreme weather events; (4) Ecosystems reduce the impact of climate-related disasters and contribute to more sustainable post-disaster recovery; (5) Human conflicts cause environmental degradation. Therefore, environmental management is essential to decrease the risk of conflict.

As highlighted above, the main task is to restore and maintain healthy marine ecosystems and to incorporate ecosystems into DRR policy, legislation, and planning. However, this task has proven difficult, possibly due to the need to integrate information and action across different economic sectors and disciplinary boundaries. Achieving ecosystem-based DRR is therefore dependent on a high degree of cross-sectorial cooperation and decentralization. In the SBR, CORALINA and the Universidad Nacional de Colombia, with the support of other institutions, made an institutional effort to start including marine ecosystems as subjects of DRR policy by generating marine ecosystem restoration protocols after hurricanes (mangroves, coral reefs, and seagrasses), in order to mitigate the impacts and risk for that natural capital of the islands (Velázquez-Calderón et al. 2022). Because the Colombian political system is highly centralized, this presents a clear obstacle. Additionally, there is a lack of understanding and inclusion of ecosystems and ES in disaster-related impact evaluations and what this means in terms of advancing climate change mitigation and adaptation efforts, and sustainable development (Walz et al. 2021). ES losses remain largely neglected in the recording and monitoring of disaster losses, as ecosystems are not considered relevant, exposed, and vulnerable assets. However, ecosystems have recently gained importance as an element for defending territories from extreme events, and people are progressively aware of human dependence on them.

Currently, San Andrés and Providencia are in the process of replanting corals and restoring mangroves. For example, currently, the goal is to restore 200 ha of coral reefs by December 2022. During this process, 320 conservation agreements were signed with Raizal families (70 in San Andrés and 250 in Providencia) to restore, recover, and rehabilitate marine and terrestrial ecosystems, including mangroves (MADS 2021).

Overall, there are three main dimensions of disaster risk analysis: hazard, exposure, and vulnerability. Most archipelago marine ecosystems are exposed to hurricane impacts and climate changes. Shallow sites or sites with little change due to tides in the paths of cyclones are highly exposed (Salazar-Vallejo 2002). Mangroves and coral reefs mitigate hazards and human exposure mainly through buffering hazards and reducing overall hazard intensity. The main ES that mitigate hazards belongs to the group of regulating services (Walz et al. 2021). For instance, ES such as the sequestration and storage of carbon removes carbon dioxide from the atmosphere, helping to regulate climate change. The service of moderation or attenuation of hurricanes (coastal protection) contributes to mitigating hazard severity by protecting coastlines from storm surges, waves, high-speed winds, and flooding. The regulation of floodwater and reduction of wind speed service prevents coastal erosion, reducing water flow velocity and thus acting as a filter for wastewater treatment. These regulating functions are hampered when disastrous events impact ecosystems. Ecosystems also have double exposure, which means that vulnerability is augmented by having to deal simultaneously with problems from the impacts of climate change and human pollution. The exposure of ecosystems to hazards can be somewhat balanced with measures to mitigate vulnerabilities, such as investments in early warning and preparedness, addressing human contamination and pollution, and the implementation of restoration programs. Part of the marine ecosystem restoration after hurricanes protocols provided for the archipelago (Velázquez-Calderón et al. 2022), provides a special chapter available online with vulnerability maps for marine ecosystems (coral reefs, mangroves, and seagrasses) at San Andrés, Providencia, and Santa Catalina, with some general thoughts that could be applied to other islands at the Caribbean (Prato et al. 2022).

The other dimension of disaster risk is vulnerability. ES can reduce people's vulnerability, but this depends on the ecosystem's health and conservation status. Ecosystem vulnerability analysis aims to identify and prioritize areas according to the level of vulnerability and includes the study of spatial exposure, temporary exposure, the intrinsic response capacity of the ecosystem, and the analysis of external interventions that can increase vulnerability. There are no integral ecosystem vulnerability studies in the archipelago regarding hurricanes. However, various studies shed light on this matter. In 2017, the Third Communication on Climate Change indicated that the archipelago was the department (Colombian administrative region) at greatest risk from climate change in Colombia, due to its high vulnerability levels, including factors like biodiversity, ecosystem provision service, land use, and threatened species (Ideam et al. 2017). A 2017 study by INVEMAR regarding coastal and marine vulnerability to climate change in the archipelago found, from 2014 and 2015 data, that 67% of coral reefs are in "regular" condition, 38% are in "good" condition, and 5% are in "alert" (INVEMAR 2017). Regarding the biotic integrity indicator evaluation, results show that from 100% of the coral areas of the Colombian Caribbean, 84% are in a "regular" state of conservation, and 16% of the coral reefs are in "alert". The results of the health status indicator and its potential for restoration in seagrasses showed a high level of sensitivity due to a very low state of conservation, and mangrove ecosystems showed a low to medium level of sensitivity. The study also analyzed the adaptive capacity of ecosystems by evaluating institutional capacity for ecosystem management. The results showed that coral reefs and seagrasses have high and very high levels, while mangroves have a medium level of adaptation. Improved long-term monitoring and restoration programs with increased cover (on different areas with permanent observation stations) that better represent ecosystem variability with a focus on some special interest areas (such as coral barrier reefs), which also include the nine islands of the archipelago, are needed to better manage and protect SBR ecosystem integrity and human well-being. Sufficient investment, encouraged by the economic benefits provided by ES (Prato and Newball 2016), is necessary for integral management strategies, that must include at least illegal fishing control, restoration efforts, reduction of main causes of ecosystem degradation, long-term monitoring programs, and economic valuation assessments for a better accounting of SBR ES benefits.

As previously stated, vulnerability levels depend on ecosystem health and, in turn, on the implementation of sustainable human activities. A recent study by CORALINA and UNAL (2021) found that the archipelago has high levels of social vulnerability. Twelve indicators were analyzed, and the results showed that 70% of San Andrés have a high level of social vulnerability and 29% medium level vulnerability. On the other hand, Providencia and Santa Catalina have 62% medium and 32% high levels of vulnerability. For the three islands, the lack of access to public services (aqueduct and sewage), the low perception of risk associated with the low pro-environmental perception, and low economic capacity were particularly relevant in the high levels of vulnerability. For instance, waste waters in the islands are discharged into the environment without prior treatment, inevitably causing contamination of coastal zones and ecosystem degradation. For this reason, effective wastewater treatment is essential to reduce not only human vulnerability but also that of ecosystems. Conservation actions and working for and with nature to mitigate and adapt to climate change, are some of our strongest allies.

8 Seaflower Ecosystem Services and Economic Valuation for Management

This chapter clearly illustrates that the natural resources within the SBR, such as coral reefs, beaches, mangroves, seagrass beds, and marine fish stocks support numerous economic activities and provide valuable contributions to human well-being. In addition to providing provisioning services such as food and water, coastal and marine resources in the SBR attract tourists, protect coastal assets from erosion, storm damage, and the harmful effects of climate change, support local livelihoods through provisioning services of fish and materials, and provide opportunities for recreation, energy creation, and carbon storage. Human activities at the local, regional, and global scales can impair the structure and function of these ecosystems, limiting their ability to deliver goods and services. To achieve the highest value from these ecosystems over time, in light of the continuing threat from climate change, these tradeoffs must be understood and managed.

Unfortunately, this is a complicated endeavor. Alternative uses of natural resources create a variety of impacts that are most often not in common units, making comparisons difficult (Schuhmann 2012). Further, many of the benefits of ecosystem conservation and the costs of degradation occur over long periods and are not revealed in easily understood metrics. Conversely, the monetary costs of conservation and the benefits of economic activities that degrade ecosystems are more likely to occur in the short term and are easily observed and understood. As a result, policymakers and the public will naturally prioritize short-term market-based outcomes such as jobs and revenues at the expense of long-term ecosystem health and function. As noted by Schuhmann (2020), the benefits provided by many ES are shared by society and cannot be sold to buyers in a way that earns a profit. As a result, there are few economic incentives for consumers and producers to engage in activities that promote healthy ecosystem function or curtail ecosystem damage.

The result of these misaligned incentives is a persistent inefficiency: valuable ecosystem goods and services are under-provided at the expense of market-based sources of well-being (Schuhmann 2020). As this chapter has documented, there is ample evidence supporting this claim for the ecosystem goods and services contained within the SBR. For example, Gavio et al. (2010) show heightened levels of biologically available nitrogen and phosphorus and pathogenic bacteria in the coastal waters of San Andrés. Sánchez et al. (2019) document significant losses in coral cover and bottom complexity (rugosity) at the Roncador and Serrana Banks. Acero et al. (2019) document notable marine fish diversity in the waters of the Roncador, Serrana, and Serranilla islands, but note a consistent absence of several commercially valuable species such as grouper and large-bodied parrotfish.

Managing the natural resource assets of the SBR in a way that allows their economic benefits and potential to support climate resilience to be fully realized requires two important steps. First, tradeoffs between market activity and ES must be made apparent to policymakers and the public. Non-market valuation research can be conducted to measure the economic value of the ecosystem goods and services within the SBR so that the costs of ecosystem loss and the benefits of environmental stewardship are clearly understood. Second, policymakers should develop and implement interventions to deliver these non-market forms of economic value and well-being. These interventions might include government provision of ES (e.g., establishing protected areas, investing in ecosystem restoration, managing sewage discharge), and creating incentives to motivate individuals to promote sustainable uses of natural resources and/or limit damage (e.g., "greening" fiscal policy by shifting taxation toward activities, goods, and services that create environmental and social costs).

Regarding the first of these steps, a deep body of research uses market and non-market valuation techniques to estimate the economic value of coastal and marine resource ES (see Schuhmann and Mahon 2015 and Heck et al. 2019 for reviews of these studies in the Caribbean), including some studies within the SBR (Castaño-Isaza et al. 2015; Wilson 2001). Two branches of this work that are especially relevant for the SBR pertain to the value of reefs, mangroves, and beaches in terms of contributions to tourism (Beharry-Borg and Scarpa 2010; Schuhmann and Mahon 2015; Burke et al. 2008; van Beukering et al. 2009), and the economic value of these same ecosystems for coastal protection (Kushner et al. 2011; van Zanten et al. 2014; Cooper et al. 2009; Milon and Scrogin 2006).

Results from this literature suggest that coastal and marine ecosystems are significant sources of economic value and that their continued degradation will result in economic losses associated with declines in visits and spending, and heightened risks to coastal real estate and infrastructure. Results also suggest unrealized opportunities for conservation funding via entry fees to Marine Protected Areas, visitor entry/exit fees, price premiums for marine recreation, and donations. These results are well known. What is perhaps less apparent is the underlying complementarity between these two commonly studied sources of value. That is, the characteristics that are most valued by tourists (wide beaches, healthy and diverse coral reefs and mangroves, and clean seawater) are precisely the same characteristics that sustain livelihoods and mitigate climate risk.

Maintaining the health and vitality of these ecosystems clearly has the potential to create "win–win" scenarios for the natural resource assets within the SBR and for the people who depend on them for their well-being. With proper management, the economic returns from tourism and fisheries can be maintained without sacrificing proper ecosystem function. However, such scenarios are unlikely to materialize under a business-as-usual approach. Without public sector interventions or collective actions to incentivize sustainable behaviors, market-based forms of well-being will continue to dominate household and business decision-making processes (Schuhmann 2020).

Economic incentives such as subsidies for sustainable behaviors, and taxes and fees for activities with the potential to damage natural resource assets, will play an important role in transitioning toward more sustainable outcomes. Policies such as user fees, entry fees, pollution taxes, and payments for ecosystem services (PES) serve to incentivize sustainable behaviors and provide revenue streams for conservation. For example, using annual data from the OECD on environmental taxes in 18 Latin American countries over the period 1994–2018, Wolde-Rufae and Mulat-Weldemeske (2022) find that environmental taxes can reduce CO_2 emissions and promote the use of renewable energy. In Belize, environmental taxes include a 1% levy paid on the arrival of vehicles and other imports, with tax revenues used to finance environmental initiatives including solid waste management, improving institutional capacity in the Department of the Environment, and environmental clean-up initiatives (Northrop et al. 2022).

In Colombia, Calderón et al. (2016) find that carbon taxes can lead to significant CO_2 reductions, but note that coupling environmental taxes with reductions in other taxes may be necessary to offset negative economic impacts. In terms of user fees, the Bonaire National Marine Park (BNMP) is one of the few Caribbean MPAs that is almost entirely financed by user fees. General tourist fees related to the environment are applied in numerous countries. These include Belize, where visitors pay an exit fee of US $56, with revenues earmarked for a conservation fund that supports the management of Belize's 103 protected areas, and the British Virgin Islands where a US $10 entry fee is charged to visitors arriving by sea or air, with funds used for environmental protection and improvement, and to addressing climate change impacts. General tourist environmental fees are also applied in the U.S. Virgin Islands where an environmental fee of US $25 per night is paid by timeshare owners, and in Aruba, where a 9.5% environmental levy is paid by non-resident guests of hotels and other accommodations. Properly applied, non-market valuation techniques can help design such policies by quantifying economic returns in monetary units. For example, estimates of tourists' willingness to pay (WTP) for changes in the quantity or quality of ecosystem goods and services within the SBR can be compared to conservation costs so that appropriate user fee systems can be designed. Estimates of the monetary costs of damage to ecosystem goods and services can be used to design policies that tax harmful behaviors.

Communicating the purpose and benefits of any new taxes or fees will be an essential component of such programs. Communicating the economic value of natural resource assets within the SBR will play an important role in this regard, including the main trends in the variety of preferences that people show and management goals for sustainability. This communication can be expressed in terms of the benefits that will be gained through enhanced conservation or the costs and losses that will be incurred following a business-as-usual approach. Findings from the non-market valuation literature suggest that WTP for losses in ecosystem goods and services often exceeds WTP for similar gains, hence the latter of these two communication strategies is likely to be more effective. It is also critical to recognize that the costs and benefits of monetary incentives will vary across stakeholder groups. As such, outreach and communication efforts should be clear in terms of distributional consequences both across stakeholders and over time. Concessions for stakeholders that are disproportionately affected may be required to mitigate opposition.

In summary, facing continuing threats from climate change and local and regional stressors, improved management of SBR natural resource assets is required to maintain the flow of economic benefits and contributions to human well-being. Measuring the non-market benefits of ecosystem goods and services, the economic costs of resource degradation through non-market valuation research, and the incorporation of those values into fiscal policy and public discourse will play a critical role in the sustainable management of the SBR. While some valuation research has been conducted in the SBR, much of this work was conducted more than two decades ago. Londoño-Díaz and Vargas-Morales (2015) review these earlier studies and note that non-market valuation studies in the SBR have tended to focus more on an informational and technical perspective than a policy-decisive one, but that valuation studies have been used to support the establishment of a MPA scheme in San Andrés, opportunities for PES, and the design of entrance fees to the Johnny Cay Regional Park. Since that time, the resources contained within the SBR have continued to degrade and the threats to economic well-being have become more apparent. Future research should be directed at understanding the economic value of the ecosystem goods and services within the SBR so that appropriate policies can be designed and/or updated to promote improved sustainability.

9 Climate Change Ecosystem-Based Adaptation for the Present and Future

Food security, coastal protection, prosperous economies, and our own well-being depend on coastal and marine ES. Respect for other living beings and biodiversity, as well as real actions for better management and investment in healthy ecosystems are key for Seaflower and its people. Climate change and hurricanes have simultaneously created significant challenges and provided important lessons regarding the interdependence between healthy ecosystems and human well-being. It is apparent that a business-as-usual approach will result in continued degradation and heightened climate risk. Real actions directed at EbA and Ecosystem-Based Living (EBL) are essential to preserve our well-being. As one sign of the National Natural Parks of Colombia states, using a Native American saying: "We do not inherit the Earth from our ancestors, we borrow it from our children." We must work together in interdisciplinary teams with communities, government, academia, and businesses on this EBL and climate change EbA.

> My name is Arnold Hudson, I'm so proud of myself to participate in the Expedición Seaflower 2021 on that expirience wos amazing worcking wthit difrent people difrent culture, my experience of doing stodies I ricognize that we can help the future and the ecosystem around the island so our kids Will fine corals and fish.
>
> Arnold Hudson, February 2022.

References

Acero PA, Tavera JJ, Polanco FA et al (2019) Fish biodiversity in three northern islands of the Seaflower biosphere reserve (Colombian Caribbean). Front Mar Sci 6:113. https://doi.org/10.3389/fmars.2019.00113

Acosta AA, Glazer RA, Ali FZ et al (2020) Science and research serving effective ocean governance in the wider Caribbean region. Report for the UNDP/GEF CLME+ Project (2015–2020). Gulf and Caribbean Fisheries Institute. Marathon, Florida USA. Technical Report No. 2

Allen Coral Atlas (2022) Imagery, maps and monitoring of the world's tropical coral reefs. https://doi.org/10.5281/zenodo.3833242

Alms V, Wolff M (2020) Identification of drivers of change of the Gulf of Nicoya ecosystem (Costa Rica). Front Mar Sci 7:707. https://doi.org/10.3389/fmars.2020.00707

Alvarez-Filip L, Dulvy NK, Gill JA et al (2009) Flattening of Caribbean coral reefs: region-wide declines in architectural complexity. Proc R Soc B 276:3019–3025. https://doi.org/10.1098/rspb.2009.0339

Bacalso RTM, Wolff M, Rosales RM et al (2016) Effort relocation of illegal fishing operations: a profitable scenario for the municipal fisheries in Danajon Bank, Central Philippines. Ecol Modell 331:5–16. https://doi.org/10.1016/j.ecolmodel.2016.01.015

Barbier E (2006) Natural barriers to natural disasters: replanting Mangroves after the Tsunami. Front Ecol Environ 4(3):124–131. https://doi.org/10.1890/1540-9295(2006)004[0124:NBTNDR]2.0.CO;2

Beharry-Borg N, Scarpa R (2010) Valuing quality changes in Caribbean coastal waters for heterogeneous beach visitors. Ecol Econ 69(5):1124–1139. https://doi.org/10.1016/j.ecolecon.2009.12.007

Bindoff NL, Cheung WW, Kairo L et al (2019) Changing ocean, marine ecosystems, and dependent communities. In: Pörtner HO, Roberts DC, Masson-Delmotte V et al (eds) IPCC special report on the ocean and cryosphere in a changing climate intergovernmental panel on climate change. IPCC, Geneva, Switzerland

Bolaños Cubillos N, Abril Howard A, Bent Hooker H et al (2015) Lista de peces conocidos del archipiélago de San Andrés, Providencia y Santa Catalina, Reserva de Biosfera Seaflower, Caribe occidental colombiano. Bol Investig Mar Cost 44(1). INVEMAR. https://doi.org/10.25268/bimc.invemar.2015.44.1.24

Burke L, Cooper E, Greenhalgh S et al (2008) Coastal capital—economic valuation of coral reefs in Tobago and St. Lucia. World Resource Institute

Burke L, Maidens J (2005) Arrecifes en peligro en el Caribe. World Resources Institute, Washington DC

Burke L, Reytar K, Spalding M et al (2011) Reefs at risk revisited. World Resources Institute, Washington DC

Burkhard B, Maes J (2017) Mapping ecosystem services. Advanced Books. https://doi.org/10.3897/ab.e12837

Calderón S, Alvarez AC, Loboguerrero AM et al (2016) Achieving CO_2 reductions in Colombia: effects of carbon taxes and abatement targets. Energy Econ 56:575–586

Carlton R, Dempsey A, Thompson L et al (2021) Global reef expedition final report. Khaled bin Sultan Living Oceans Foundation, Annapolis, MD, vol 15. https://www.livingoceansfoundation.org/wp-content/uploads/2021/10/GRE-Final-Report.pdf

Castellanos-Galindo GA, Casella E, Mejía-Rentería JC et al (2019) Habitat mapping of remote coasts: evaluating the usefulness of lightweight unmanned aerial vehicles for conservation and monitoring. Biol Conserv 239:108282. https://doi.org/10.1016/j.biocon.2019.108282

Castaño D, Morales-de-Anda D, Prato J et al (2021) Reef structural complexity influences fish community metrics on a remote oceanic Island: Serranilla Island, Seaflower Biosphere Reserve, Colombia. Oceans 2(3):611–623. https://doi.org/10.3390/oceans2030034

Castaño-Isaza J, Newball R, Roach B et al (2015) Valuing beaches to develop payment for ecosystem services schemes in Colombia's Seaflower marine protected area. Ecosyst Serv 11:22–31

Chasqui L, Rincón-Díaz N, Vanegas MJ (2020) Abundancia del pez león invasor Pterois volitans en los arrecifes de coral costeros del Caribe colombiano. Bol Investig Mar Costeras 49(1) https://doi.org/10.25268/bimc.invemar.2020.49.1.779

Christensen V, Walters C (2004) Ecopath with Ecosim: methods, capabilities and limitations. Ecol Modell 172:109–139. https://doi.org/10.1016/j.ecolmodel.2003.09.003

Cooper E, Burke L, Bood N (2009) Coastal capital: Belize. The economic contribution of Belize's coral reefs and Mangroves. WRI Working Paper. World Resources Institute, Washington DC, USA

Coralina-Invemar (2012) Gómez-López, DI, Segura-Quintero C, Sierra-Correa PC, Garay-Tinoco J. Atlas de la Reserva de Biósfera Seaflower. Archipiélago de San Andrés, Providencia y Santa Catalina. Instituto de Investigaciones Marinas y Costeras "José Benito Vives De Andréis" -INVEMAR- y Corporación para el Desarrollo Sostenible del Archipiélago de San Andrés, Providencia y Santa Catalina -CORALINA-. Serie de Publicaciones Especiales de INVEMAR, 28. Santa Marta, Colombia

CORALINA & UNAL (2021) Análisis de vulnerabilidad social de las Islas de San Andrés, Providencia y Santa Catalina

Costanza R, d'Arge R, de Groot R et al (1997) The value of the world's ecosystem services and natural capital. Nature 387(6630):253–260. https://doi.org/10.1038/387253a0

Cramer KL, Jackson JB, Donovan MK et al (2020) Widespread loss of Caribbean acroporid corals was underway before coral bleaching and disease outbreaks. Sci Adv 6(17):eaax9395. https://doi.org/10.1126/sciadv.aax9395

Díaz JM, Garzón-Ferreira J, Zea S (1995) Los arrecifes coralinos de la Isla de San Andrés, Colombia: estado actual y perspectivas para su conservación. Academia Colombiana de Ciencias Exactas, Físicas y Naturales, Colección Jorge Alvarez Lleras, 7, Santafé de Bogotá, Colombia

Díaz JM, Barrios LM, Cendales MH et al (2000) Áreas coralinas de Colombia. INVEMAR, Serie Publicaciones Especiales, 5, Santa Marta, Colombia

Donato DC, Kauffman JB, Murdiyarso D et al (2011) Mangroves among the most carbon-rich forests in the tropics. Nat Geosci 4:293–297. https://doi.org/10.1038/ngeo1123

Edwards CB, Friedlander AM, Green AG et al (2014) Global assessment of the status of coral reef herbivorous fishes: evidence for fishing effects. Proc R Soc B 281(1774):20131835. https://doi.org/10.1098/rspb.2013.1835

Ellison AM, Felson AJ, Friess DA (2020) Mangrove rehabilitation and restoration as experimental adaptive management. Front Mar Sci 7:327. https://doi.org/10.3389/fmars.2020.00327

FAO (2018) The state of world fisheries and aquaculture—Meeting the Sustainable Development Goals. FAO. Rome, Italy

Feller IC, Lovelock CE, Berger U et al (2010) Biocomplexity in mangrove ecosystems. Annu Rev Mar Sci 2:395–417. https://doi.org/10.1146/annurev.marine.010908.163809

Ferrari R, Gonzalez-Rivero M, Ortiz JC et al (2012) Interaction of herbivory and seasonality on the dynamics of Caribbean macroalgae. Coral Reefs 31:683–692. https://doi.org/10.1007/s00338-012-0889-9

Ferrario F, Beck MW, Storlazzi CD et al (2014) The effectiveness of coral reefs for coastal hazard risk reduction and adaptation. Nat Commun 5(1). https://doi.org/10.1038/ncomms4794

Ferrol-Schulte D, Wolff M, Ferse S et al (2013) Sustainable livelihoods approach in tropical coastal and marine social–ecological systems: a review. Mar Policy 42:253–258. https://doi.org/10.1016/j.marpol.2013.03.007

Franklin G, Mariño-Tapia I, Torres-Freyermuth A (2013) Effects of reef roughness on wave setup and surf zone currents. J Coast Res 165:2005–2010. https://doi.org/10.2112/SI65-339.1

Friedlander A, Nowlis JS, Sanchez JA et al (2003) Designing effective marine protected areas in seaflower biosphere reserve, Colombia, based on biological and sociological information. Conserv Biol 17(6):1769–1784. https://doi.org/10.1111/j.1523-1739.2003.00338.x

Frisch AJ, Ireland M, Rizzari JR et al (2016) Reassessing the trophic role of reef sharks as apex predators on coral reefs. Coral Reefs 35(2):459–472. https://doi.org/10.1007/s00338-016-1415-2

Gardner TA, Côté IM, Gill JA et al (2003) Long-term region-wide declines in Caribbean corals. Science 301(5635):958–960. https://doi.org/10.1126/science.1086050

Gavio B, Palmer-Cantillo S, Mancera JE (2010) Historical analysis (2000–2005) of the coastal water quality in San Andrés Island, SeaFlower biosphere reserve, Caribbean Colombia. Mar Pollut Bull 60(7):1018–1030. https://doi.org/10.1016/j.marpolbul.2010.01.025

Garzón-Ferreira J, Kielman M (1995) Extensive mortality of corals in the Colombian Caribbean during the last two decades. Oceanogr Lit Rev 9(42):779

Gil-Agudelo DL, Navas Camacho R, Rodríguez Ramírez A et al (2009) Enfermedades coralinas y su investigación en los arrecifes colombianos. Bol Invest Mar Cost 38(2):189–224

González-Rivero M, Harborne AR, Herrera-Reveles A et al (2017) Linking fishes to multiple metrics of coral reef structural complexity using three-dimensional technology. Sci Rep 7(1):1–15. https://doi.org/10.1038/s41598-017-14272-5

Graham NAJ, Nash KL (2013) The importance of structural complexity in coral reef ecosystems. Coral Reefs 32(2):315–326. https://doi.org/10.1007/s00338-012-0984-y

Guarderas AP, Hacker SD, Lubchenco J (2008) Current status of marine protected areas in Latin America and the Caribbean. Conserv Biol 22(6):1630–1640. https://doi.org/10.1111/j.1523-1739.2008.01023.x

Guzmán-Hernández V, del Monte-Luna P, López-Castro MC et al (2022) Recuperación de poblaciones de tortuga verde y sus interacciones con la duna costera como línea base para una restauración ecológica integral. Acta Bot Mex 129:e1954. https://doi.org/10.21829/abm129.2022.1954

Halpern BS, Walbridge S, Selkoe KA et al (2008) A global map of human impact on marine ecosystems. Science 319:948–952. https://doi.org/10.1126/science.1149345

Hauser S, Meixler MS, Laba M (2015) Quantification of impacts and ecosystem services loss in New Jersey coastal wetlands due to hurricane sandy storm surge. Wetlands 35:1137–1148. https://doi.org/10.1007/s13157-015-0701-z

Hawkins JP, Roberts CM (2004) Effects of artisanal fishing on Caribbean coral reefs. Conserv Biol 18(1):215–226. https://doi.org/10.1111/j.1523-1739.2004.00328.x

Heck N, Narayan S, Beck MW (2019) Benefits of Mangroves and Coral Reefs in the Caribbean. Policy Brief. The Nature Conservancy

Hochard HP, Hamilton S, Barbier EB (2019) Mangroves shelter coastal economic activity from cyclones. Proc Natl Acad Sci USA 116(25):12232–12237. https://doi.org/10.1073/pnas.1820067116

Hoegh-Guldberg O, Mumby PJ, Hooten AJ et al (2007) Coral reefs under rapid climate change and ocean acidification. Science 318(5857):1737–1742. https://doi.org/10.1126/science.1152509

Ideam, PNUD, MADS et al (2017) Resumen ejecutivo Tercera Comunicación Nacional De Colombia a La Convención Marco De Las Naciones Unidas Sobre Cambio Climático (CMNUCC). Tercera Comunicación Nacional de Cambio Climático. IDEAM, PNUD, MADS, DNP, CANCILLERÍA, FMAM. Bogotá D.C., Colombia

Invemar (2017) Análisis de vulnerabilidad marino costera e insular ante el cambio climático para Colombia como insumo para la Tercera Comunicación Nacional de Cambio Climático. Santa Marta, Colombia

Invemar (2021) Expedición cangrejo negro - providencia 2021: Respuesta a los impactos del huracán iota en los ecosistemas marino costeros, recomendaciones al proceso de restauración y primeras acciones implementadas. Santa Marta, Colombia

IPCC (2014) Climate change 2014: synthesis report. Contribution of Working Groups I, II and III to the Fifth Assessment Report of the Intergovernmental Panel on Climate Change [Core Writing Team, Pachauri RK, Meyer LA (eds)]. IPCC, Geneva, Switzerland

IPCC (2021) Summary for policymakers. In: Climate change 2021: the physical science basis. Contribution of Working Group I to the Sixth Assessment Report of the Intergovernmental Panel on Climate Change [Masson-Delmotte VP, Zhai A, Pirani SL et al (eds)]. Cambridge University Press, Cambridge, UK

IUCN (2022) Five reasons why ecosystems are central to disaster risk reduction. https://www.iucn.org/theme/ecosystem-management/our-work/environment-and-disasters/about-ecosystem-based-disaster-risk-reduction-eco-drr/five-reasons-why-ecosystems-are-central-disaster-risk-reduction

Jackson J, Kirby M, Berger W et al (2001) Historical overfishing and the recent collapse of coastal ecosystems. Science 293(5530):629–637. https://doi.org/10.1126/science.1059199

Kluger L, Kochalski S, Müller M et al (2015) Towards an holistic analysis of social-ecological systems (SES) in the marine realm, YOUMARES 6. Conference Paper 107–121. https://doi.org/10.13140/RG.2.1.4780.6165

Komyakova V, Munday PL, Jones GP (2013) Relative importance of coral cover, habitat complexity and diversity in determining the structure of reef fish communities. PLoS ONE 8(12):e83178. https://doi.org/10.1371/journal.pone.0083178

Komarkova J, Jech J (2020) Processing UAV based RGB data to identify land cover with focus on small water body comparison of methods. In: 2020 15th Iberian conference on information systems and technologies (CISTI), pp 1–6. https://doi.org/10.23919/CISTI49556.2020.9141170

Kroodsma DA, Mayorga J, Hochberg T et al (2018) Tracking the global footprint of fisheries. Science 359(6378):904–908. https://doi.org/10.1126/science.aao5646

Kroon FJ, Motti CE, Jensen LH et al (2018) Classification of marine microdebris: a review and case study on fish from the great barrier reef, Australia. Sci Rep 8(1):1–15. https://doi.org/10.1038/s41598-018-34590-6

Kushner B, Edwards P, Burke L et al (2011) Coastal capital: Jamaica. In: Coral reefs, beach erosion and impacts to tourism in Jamaica. Working Paper. World Resources Institute, Washington DC, USA

Lamb JB, Willis BL, Fiorenza EA et al (2018) Plastic waste associated with disease on coral reefs. Science 359(6374):460–462. https://doi.org/10.1126/science.aar3320

Lagomasino D, Fatoyinbo T, Castañeda-Moya E et al (2021) Storm surge and ponding explain mangrove dieback in southwest Florida following Hurricane Irma. Nat Commun 12:4003. https://doi.org/10.1038/s41467-021-24253-y

Leal M, Spalding MD (2022) The state of the world's Mangroves 2022. The Global Mangrove Alliance

Lee SY, Primavera JH, Dahdouh-Guebas F et al (2014) Ecological role and services of tropical mangrove ecosystems: a reassessment: reassessment of mangrove ecosystem services. Glob Ecol Biogeogr 23:726–743. https://doi.org/10.1111/geb.12155

Li S, Yu K, Chen T et al (2011) Assessment of coral bleaching using symbiotic zooxanthellae density and satellite remote sensing data in the Nansha Islands, South China Sea. Chin Sci Bull 56(10):1031–1037. https://doi.org/10.1007/s11434-011-4390-6

Londoño-Díaz L, Vargas-Morales M (2015) An insight into the economic value of reef environments through the literature: the case of the seaflower biosphere reserve. Bol Investig Mar Cost 44(1):93–116

López Rodríguez A, García M, Sierra-Correa PC et al (2009) Ordenamiento Ambiental de los manglares del Archipiélago San Andrés, Providencia y Santa Catalina. Serie de documentos generales 30

Lowe RJ, Falter JL, Bandet MD et al (2005) Spectral wave dissipation over a barrier reef. J Geophys Res 110(C4). https://doi.org/10.1029/2004JC002711

MADS, IDEAM, IAvH et al (2015) Mapa de ecosistemas continentales, costeros y marinos de Colombia versión 1.0 a. Ministerio de Ambiente y Desarrollo Sostenible et al. http://www.ideam.gov.co/web/ecosistemas/mapa-ecosistemas-continentales-costeros-marinos

MADS (2021) Así avanza la recuperación de San Andrés, Providencia y Santa Catalina. https://www.minambiente.gov.co/asuntos-marinos-costeros-y-recursos-acuaticos/avanza-la-recuperacion-de-san-andres-providencia-y-santa-catalina/

Mancera Pineda JE, Poveda AP, Gavio B (2019) Cambios en la cobertura de playas y manglares en la isla de San Andrés a lo largo de siete décadas: 1944–2010. In: Campos NH, Acero Pizarro A (eds) Ciencias del mar, una mirada desde la Universidad Nacional de Colombia. Colección 20 años de presencia—Sede Caribe. Editorial Universidad Nacional de Colombia. Bogotá, Colombia

McLeod E, Chmura GL, Bouillon S et al (2011) A blueprint for blue carbon: toward an improved understanding of the role of vegetated coastal habitats in sequestering CO_2. Front Ecol Environ 9:552–560. https://doi.org/10.1890/110004

Matera J (2016) Livelihood diversification and institutional (dis-)trust: artisanal fishing communities under resource management programs in Providencia and Santa Catalina, Colombia. Mar Policy 67:22–29. https://doi.org/10.1016/j.marpol.2016.01.021

Milon JW, Scrogin D (2006) Latent preferences and valuation of wetland ecosystem restoration. Ecol Econ 56(2):162–175. https://doi.org/10.1016/j.ecolecon.2005.01.009

Monismith SG (2007) Hydrodynamics of Coral Reefs. Annu Rev Fluid Mech 39:37–55. https://doi.org/10.1146/annurev.fluid.38.050304.092125

Monismith SG, Rogers JS, Koweek D et al (2015) Frictional wave dissipation on a remarkably rough reef. Geophys Res Lett 42:4063–4071. https://doi.org/10.1002/2015GL063804

Morris RL, Konlechner TM, Ghisalberti M et al (2018) From grey to green: efficacy of eco-engineering solutions for nature-based coastal defence. Glob Chang Biol 24(5):1827–1842. https://doi.org/10.1111/gcb.14063

Mumby PJ, Green EP, Edwards AJ et al (1997) Coral reef habitat mapping: how much detail can remote sensing provide? Mar Biol 130(2):193–202. https://doi.org/10.1007/s002270050238

Mumby PJ, Flower J, Chollett I et al (2014) Towards reef resilience and sustainable livelihoods: a handbook for Caribbean coral reef managers. University of Exeter, Exeter, UK

Nagelkerken I, Sheaves M, Baker R et al (2015) The seascape nursery: a novel spatial approach to identify and manage nurseries for coastal marine fauna. Fish Fish (oxf) 16:362–371. https://doi.org/10.1111/faf.12057

Nagendra H, Rocchini D (2008) High resolution satellite imagery for tropical biodiversity studies: the devil is in the detail. Biodiv Conserv 17:3431–3442. https://doi.org/10.1007/s10531-008-9479-0

Navas-Camacho R, Acosta-Chaparro A, González-Corredor JD (2019) 20 years (1998–2017) of coral formations monitoring in San Andrés and Providencia. General Publications Series, 106. INVEMAR-CORALINA, Santa Marta, Colombia

Nellemann C, Corcoran E, Duarte CM et al (eds) (2009) Blue carbon. A rapid response assessment. United Nations Environment Programme, GRID-Arendal

Northrop E, Schuhmann P, Burke L et al (2022) Opportunities for transforming coastal and marine tourism: towards sustainability, regeneration and resilience, high level panel for a sustainable ocean economy. High Level Panel for a Sustainable Ocean Economy, Washington DC, USA

Ortiz Royero JC, Plazas Moreno JM, Lizano O (2015) Evaluation of extreme waves associated with cyclonic activity on San Andres Island in the Caribbean Sea since 1900. J Coast Res 31(3):557–568. https://doi.org/10.2112/JCOASTRES-D-14-00072.1

Pauly D, Christensen V, Dalsgaard J et al (1998) Fishing down marine food webs. Science 279:860–863. https://doi.org/10.1126/science.279.5352.860

Plass-Johnson JG, Ferse SC, Jompa J et al (2015) Fish herbivory as key ecological function in a heavily degraded coral reef system. Limnol Oceanogr 60(4):1382–1391. https://doi.org/10.1002/lno.10105

Pittman SJ, Costa BM, Battista TA (2009) Using Lidar bathymetry and boosted regression trees to predict the diversity and abundance of fish and corals. J Coast Res 10053:27–38. https://doi.org/10.2112/SI53-004.1

Polovina J (1984) An overview on the ECOPATH model. Fishbyte 4–7

Prada M, Castro E, Puello E et al (2007) Threats to the grouper population due to fishing during reproductive seasons in the San Andres and Providencia Archipelago, Colombia. Proc Gulf Caribbean Fish Inst 58:270–275

Prahl HV (1983) Blanqueo masivo y muerte de los corales en la isla de Gorgona, Pacífico colombiano. Cespedesia 12:125–129

Prato J, Newball R (2016) Aproximación a la valoración económica ambiental del departamento Archipiélago de San Andrés, Providencia y Santa Catalina—Reserva de la Biósfera Seaflower. Secretaría Ejecutiva de la Comisión Colombiana del Océano- SECCO, Corporación para el desarrollo sostenible del Archipiélago de San Andrés, Providencia y Santa Catalina -CORALINA

Prato J (2018) Relationships between coral reef complexity and ecosystem services at Caribbean oceanic islands, Seaflower Biosphere Reserve, Colombia. Doctoral thesis project. Hermes registration 45729. Universidad Nacional de Colombia—Sede Caribe. Approved by the postgraduate committee, Dec 2018

Prato J, Reyna J (2015) Aproximación a la valoración económica de la Zona Marina y Costera del Caribe Colombiano. Secretaría Ejecutiva de la Comisión Colombiana del Océano- SECCO, Bogotá, Colombia

Prato J, Santos-Martínez A, Castaño D et al (2020) Natural shields for Caribbean insular territories: Wave and wind attenuation by coral reef barriers and mangroves at San Andrés Island, Seaflower Biosphere Reserve, Colombian Caribbean. GCFI 73, book of abstracts. Gulf and Caribbean Fisheries Institute

Prato J, Mejía-Rentería JC, Echeverry P et al (2022) Mapas de áreas de los ecosistemas susceptibles a ser impactados por huracanes—Islas de San Andrés, Providencia y Santa Catalina. San Andrés Isla

Quataert E, Storlazzi C, van Rooijen A et al (2015) The influence of coral reefs and climate change on wave-driven flooding of tropical coastlines. Geophys Res Lett 42(15):6407–6415. https://doi.org/10.1002/2015GL064861

Ranganathan J, Raudsep-Hearne C, Lucas N et al (2008) Ecosystem services: a guide for decision makers. World Resources Institute, Washington DC, USA

Reguero BG, Beck MW, Agostini VN et al (2018) Coral reefs for coastal protection: a new methodological approach and engineering case study in Grenada. J Environ Manag 210:146–161. https://doi.org/10.1016/j.jenvman.2018.01.024

Reguero BG, Storlazzi CD, Gibbs AE et al (2021) The value of US coral reefs for flood risk reduction. Nat Sustain 4(8):688–698. https://doi.org/10.1038/s41893-021-00706-6

Riascos RH (1999) Caracterización de la avifauna y herpetofauna asociada a los bosques de manglares de la isla de San Andrés. SENA-SECAB-CORALINA

Robertson DR, Van Tassell J (2019) Shorefishes of the Greater Caribbean: online information system. Version 2.0. Smithsonian Tropical Research Institute

Rodríguez-Ramírez A, Reyes-Nivia MC, Zea S et al (2010) Recent dynamics and condition of coral reefs in the Colombian Caribbean. Rev Biol Trop 58(1):107–113. https://doi.org/10.15517/rbt.v58i1.20027

Romagnoni G, Mackinson S, Hong G et al (2015) The ecospace model applied to the North Sea: evaluating spatial predictions with fish biomass and fishing effort data. Ecol Modell 300:50–60. https://doi.org/10.1016/j.ecolmodel.2014.12.016

Rogers A, Blanchard JL, Mumby PJ (2014) Vulnerability of coral reef fisheries to a loss of structural complexity. Curr Biol 24(9):1000–1005. https://doi.org/10.1016/j.cub.2014.03.026

Rogers JS, Monismith SG, Koweek DA et al (2016) Wave dynamics of a Pacific Atoll with high frictional effects. J Geophys Res Oceans 121(1):350–367. https://doi.org/10.1002/2015JC011170

Rueda M, Marmol D, Viloria E et al (2010) Identificación, ubicación y extensión de caladeros de pesca artesanal e industrial en el territorio marino-costero de Colombia. INVEMAR, INCODER, AGENCIA NACIONAL DE HIDROCARBUROS-ANH. Santa Marta, Colombia

Salas S, Chuenpagdee R, Charles AT et al (eds) (2011) Coastal fisheries of Latin America and the Caribbean. FAO Fisheries and Aquaculture Technical Paper, 544. Food and Agriculture Organization of the United Nations, Rome, Italy

Salazar-Vallejo SI (2002) Huracanes y biodiversidad costera tropical. Rev Biol Trop 50:415–428

Sánchez JA (2021) ¿Por qué dependemos de la biodiversidad? Universidad de los Andes, Intermedio Editores. Bogotá, Colombia

Sánchez JA, Gómez-Corrales M, Gutierrez-Cala L et al (2019) Steady decline of corals and other benthic organisms in the Seaflower biosphere reserve (Southwestern Caribbean). Front Mar Sci 6:73. https://doi.org/10.3389/fmars.2019.00073

Sanchez JA, Pizarro V, Acosta-De-Sanchez AR et al (2005) Evaluating coral reef benthic communities in remote atolls (Quitasueño, Serrana, and Roncador Banks) to recommend marine-protected areas for the Seaflower biosphere reserve. Atoll Res Bull 531:1–66. https://doi.org/10.5479/si.00775630.531.1

Sánchez-Núñez DA, Bernal G, Mancera-Pineda JE (2019) The relative role of mangroves on wave erosion mitigation and sediment properties. Estuaries Coast 42:2124–2138. https://doi.org/10.1007/s12237-019-00628-9

Sánchez-Núñez DA, Mancera-Pineda JE, Osorio A (2020) From local-to global-scale control factors of wave attenuation in mangrove wave attenuation. Estuar Coast Shelf Sci 245:106926. https://doi.org/10.1016/j.ecss.2020.106926

Sánchez-Jiménez A, Fujitani M, MacMillan D et al (2019) Connecting a trophic model and local ecological knowledge to improve fisheries management: the case of Gulf of Nicoya. Costa Rica. Front Mar Sci 6:126. https://doi.org/10.3389/fmars.2019.00126

Santos Martínez A, Hinojosa S, Sierra-Rozo O (2009) Proceso y avance hacia la sostenibilidad ambiental: La Reserva de la Biosfera Seaflower, en el Caribe colombiano. Cuadernos de Caribe 13:7–23. https://repositorio.unal.edu.co/handle/unal/73998

Santos-Martínez A, Rojas Archbold A, García Escobar MI et al (2020a) Dinámica de la Pesca Artesanal y Propuestas de Manejo Sustentable, Zona Providencia y Santa Catalina, Reserva de Biosfera Seaflower Caribe Colombiano. GCFI 72:215–228. Gulf and Caribbean Fisheries Institute

Santos-Martínez A, Medina JH, Rojas Archbold A (2020b) El reto del manejo sustentable de la pesca artesanal en la Reserva de Biosfera Seaflower. UN Periódico, Universidad Nacional de Colombia, Voces y lecciones desde el Archipiélago

Santos-Martínez A, Prato J (2020) Barreras de coral y manglares protegen de huracanes y tsunamis. UN Periódico, Universidad Nacional de Colombia, Voces y lecciones desde el Archipiélago

Santos-Martínez A, Velásquez C (eds) (2009) Gestión y Manejo de Riesgo Frente a Huracanes. Universidad Nacional de Colombia Sede Caribe, San Andrés, Cargraphics—Carvajal, Bogotá, Colombia

Santos-Martínez A, Mancera Pineda JE, Castro González E et al (2013) Propuesta para el manejo pesquero de la zona del sur del área marina protegida en la Reserva de Biosfera Seaflower—Archipiélago de San Andrés, Providencia y Santa Catalina, Caribe Colombiano. Gobernación Departamento de Archipiélago de San Andrés, Providencia y Santa Catalina y Universidad Nacional de Colombia—Sede Caribe. Editorial Unibiblos, Bogotá, Colombia

Schuhmann PW (2012) The valuation of marine ecosystem goods and services in the wider Caribbean region. CERMES Technical Report (63):57. Center for Resource Management and Environmental Studies, University of the West Indies, Barbados

Schuhmann PW, Mahon R (2015) The valuation of marine ecosystem goods and services in the Caribbean: a literature review and framework for future valuation efforts. Ecosyst Serv 11:56–66. https://doi.org/10.1016/j.ecoser.2014.07.013

Schuhmann PW (2020) Valuation of ecosystem services as a basis for investment in blue economies. In: Clegg P, Mahon R, McConney P et al (eds) The Caribbean Blue Economy. Routledge, London, UK, pp 78–91

Sheppard C, Dixon DJ, Gourlay M et al (2005) Coral mortality increases wave energy reaching shores protected by reef flats: examples from the Seychelles. Estuar Coast Shelf Sci 64(2–3):223–234. https://doi.org/10.1016/j.ecss.2005.02.016

Spalding M, Kainuma M, Collins L (2010) World atlas of mangroves. Earthscan, London, UK

Spalding MD, Ruffo S, Lacambra C et al (2014) The role of ecosystems in coastal protection: adapting to climate change and coastal hazards. Ocean Coast Manag 90:50–57. https://doi.org/10.1016/j.ocecoaman.2013.09.007

Stewart SR (2021) Hurricane Iota (AL312020). National hurricane center. Tropical Cyclone Report. NOAA-NWS. 18 May 2021

Swartz W, Sala E, Tracey S et al (2010) The spatial expansion and ecological footprint of fisheries (1950 to present). PLoS ONE 5(12):e15143. https://doi.org/10.1371/journal.pone.0015143

Taylor MH, Wolff M, Mendo J et al (2008a) Changes in trophic flow structure of Independencia Bay, Peru over an ENSO cycle. Prog Oceanogr 79:336–351. https://doi.org/10.1016/j.pocean.2008.10.006

Taylor E, Prada M, Peñaloza G et al (2008b) Evaluación, restauración y recuperación ambiental y de sistemas productivos de las islas de Old Providence y Santa Catalina después del paso del huracán Beta. In: Santos-Martínez A, Velásquez C (eds) Gestión del riesgo y manejo de crisis frente a huracanes guía de preparación. Universidad Nacional de Colombia—Sede Caribe, San Andrés, Colombia

van Beukering PJH, Sarkis S, McKenzie E et al (2009) Total economic value of Bermuda's coral reefs: valuation of ecosystem services. Technical Report, Department of Conservation Services, Government of Bermuda, Bermuda

van Zanten BT, Van Beukering PJH, Wagtendonk AJ (2014) Coastal protection by coral reefs: a framework for spatial assessment and economic valuation. Ocean Coast Manag 96:94–103

Velásquez C, Santos-Martínez A (2020) Lecciones para la gestión del riesgo en el Archipiélago. UN Periódico, Universidad Nacional de Colombia, Voces y lecciones desde el Archipiélago

Velásquez-Calderón C, Prato J, Santos-Martínez A (2022) Documento síntesis protocolos de restauración ecosistemas frente a huracanes (procedimientos restauración manglares, pastos marinos. San Andrés, Colombia. https://coralina.gov.co/planes/protocolo-de-evaluacion-y-restauracion-de-ecosistemas-marinos-frente-a-huracanes

Vides, M, Alonso D, Castro E et al (eds) (2016) Biodiversidad del mar de los siete colores. Instituto de Investigaciones Marinas y Costeras—INVEMAR y Corporación para el Desarrollo Sostenible del Archipiélago de San Andrés, Providencia y Santa Catalina—CORALINA. Serie de Publicaciones Generales del INVEMAR, 84, Santa Marta, Colombia

Waite R, Burke L, Gray E et al (2014) Coastal capital: ecosystem valuation for decision making in the Caribbean. World Resources Institute, Washington DC, USA

Walz Y, Janzen S, Narvaez L et al (2021) Disaster-related losses of ecosystems and their services. Why and how do losses matter for disaster risk reduction? Int J Disaster Risk Reduct 63:102425. https://doi.org/10.1016/j.ijdrr.2021.102425

Wilson R (2001) Economic valuation of the non-market values of mangroves of San Andres Island, Colombia, and recommendations for management. Master's thesis, Heriot Watt University, Edinburgh, UK

Wilson SK, Fisher R, Pratchett MS et al (2010) Habitat degradation and fishing effects on the size structure of coral reef fish communities. Ecol Appl 20(2):442–451. https://doi.org/10.1890/08-2205.1

Wolde-Rufael Y, Mulat-Weldemeskel E (2022) The moderating role of environmental tax and renewable energy in CO_2 emissions in Latin America and Caribbean countries: evidence from method of moments quantile regression. Environ Chall 6:100412. https://doi.org/10.1016/j.envc.2021.100412

Wolff M, Ruiz D, Taylor M (2012) El Niño induced changes to the Bolivar Channel ecosystem (Galapagos): comparing model simulations with historical biomass time series. Mar Ecol Prog Ser 448:7–22. https://doi.org/10.3354/meps09542

Worthington TA, zu Ermgassen PSE, Friess DA et al (2020) A global biophysical typology of mangroves and its relevance for ecosystem structure and deforestation. Sci Rep 10:14652. https://doi.org/10.1038/s41598-020-71194-5

Zea SJ, Geister J, Garzón-Ferreira JM et al (1998) Biotic changes in the reef complex of San Andrés Island (Southwestern Caribbean Sea, Colombia) occurring over nearly three decades. Atoll Res Bull 456:1–30. https://doi.org/10.5479/si.00775630.456.1

Zhang K, Liu H, Li Y et al (2012) The role of mangroves in attenuating storm surges. Estuar Coast Shelf Sci 102–103:11–23. https://doi.org/10.1016/j.ecss.2012.02.021

Open Access This chapter is licensed under the terms of the Creative Commons Attribution 4.0 International License (http://creativecommons.org/licenses/by/4.0/), which permits use, sharing, adaptation, distribution and reproduction in any medium or format, as long as you give appropriate credit to the original author(s) and the source, provide a link to the Creative Commons license and indicate if changes were made.

The images or other third party material in this chapter are included in the chapter's Creative Commons license, unless indicated otherwise in a credit line to the material. If material is not included in the chapter's Creative Commons license and your intended use is not permitted by statutory regulation or exceeds the permitted use, you will need to obtain permission directly from the copyright holder.

Climate Change Effects on Seaflower Biosphere Reserve Fishery Resources

Carolina Sofia Velásquez-Calderón ⓘ, Adriana Santos-Martínez, Anthony Rojas-Archbold, and Julián Prato

Abstract Climate Change (CC) is a global phenomenon with differentiated impacts. Its effects are felt in marine and terrestrial ecosystems and organisms, and in the most vulnerable economies and societies. CC is altering the ocean's chemistry, initiating cascading socioenvironmental impacts. The fisheries sector is the most affected. In the Western Caribbean's Archipelago of San Andrés, Providencia, and Santa Catalina, identified as having the highest climatic risk, these impacts are pronounced. This study comprehensively reviews existing knowledge on climate change effects on fishery resources and incorporates fishers' perceptions through two rounds of surveys in 2019 and 2022. The findings reveal significant consequences for fishery resources, including alterations in biological properties and species distribution, loss of critical coastal fish breeding habitats, reduced fisheries productivity, and increased local and cross-border conflicts over fish resources. Especially, after the destructive impact of Hurricane Iota (2020), fishers shifted their hazard perception, elevating hurricanes as a significant threat alongside drought. These evolving perceptions emphasize the need for comprehensive policy strategies to address multiple hazards and their interactions, aligning with fishers' priorities and enhancing the resilience of the fishing sector. This research underscores the urgency of ecosystem-based and co-management policies, alternatives for artisanal fishers, and heightened climate risk perception.

C. S. Velásquez-Calderón (✉)
Assistant Professor, Department of Geography & African American Studies, Florida State University, Tallahassee, FL, USA
e-mail: csv23@fsu.edu

A. Santos-Martínez
Universidad Nacional de Colombia - Caribbean Campus, San Andrés isla, Colombia
e-mail: asantosma@unal.edu.co

A. Rojas-Archbold
Associate Professor, The National Authority for Aquaculture and Fisheries (AUNAP) of Colombia, San Andrés isla, Colombia

J. Prato
Researcher, Universidad Nacional de Colombia - Caribbean Campus, San Andrés isla, Colombia
e-mail: jprato@unal.edu.co

© The Author(s) 2025
J. E. Mancera Pineda et al. (eds.), *Climate Change Adaptation and Mitigation in the Seaflower Biosphere Reserve*, Disaster Risk Reduction, https://doi.org/10.1007/978-981-97-6663-5_9

Keywords Marine fishery resources · Climate risk management · Fisheries management · Fishers' perception · Seaflower biosphere reserve

1 Introduction

The IPCC (2018, 2021) has indicated that global warming is likely to reach 1.5 °C or even 2 °C in the coming decades and has assured that carbon dioxide (CO_2) emissions should fall by 45% by 2030 if we want to avoid the most catastrophic effects of global warming. Climate change effects are being felt in marine and terrestrial organisms and ecosystems, and the most vulnerable economies and societies. The Caribbean region has been identified as one of the areas with the most significant risk, as it is increasingly affected by increases in sea and land temperatures, storm surges, changing precipitation patterns, sea level rise, coral bleaching, impacts of intense tropical cyclones, and invasive species. The Caribbean region will experience 0.5–1.5 °C of warming compared to the 1971–2000 baseline. Impacts of ocean warming at 1.5 °C mean an overall high risk to the fisheries sector, implying moderate risk for seagrasses, very high risk of severe impacts for coral reefs, high risk for pteropods, and very high risk for bivalves and finfish (Hoegh-Guldberg et al. 2018). Furthermore, there will be multiple cascade socioenvironmental impacts on marine ecosystems and fisheries in the Caribbean region (Oxenford and Monnereau 2018).

Since the 1990s, a large volume of literature has emerged on the impacts of climate change on marine ecosystems. Recent research on this topic includes Cheung et al. (2010), Nurse (2011), Barange et al. (2014), FAO (2009, 2014, 2016), Gordon et al. (2018), Daw et al. (2009), and more specifically on fishery resources in the Caribbean Sea, Boavida-Portugal et al. (2018), Bonebrake et al. (2018), Caputi et al. (2013), and Oxenford and Monnereau (2017, 2018) highlighting numerous interrelated impacts on commercially important fishery species, including effects on distribution, abundance, seasonality, physiology, life processes, and indirect effects arising from habitat deterioration and socioeconomic implications. However, research on the effects of climate change on Caribbean marine species is, in general, considerably scarcer than in other regions.

Additionally, there is a lack of data related to island-specific effects and the integration of those into strategic planning and public policy (OECD 2021). Furthermore, there is a need for local island studies on fisheries that integrate the expert point of view and the fisher community's perspective (Kettle et al. 2014). It is critical to consider different perceptions in order to, from the ground up, identify the right problem and formulate and implement participatory solutions.

Climate adaptation strategies are unlikely to be effective without understanding fishers' perceptions (Mulyasari et al. 2018). According to Acosta et al. (2021), science should include various knowledge types across sectors, including community and traditional knowledge. Studies integrating local perspectives, such as bottom-up approaches, are increasingly valuable (Mastrandrea et al. 2010; Kettle et al. 2014; Monirul et al. 2017).

The relevance of understanding public perceptions of climate change effects on fishery resources is in its infancy. Until now, studies have explored fishers' perceptions of climate change mainly in the Pacific and Asian regions. For example, Mulyasari et al. (2018) studied fishermen's perceptions in Bengkulu Province, Indonesia, finding that Bengkulu fishermen are less aware of climate change and that there is, therefore, less implementation of climate change adaptation strategies. Diouf et al. (2020) analyze Senegalese fishers' perception of climate variability and change and their attitude towards weather forecasts. The results show that the fishing communities in Senegal are aware of and willing to act to adapt to the effects of climate change. However, the study also found that access to weather forecasts was limited and the authors state that if all fishers could access weather forecasts, at least 83% of them would decide to postpone sea fishing activities in certain circumstances. The study highlights that the problem in taking adaptation measures is based on weaknesses in early warning systems. Jyun-Long (2020) selected Keelung City, New Taipei City, and Yilan County as case studies to analyze fishers' perceptions of climate change in northeastern Taiwan. The author found that three variables significantly and positively affected fishers' willingness to adapt fishing behavior: experience in fishing, recognition of impacts on marine physical environments, and preferences of risk control measures. To fully understand the challenges of climate change for marine fishery resources and to identify opportunities to mitigate them, the inclusion of fishers' perception is essential.

The Archipelago of San Andrés, Providencia, and Santa Catalina (hereafter, the archipelago) is located in the western Caribbean region, and it has been classified as having one of the highest levels of climatic risk. Because of its biological diversity, cultural values, and natural ecosystem, UNESCO declared the archipelago as the Seaflower Biosphere Reserve in November 2000. The islands' primary sources of income are tourism and commerce, followed by fishing. Fishing is key to the economy, as well as to cultural identity and food security. According to data from the Fishing and Aquaculture Registry and UNAL (2019), in 2018 there were approximately 887 artisanal fishers in the archipelago, with 624 located in San Andrés, and 263 in Providencia and Santa Catalina. Fishers undertake artisanal fishing activities in the archipelago in the following areas: Outside Bank, Southend Bank, Bolívar Cay, Albuquerque, Far Bank, and Serrana, Quitasueño, Serranilla, and Bajo Nuevo islands (Llanos 2015). Catches are made using traditional fishing techniques, especially hand lines. Some of the commercially significant fish and shellfish species in the archipelago include *Panulirus argus* (lobster), *Eustrombus gigas* (queen conch), *Lutjanus spp.* (Blackfin snappers), *Sphyraena barracuda* (barracuda), *Ocyurus chrysurus* (yellowtail), *Etelis oculatus* (mandilos), *Elagatis bipinnulata* (Ocean Yellowtail), *Euthynnus alletteratus* (bonito), *Apsilus dentatus* (black snapper), *Epinephelus spp.* (Rockfish), and *Mycteroperca spp.* (Yellowfin Grouper, Black Grouper). These species form the backbone of the local fishing industry and play a vital role in the archipelago's food supply.

Local experts have warned that marine ecosystems have been experiencing dramatic deterioration processes in recent decades, specifically, pollution, the mechanical destruction of reefs, the effects of coral bleaching episodes, the loss of structural complexity in coral reefs, and recently (2022), the presence and spreading of the Stony Coral Tissue Loss Disease (Navas-Camacho et al. 2019; CORALINA-INVEMAR 2017; CORALINA 2022). In this sense, there is a pending task in understanding and consolidating the impacts of climate change on Caribbean fishery resources and marine ecosystems. This chapter addresses this research priority and gap, and aims to provide information on the fishers' perceptions of climate change impacts on fishery resources and proposed local adaptation strategies. It uses data from 27 surveys in 2019 and 22 in 2022. The research questions are: (i) What are the potential impacts of climate change on fishery resources and fishing activities? (ii) What are the impacts perceived by fishers? Moreover, (iii) How do fishers think climate change adaptation should be implemented?

The chapter is structured as follows. First, it describes the methodology for collecting and analyzing data about fishers' perceptions and the integration of fishers' knowledge into adaptation recommendations. Second, it provides an overview of the literature on climate change impacts on fishers and fishery resources. Third, the findings are divided into two parts: climate change impact perceptions, and proposed climate change adaptation strategies for the archipelago. Finally, the conclusion draws attention to the benefits of combining and contrasting the fishers' survey results with scientific data to provide a seascape perspective of the fisheries environment. The direct benefit of this research is an improved understanding of the future risk scenario for fishery resources. An indirect benefit is that it empowers fishers to be actively involved in developing management options for the fisheries sector.

2 Methods

The general objective was to analyze the different effects of climate change on the archipelago's primary fishery resources. The methodology used was a systematic literature review mixed with data from the knowledge of local fishers and key fisheries actors. The steps followed were: (1) search for scientific articles in the library catalog of various universities, Google Scholar, and multidisciplinary databases, (2) evaluate and select literature at the intersection of the search terms Climate Change, Fisheries, Caribbean, San Andrés Island, and Seaflower Biosphere Reserve. After that, the key information about specific fishery species was identified. Finally, (3) once the documentation was identified, information processing began through the following guiding questions: What are the key impacts? What are the conclusions and results of the investigations? How can this research be applied to the literature review?

The initial dataset in the literature review included around 104,000 publication results. In the first screening of the dataset, the authors read the titles and abstracts of the publications to determine which publications would be included in the final analysis. Based on this, approximately 40 articles were retained for final review. The second phase of a more specific search was carried out around the effects of climate change on strategic coastal marine ecosystems (mangroves, seagrasses, and coral reefs) and a final search for information on the archipelago's most important commercial species, such as the queen conch and spiny lobster.

To include fishers' voices, this chapter describes and analyzes how they framed climate change and its impacts. Forty-nine surveys were conducted, 27 in 2019 and 22 in 2022. The survey sample was a combination of randomly directed and targeted. Before data collection started, a structured survey questionnaire was tested with two respondents to ensure the adequacy of the questions and the information obtained. Data were collected using face-to-face and electronic surveys through Google Forms in November 2019 and June 2022, respectively. The survey sought information on the impacts of climate change on fishery resources, perceptions of climate change understandings, and adaptation strategies. Statistical analyses such as descriptive analysis were conducted to compare participants' perceptions about climate change understandings and impacts. We went from deductive to inductive analysis, where umbrella themes were organized to identify adaptation strategies to climate change. The individuals surveyed for this study were fishers and practitioners with extensive experience and knowledge of marine resources. However, the information provided by these participants reflects their perceptions of climate change and should not be understood as representing the views of all stakeholders engaged in fisheries activities in the archipelago. We summarized the results of these surveys and focused on crucial similarities and dissimilarities between the scientific literature and local knowledge concerning climate change's effects on fisheries.

3 A Review of Climate Change Impacts on Marine Ecosystems and Fishery Resources in the Caribbean Region

According to IPCC reports (2018, 2021), climate change impacts are amplified for small tropical Caribbean islands. There are four main hazards for the Colombian Caribbean Sea (INVEMAR 2017): (1) sea level rise, which will be 81–90 mm by 2040, 171–200 mm by 2070, and 301–350 mm by 2100; (2) changes in sea surface temperature whose projections under the RCP 4.5 scenario for the 2041–2070 period increment between 27.5 and 28 °C, and the 2071–2100 scenario between 28 and 28.5 °C; (3) marine acidification whose pH trend concerning atmospheric content is towards a decrease in the Caribbean: a decrease of 0.102 (RCP4.5) and 0.159 (RCP6.0) is projected by 2100, in turn generating lower disposal of $CaCO_3$ in the form of aragonite; and (4) extreme events such as tropical cyclones will be more frequent and more intense, and are influenced by climate variability.

According to Monnereau and Oxenford (2017) and Monnereau et al. (2021), considering the high level of exposure to climate change, the high socioeconomic fragility of the fishing sector, and the low adaptive capacity of many small islands in the Caribbean region, the impact of climate change will be high. The Caribbean Sea is considered the second largest sea in the world, containing 7.64% of the world's coral reefs and supporting over 500 fish species. The most important fish groups caught in commercial quantities are medium-sized pelagic fish, shrimps, lobster, and benthic mollusks (Smikle et al. 2010).

Coastal ecosystems provide essential services such as habitat and food provision, and carbon sequestration. For instance, most of the carbon stored by these ecosystems is found in the sediments: from 95 to 99% for coastal lagoons and seagrass, from 50 to 90% for mangroves, and the remaining carbon is in the biomass (Prato and Newball 2015). Coastal ecosystems are widely recognized for protecting the coastline by 70–90% against storms or tropical cyclones (Rodríguez 2015). Specifically, coral reefs dissipate wave energy by 97% on a global average (Ferrario et al. 2014). Coral reefs, seagrasses, and mangroves also serve as essential habitats and food provisions for local fishery resources (OECD 2021). McAllister (1988) found that healthy reefs in excellent conditions could provide around 18 tons of fish per km^2, which can decrease to 13 tons/km^2 for acceptable conditions and drop to 8 tons/km^2 for reefs in poor conditions. Mangrove-related fisheries could contribute USD 750 to USD 16,750 per hectare as highlighted by Rönnbäck (1999). This underscores that those economic benefits could strongly lessen or even disappear with each hectare of mangrove lost (Table 1). Among the primary marine ecosystems facing severe global pressures are coral reefs, mangroves, and seagrasses (Burke et al. 2011). Coral reefs are vulnerable to rising temperatures, leading to phenomena like coral bleaching, as observed by Hughes et al. (2017). Furthermore, ocean acidification negatively impacts coral growth and calcification (Doney et al. 2009), while tropical cyclones can rapidly destroy extensive areas of coral reefs, mangroves, and seagrasses, as documented by Cheal et al. (2017). Climate change exacerbates these challenges, driving up sea temperatures and the frequency and intensity of tropical cyclones (IPCC 2021), causing cover and extension losses for marine ecosystems and socioenvironmental impacts on fisheries. In addition, in recent decades, climate change has significantly impacted coral reef health, for example, the rapid spreading of Stony Coral Tissue Loss Disease, as coral reefs are more susceptible to diseases.

Optimal environments for coral development occur in regions with temperature ranges between 25 and 29 °C (Buddemeier et al. 2004). In Colombia, the temperatures recorded in coral reefs are between 24 and 28 °C on average per year under normal conditions. However, temperatures ranging between 26 and 29.5 °C have been reported in the archipelago (Guzman-Amaya et al. 2010). For the period 2041–2070, the mean SST is expected to increase between 0.9 and 1 °C (INVEMAR 2017). A temperature increase of 1–2 °C above the annual maximum value in an area can cause heat stress in corals (Mumby and van Woesik 2014; Kim et al. 2000; Oxenford and Monnereau 2017). If the temperature increases and persists for several weeks (three to four weeks), there is a higher incidence of diseases and massive coral bleaching events (Navas-Camacho et al. 2019; Hoegh-Guldberg et al. 2018; Grimsditch and Salm 2005; Oxenford and Monnereau 2017). Van Hooidonk et al. (2015) have indicated that bleaching events are likely to become the key driver of reef decline. In San Andrés, coral bleaching events have occurred intensely and recurrently over the past three decades, in 1985, 1995, 2005, and 2017 (Gómez-Campo et al. 2011). However, their impacts have been little documented. A coral reef grows through calcification

Table 1 Summary of climate change effects on fishery resources in the western Caribbean region

	Fishery resources	Level of affectation	Climate change impacts
Reef-associated shallow shelf species and coastal benthic species	*Haemulidae (grunts)*, *Lutjanus spp*, *Etelis oculatus*, *Mycteroperca spp. Serranidae Lobatus gigas (queen conch)*, *Panulirus argus (lobster)*, *Lenguados (flounders)*	Highly affected Highly vulnerable fish group	Decreased health, productivity, and overall abundance Reef fish will have lower fecundity and smaller eggs and larvae being produced Decreased health, productivity, and overall abundance Reef fish will have lower fecundity with smaller eggs and larvae being produced Reduced calcification in corals, mollusks, echinoderms, and larval fishes is more likely to be observed in the medium term Spiny lobster migration is influenced by the strength of local currents, which are likely to be affected over the long term by climate change
Deep slope species	Snappers, groupers Deep water *Lutjanus spp*, (red, black, yellowfin, yellowmouth), *Carangidae* (carangidos) (mackerel) (Jureles) (jacks, amberjacks, blackjacks)	Medium affected Slightly less severe than the coastal reef-associated species group Highly vulnerable fish group	Decreased recruitment of the species to its adult stage Adults will move deeper, avoiding warm waters, and the timing and location of spawning aggregations will likely be negatively affected

(continued)

Table 1 (continued)

	Fishery resources	Level of affectation	Climate change impacts
Oceanic pelagic fish	*Dolphinfish, wahoo, tunas Makaira spp. Thunnus atlanticus, Acanthocybium solandri, Coryphaena* hippurus and *Katsuwonus pelamis. Thunnus Obesus, Acanthocybium solandri, Coryphaena hippurus* and *Katsuwonus pelamis*	Less affected in the short term	Less abundant oceanic pelagic species Shifted tuna distribution Changes in horizontal and/or vertical distribution to escape unfavorable conditions Reproductive changes Possible affectations in their spawning stage Reductions in productivity of the oceanic pelagic species are expected over the medium to long term

processes that must be greater than erosion rates (Mallela and Perry 2007); the rise in sea level may increase wave energy, and thus corals will experience a higher level of energy, affecting coral growth and effectiveness in attenuating the energy of the waves. Consequently, the coastlines will be more eroded, and the impact of tropical storms and hurricanes will be more significant on the coasts.

4 Effects on Fishery Resources in Seaflower Biosphere Reserve

Historically, the oceans have acted as vital buffers against climate change by absorbing carbon dioxide (CO_2). However, the current state of climate change is altering ocean chemistry, triggering a series of interconnected and cumulative impacts within the fisheries sector. As a result, multiple adverse consequences have been observed, including reductions in fishing yields, economic disruption due to the destruction of settlements and infrastructure, economic deterioration, failure of livelihoods from fishing, biodiversity loss in traditional fishing areas, reduction in the habitability of reef islands leading to fish displacement or migration elsewhere, and loss of ecosystems and biodiversity (Lotze 2021; OECD 2021).

Oxenford and Monneareou (2017) contend that commercial Caribbean fisheries species are particularly vulnerable due to their already high thermal tolerance in the western Central Atlantic and the geographical constraints imposed by land barriers in the Gulf of Mexico and the Caribbean Sea. These species are at heightened risk for two primary reasons. Firstly, their high thermal tolerance has allowed them to adapt to the naturally warm waters of the western Central Atlantic. However, as the climate changes, rising sea temperatures surpass the limits of their thermal tolerance, leading to various issues, including stress, reduced reproductive success, and elevated mortality rates among these species. Once a tipping point is crossed, these consequences can persist over extended periods and may become irreversible. Secondly, the Gulf of Mexico and the Caribbean Sea, encircled by landmasses, impede the movement of marine species. As temperatures within these confined waters continue to rise, these species encounter limited opportunities for migration to cooler regions. In contrast to species in more open marine environments, which can migrate poleward to seek suitable temperatures, those within the Gulf of Mexico and Caribbean Sea can become trapped in increasingly unfavorable conditions. This combination of factors makes commercial Caribbean fisheries species highly susceptible to the impacts of climate change.

Changes will not be felt homogeneously throughout the Caribbean region and in different marine species. The magnitude of the impacts will depend on greenhouse gas emission levels which, in turn, depend on the effort and initiatives carried out by society to reduce these emissions, and the awareness of the cause and effects of climate change. Furthermore, factors such as the ecological characteristics of marine organisms, their geographic locations, the level of development in specific areas, their vulnerability, and the choice of adaptation and mitigation strategies all play pivotal roles in either exacerbating or mitigating the effects of climate change.

According to Cheung et al. (2010) and FAO (2016), the main impacts on primary fishery resources are changes in the distribution of some marine species towards the poles and deeper waters, a reduction in the productivity of most marine organisms, and in general a reduction in the size of fish (Sheridan and Bickford 2011; Cheung et al. 2010; Daw et al. 2009) and shifting baselines (Jackson et al. 2011). According to Cheung (2018), the impacts on fishery resources can be grouped as follows:

1. Changes in body size, reproduction, primary productivity, and habitats.
2. Changes in marine organism growth, abundance, and distribution.
3. Changes in community structure, trophic interaction, and biodiversity directly relate to socioeconomic changes in fisheries captures, fisheries economics, and fisheries management.
4. Marine ecosystem degradation.

Essentially, catches of commercial marine species might be reduced (Barange et al. 2014; Cheung et al. 2010; FAO 2018) meaning fishing operations, capture, landings of fish, and fishers' livelihoods will be seriously affected (Fig. 1).

Climate change affects, directly and indirectly, the Caribbean fishery resources in reef fishes, conch, spiny lobster, coastal pelagic, oceanic pelagic, and deep slope fishes. Direct effects are related to physiology and life processes, for example, rates of growth and development, reproduction, and longevity. The indirect effects are related to significant impacts on marine-coastal ecosystems that affect nursery areas, living space, refuge, and predator–prey relationships, as well as the physical and biological oceanographic changes that affect the survival, dispersal, and settlement of the early stages of the life history, and influence migration and distribution shifts (Monnereau and Oxenford 2017; Cheung 2018). The effects on fishing resources are organized into three main groups, according to types of habitats and fishing techniques: reef-associated shallow shelf, deep slope species, and oceanic pelagic fish. The following paragraphs and Table 1 use Monnereau and Oxenford's (2017) classification and analysis to describe species in each category and their impacts.

(1) **Reef-associated shallow shelf**: species that depend on critical reef habitats, including rocky shores, rock reefs, and coral reefs. Some of the most prominent families include *groupers (Serranidae), snappers (Lutjanidae), grunts (Haemulidae), parrotfishes (Labridae),* and *Lutjanus (Lutjanidae)*. Species have a biphasic life cycle involving an early pelagic life stage and a specific benthic nursery habitat for development from juvenile to adult stages.

   This group includes demersal, benthic fish, and shellfish. Juveniles are generally associated with brackish mangroves, and adults live in shallow, soft-bottomed, muddy, or sandy continental shelf areas, for example, lobsters and queen conch. Impacts include the following. (1) Acidification directly affects the exoskeleton of the Caribbean spiny lobster, especially during the molting process, delaying larval development and growth size for the market. (2) The risk of predation may increase, causing the overall abundance of adult lobsters to be reduced. (3) Lobster populations will seek deeper waters with lower temperatures, and lobster production will be very low during this period. (4) Increased spread of diseases, bacteria, viruses, and fungi that grow better at warmer temperatures (Kough et al. 2014). Authors that study the impacts of climate effects on lobsters include Briones-Fourzan and Lozano-Alvarez (2015), Caputi et al. (2013), and Lestang et al. (2012). The study carried out by Aldana and Manzano (2017) on the effect of climate change on the queen conch shows that temperature has a negative effect on the reproductive cycle and indicates that one effect is the reduction of the survival rate of the snail by 25% and of calcification by 50%.

(2) **Deep slope species**: this group relates directly to coral reefs, mangroves, and seagrasses. Most of the fish species in this group have a biphasic life cycle. As a result, these large deep-sea species tend to grow more slowly, mature at a higher age, and live longer than their shallow-reef counterparts.

(3) **Oceanic pelagic fish**: this group includes the more offshore, open water, highly migratory, epipelagic (surface) species (e.g., flying fish, dolphinfish, wahoo). Examples include larger fish such as mackerel, swordfish, and tuna. Species in this group are expected to respond by changing their horizontal and vertical distribution to escape unfavorable conditions.

In general, marine organisms have multiple levels of responses to climate change, for example, changes in the distribution and composition of plankton and the timing of phytoplankton blooms, ocean species range shifts, increased incidence of marine diseases, changes in physiology and fish behavior, altered timing and duration of spawning and migrations, disruption of food webs, and changes in the populations' genetic structure and productivity. The fish group that will be most affected by climate change-associated hazards are the reef-associated shallow shelf and benthic species, not only due to the species' ecology but also because of habitat deterioration. In general, the few published studies on the direct impacts of climate change on Caribbean-specific species like queen conch, spiny lobster, and red snappers, among others, demonstrate a significant research gap for the Caribbean region and especially for the archipelago.

To summarize and consolidate the various climate change impacts on marine ecosystems, fishery resources, and the fisheries sector stated in the literature, a progression graph is developed that shows the climate drivers, changes in fishery resources, fishers' fragility, and the adverse consequences (Fig. 1).

5 Seaflower Fishers in the Context of Climate Change

From the seventeenth century to the present, marine ecosystems have played a vital role in sustaining local food sources for the archipelago. Artisanal fishing activities have historically provided fishery resources, including fish, lobster, queen conch, and more (INVEMAR-ANH 2012; Prato and Newball 2015; Velásquez 2019). These activities are primarily carried out by the Raizal community, utilizing small boats of approximately 40 ft. or 12 m in length, with a capacity of under 5 tons (Santos-Martínez et al. 2019).

However, the fisheries in Seaflower face many challenges, mirroring global issues in the sector. These challenges include overfishing, habitat loss, illegal fishing, data uncertainty for population assessments, environmental impacts, and political factors affecting fisheries management (Santos-Martínez et al. 2019).

Amid these challenges, changes in the artisanal fishing landscape have been observed in recent years. The number of artisanal fishers in the area has undergone notable reductions over the past five years. In 2015, San Andrés had 1,408 artisanal fishers, with Providencia having 472. However, by 2017, these numbers had declined, with San Andrés at 585 and Providencia at 249. Furthermore, in 2019, the count stood at 624 registered artisanal fishers in San Andrés and 263 in Providencia, indicating a general decrease in recent years (Gobernación departamental de San Andrés, Providencia y Santa Catalina 2019).

Additionally, climate change's negative impacts and climate variability affect fishers' livelihoods in the archipelago. This is especially true as the archipelago is exposed to the impact of both climate change-related stressors and human activities. According to Bejarano (2016) and Ortiz (2016), fishers have experienced a prolonged impact since 2012. The International Court of Justice-ICJ judgment considerably reduced the daily catch. In 2012, the ICJ confirmed Colombia's sovereignty over the disputed islands while granting Nicaragua a significant maritime territory of approximately 75,000 km^2 extending Nicaragua's Continental Shelf (CS) and Exclusive Economic Zone (EEZ). However, this decision adversely affected the spatial integrity

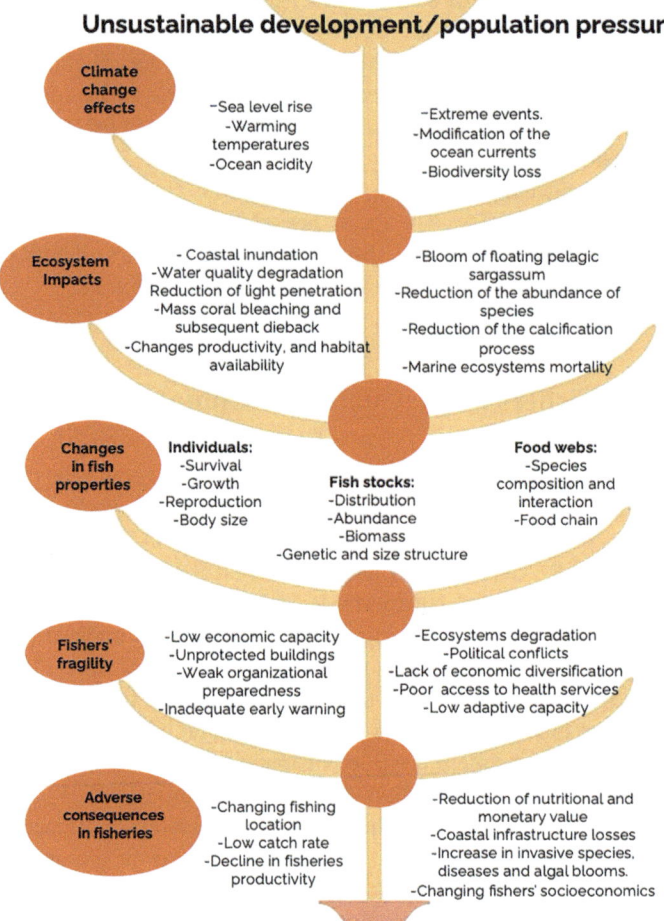

Fig. 1 This progression graph serves as an illustrative summary of the intricate interplay between climate change, ecosystems, and fishery resources within the fisheries sector. It provides a visual representation of the multifaceted effects of climate change on these vital components. Moreover, by considering the vulnerabilities of the fishing community, it elucidates how these environmental shifts intensify risks within the fisheries, culminating with a description of the challenges faced in this dynamic sector

of the Archipelago, the RB Seaflower, its Marine Protected Areas, the ancestral territory of the Raizal People, and their cultural, environmental, and food security practices tied to fishing (ICJ 2012; Ortiz 2023). For instance, in 2012, the average fishing catch reached 1,000 pounds a week, while in 2016, fishers got only 250 pounds of fish a week. Since 2012, fishers do not have access to traditional fishing areas since they are no longer under Colombian jurisdiction. Indeed, the gradual reduction of fishery resources, climate change effects, and high tourism demand have made it more difficult for fishers to supply local fish than foreign freshwater fish. Few studies have explored the social and political impacts on the small-scale fisheries of San Andrés.

On November 16, 2020, Hurricane Iota hit the archipelago, mainly the islands of Providencia and Santa Catalina, causing severe impacts in the fishing sector. There were severe impacts on mangroves, seagrasses, and coral reefs. For instance, in Providencia, observations include the defoliation of 90% of mangroves, a significant decrease in the biomass of seagrasses, and significant damages to coral reefs, affecting food provision, natural barriers, and carbon sequestration ecosystem services (ES). Moreover, the entire fishing fleet and fishing infrastructure were lost, and more than 350 fishers were affected. There were losses and damages to more than 170 boats and 188 motors, fishing gear, refrigerators, and infrastructure for fishery product collection, conservation, and commercialization (Guzman 2021). In response to this situation, priority actions were established through collaboration between the Ministry of Agriculture and Rural Development, the National Aquaculture and Fisheries Authority, and the fishing organizations in Providencia, aiming to facilitate the recovery of the sector and strengthen the associative and entrepreneurial processes of fishing organizations.

Overall, San Andrés, Providencia, and Santa Catalina, all part of the archipelago, exhibit differences in geographic, environmental, and economic factors. Hurricane Iota, for instance, inflicted far more devastating damage on Providencia than on San Andrés, shaping the islands' responses to climate-related challenges. Additionally, the islands differ significantly regarding tourism dynamics and resource management. With its developed tourism infrastructure, San Andrés experiences higher tourism demand and, consequently, it faces more pronounced challenges related to resource depletion and environmental sustainability. Providencia and Santa Catalina, on the other hand, have maintained a more eco-conscious approach to tourism, emphasizing conservation efforts and preserving their natural assets. These differences in economic and environmental strategies highlight the unique paths each island has taken in balancing fisheries and tourism growth.

The Seaflower Biosphere Reserve's fishers are highly vulnerable. Results from the socioeconomic characterization survey carried out with 636 artisanal fishers by the Secretariat for Agriculture in 2014, give us an approximation of their social fragility, in which 87.30% of participants were not affiliated with a retirement pension fund. For most of them, their economic income is not enough to pay for health insurance and a pension fund. 82.29% are heads of the family, 30% of these are married, 28% live in free union and almost 15% are single. This shows the relevance of fisheries work in supporting their families. 22.2% of fishers earn between 301,000 Colombian Pesos (COP) (US $155) and COP 600,000 (US $309) per month to support an average of 4 people. The fishers' vulnerability decreases as their socioeconomic situation improves.

Accordingly, taking preventive and adaptive local actions is the main way to face global changes. Adaptation is a two-step estimation process: first, perceiving the change and, second, deciding whether to adapt by adopting a particular measure

(Mulyasari et al. 2018). In this sense, implementing local future management interventions in fishing activities depends on fishers' vulnerability and perceptions. Unfortunately, there have been few studies in the archipelago documenting islanders' perceptions: Velásquez (2011) studied farmers' perception of hurricanes, Ruiz de la Cruz (2016) analyzed the perception of the deterioration of the landscape and natural resources related to the tourism sector, and Correa (2012) studied local knowledge and beliefs about the climate on the islands of Providencia and Santa Catalina. In other words, there is a knowledge gap concerning the perception of climate change impacts on fishery resources.

Fishers' perceptions influence resource extraction patterns and the protective and adaptation measures taken (Acosta et al. 2021). According to Mulyasari et al. (2018), perceptions influence the readiness and willingness of fishers to adapt and adjust to climate change. These perceptions are based on beliefs, knowledge, and past experiences, and are particularly important for the archipelago where there is weak fisheries governance. Accordingly, this study conducted forty-nine surveys, 27 in 2019 and 22 in 2022, to learn more about fishers' perceptions.

Findings show that fishers have noticed changes in the climate and fishery resources, based on climate-related manifestations. Participants perceive changes in dry and rainy seasons, increased coastal erosion, changes in the ocean and earth temperatures, aquifer salinization, and increased hurricane frequency. Participants in 2019 said that the most critical climate change stressor in recent years has been droughts (85%), followed by changes in the sea surface temperature and hurricanes. Precipitation reduction magnifies the effects of the rise in salinity, which has repercussions on the quantity and quality of fishery resources (Diouf et al. 2020). Additionally, the areas of rapid coastal erosion on San Andrés, Providencia, and Santa Catalina are linked not only to sea level rise, but also to the extraction of sand from the beach for building houses and roads which, in turn, is affecting fishing docks and houses.

Analyzing fishers' perceptions before and after Hurricane Iota offers valuable insights into their experiences and evolving environmental awareness. Prior to the hurricane (2019 survey), fishers primarily perceived drought as the main hazard, influenced by recurrent dry conditions in the region. However, after the destructive impact of Hurricane Iota (2022 survey), there was a noticeable shift in hazard perception. While drought continued to be a concern, the hurricane's devastating effects elevated the perception of hurricanes as a significant hazard. This heightened awareness of multiple hazards profoundly impacted their priorities and risk perception, causing them to view drought and hurricanes as substantial threats to their livelihoods and communities. Policymakers and disaster management agencies must consider these evolving perceptions when crafting resilience and preparedness strategies that address a broader spectrum of hazards and their interactions.

Participants perceived that the main problem facing the fisheries sector is the capture and commercialization of small-sized juveniles and ovate females (55.5%), the use of prohibited equipment, such as diving tanks that affect the renewal of fishery resources, and 44.5% mentioned fish overexploitation. Participants did not prioritize climate change as the main current problem.

Overall, the survey results (2019 and 2022) show that 100% of participants think that the climate is changing, and that the fisheries sector is currently being affected by these changes. Participants mentioned that the various changes directly affect household food security and infrastructure. They also argued that this makes it difficult to plan their fishing activities: "there is a reduction of the fishing resource in the areas traditionally used for fishing, because when the water warms up, fish migrate."

Results show that participants are aware of human responsibility for global warming. For example, 51% of participants think that natural and social processes cause climate change, 30.6% think climate change is caused only by social factors, and 16.3% think only natural factors cause climate changes. These results, however, indicate the need to continue implementing risk communication strategies explaining the causes, risks, and uncertainties surrounding potential changes over the coming years, decades, and centuries (Pidgeon and Fischhoff 2011).

Participants listed the leading climate change effects as follows: decreased fish population, the disappearance of species, fish migration, and changes in the reproduction cycle. Most participants (82%) in 2022 perceived that all fishery resources, including reef-associated shallow shelf species, deep slope species, queen conch, and spiny lobster, were being affected by climate change. In 2019, participants perceived that the primary fishery resources affected by climate change were reef-associated shallow shelf species (45%), followed by queen conch and spiny lobster (37%). Likewise, half of the participants in 2022 perceived that all marine ecosystems, including coral reefs, soft sandy bottoms, seagrasses, and mangroves, have been affected by climate change. Some participants (31.8%) highlighted that the coral reef ecosystem is the one with the most significant impacts, followed by mangrove ecosystems (22.7%).

In response to the question, what are the potential impacts of climate change on fishery resources and fishing activities? Participants answered (including 2019 and 2022 participants):

- Large shoal fish are migrating to deeper waters.
- Shallow water fish are now only available in some seasons.
- There is a reduction of the fishing resources in the traditionally used areas because fish migrate when the water warms up.
- Alteration of the reproductive cycles does not allow the fish population to increase, and the catch will decrease.
- It would affect the growth and feed of lobster juveniles, and the queen conch could lose its shell in a few decades.
- There has been a change in the sea level.
- There is an increase in diseases and invasive species in marine ecosystems.
- Acidification and warming of the sea will affect coastal marine ecosystems and species.
- Marine species extinction.
- Climate change is a threat to the food sovereignty of the islands.
- Mortality of coral reefs, more hurricanes, cloudy water, and an increase in ocean currents.

- Fishers will not be able to carry out their work, and there will be conflicts between fishermen over the fishing resources that are becoming scarce.

Participants perceive that local institutions are not prepared to face climate change impacts and its associated stressors, and they make the following recommendations:

- Maintain only artisanal fishing and stop industrial fishing in the archipelago.
- Change the development model to give marine organisms more time to grow.
- Implement regulations, sanctions, and control of illegal fishing catches.
- Promote alternative energy solutions.
- Implement a systemic approach incorporating synergies in mitigation, adaptation, and sustainable food production.
- Integrate local and scientific knowledge in public fishing policy and strengthen marine protected areas.
- Training and education programs to sensitize islanders and tourists concerning the protection of fishery resources.
- Promote international agreements on climate change, desertification, biodiversity, and fishing.

Overall, local perspectives are in line with general scientific results. Fishers recognize the climate stressors and consequences and identify management actions. However, participants did not prioritize climate change as the main current problem for the fisheries sector. Fishers perceive the general effects of climate change and the relationship between marine ecosystem deterioration and marine organism changes. However, participants did not point out specific impacts on commercial fishery resources. Results indicate the necessity to move from general to specific and detailed information for each fishery resources group in the archipelago. It is also necessary to raise awareness about the socioenvironmental cascade impacts in the fishing sector and the interaction between hazards and vulnerabilities in impact magnitude. Generally, new observations made by fishers are related to the intensification of transboundary and local social conflicts between fishers and the illegal fisheries practices related to climate change effects and exacerbation. Moreover, fishers argue that the catching and commercialization of small-sized juveniles, ovate females, and the use of prohibited equipment such as diving tanks are affecting the renewal of fishery resources. Fishers did not mention neurological and genetic fish changes resulting from ocean changes.

There is a growing perception that climate change is caused by a combination of natural and anthropic factors. However, there is still a belief that climate change is exclusively a natural phenomenon. There is not enough clarity on how each fisher can contribute to reducing impacts and adapting to climate change. Fishers also firmly believe that industrial fishing should not be allowed in the archipelago. Fishers have cared for the environment, and marine protected areas are primarily considered in terms of both biodiversity conservation and fisheries management, and, most importantly, as a strategy to face and adapt to climate changes. Recommendations related to improving monitoring programs, data collection, and risk assessments were not mentioned.

6 Integrating Scientific and Local Knowledge to Inform Adaptation Strategies

Based on the general review of the threats and their effects on the strategic marine-coastal ecosystems and primary fishing resources of the Caribbean, and considering the recommendations made by fishers and the different researchers cited in previous sections, as well as international organizations like FAO and national ones like the National Planning Department and the Ministry of Agriculture and Fisheries, five guidelines for adaptation to climate change are listed below.

6.1 A Participatory Ecosystem Approach for the Co-management of Shared Natural Resources

An ecosystem approach promotes a holistic, integrated, and participatory vision that seeks to achieve the sustainability of the fishing sector (FAO 2014). It seeks to implement comprehensive strategies that anticipate future changes, evaluate the consequences, and develop responses according to the local reality and based on ecosystem connectivity. The effects of climate change on ecosystems make it essential to promote the conservation of biodiversity, support the generation of knowledge of the species captured throughout the year, and monitor changes occurring annually through the implementation of information systems. Additionally, it is necessary to implement a co-management strategy in which fishers take a leading role in managing the activity and conservation of ecosystems. There is a shared responsibility between fishers and the local government. For example, fishers take responsibility for the state of fishery resources by creating, reviewing, and improving rules and adaptation strategies. This approach legitimizes fishers' knowledge, and places value on traditional ecological knowledge, seeking greater stability in fishing activity.

6.2 Alternative Economic Activities for Artisanal Fishers

Fishing resources are already presenting problems of population decline due to the significant impact caused by human activities such as overexploitation, deterioration, and contamination of coastal ecosystems. Climate change is aggravating this situation and projecting a dark scenario for small tropical islands. Therefore, artisanal fishers need to strengthen their capacity to generate additional income through alternative economic activities, for example, ecotourism, handicrafts, and agriculture, as well as participating in government programs to strengthen fishers' entrepreneurship abilities. Likewise, this guideline includes promoting sustainable fishing methods and practices that allow sustainable resource extraction and strengthening fisher's associations.

6.3 Knowledge and Risk Communication to Raise Awareness and Inform Fishers About Climate Change Cascade Effects

Uncertainty regarding the effects of climate change on fishery resources is high and requires extensive research on marine species. The purpose is to have and share accurate and up-to-date information on fishing exploitation, ecosystem connectivity and conservation, threats, and vulnerabilities of marine species, ecosystems, and fishers. Making risk information accessible in an easy-to-understand way is important for facilitating the application of risk information by fishers.

6.4 Promote Vulnerability Studies to Understand Root Causes and Stimulate Adaptation Actions

Knowing the level of vulnerability of fishers, fishery resources, and marine-coastal ecosystems is essential to have high-quality information on climate risks in the archipelago. Vulnerability leads us to find and analyze the factors that make a system, organization, community, or sector susceptible and, from there, to work on those factors that would not be able to face new situations. Risk management, with its conceptual components of hazard and vulnerability, is proposed as the ideal way to reduce climatic risk, which implies acting on the causes that produce them and understanding risk perception (UNDP 2007).

6.5 Strengthening the Seaflower Marine Protected Area

A marine protected area is the best planning tool to guarantee the conservation of natural resources, preserve marine species, and guide human activity through environmental zoning (Santos-Martínez et al. 2012). The strategy guarantees:

- Protecting priority areas, key species, and the most vulnerable communities.
- Reducing conflicts of interest regarding the use of resources.
- Generating tools so that the use of resources is sustainable.

Researchers such as Pauly and Cheung (2017) agree that extensive conservation areas are required to increase fish populations' resilience, along with reviewing surveillance systems and implementing fishing activity controls.

7 Conclusions

This chapter first reviewed existing knowledge of the multiple impacts of climate change on commercial fishery resources in the western Caribbean region. Second, to promote the integration of local and scientific knowledge, it assessed fishers' perceptions of climate change impacts on fishery resources.

There is extensive information available on climate change hazard projections and their possible effects, but little information on marine species levels in the Caribbean region. The same situation is true for marine ecosystems, where the coral reef is the most studied. Key knowledge gaps exist in our understanding of the implications of climate and ocean chemistry changes for marine fisheries in the archipelago, particularly on the social and economic responses of the fishing sector to climate change. Despite the increasing interest in the impacts of climate change in the archipelago, economic vulnerability and social-ecological adaptation strategies are still largely unknown. There is a need for specific studies on biophysical variables and species, long-term impacts of ocean acidification on species, including the relationship to disease outbreaks, and cascade effects related to the interconnectivity of coastal ecosystems (Mycoo et al. 2022; McField 2017; Wilson 2017). However, these knowledge gaps should not delay the implementation of climate change adaptation strategies.

The analysis of the fishers' perceptions about climate change and its impacts on fisheries, revealed that fishers have already noticed changes in temperature and rainfall trends. Findings showed that the fishers in the archipelago are aware of and are willing to act to adapt to the effects of climate change. However, participants did not prioritize climate change as the main current problem. Fishers' perceptions are largely aligned with the consequences of climate change as described and analyzed by scientists. Climate change impacts have adversely affected the fisheries sector in multiple ways: loss of coastal fish breeding habitats such as coral reefs, soft sandy bottoms, seagrasses, and mangroves; the migration of large schools of fish to deeper waters; a decline in fish and shellfish productivity in traditional fishing areas; an increase in invasive species, diseases, and algal blooms; changes in biological properties in marine organisms; queen conch and spiny lobster present reduced calcification and overall abundance; reduced fisheries productivity, reduced fishing operations, and higher adaptation costs; increased risk to food security; and increased local and cross border conflicts over fishery resources.

Finally, as a way to integrate the knowledge of scientists and fishers, we proposed five guidelines for adaptation to climate change for the archipelago: a participatory ecosystem approach for the co-management of shared natural resources; alternative economic activities for artisanal fishers; knowledge and risk communication to raise awareness and inform fishers about cascade climate change effects; promote vulnerability studies to understand root causes and stimulate adaptation actions; and strengthening the Seaflower Marine Protected Area.

Acknowledgements We gratefully acknowledge the support provided by the Agriculture and Fishery Secretariat of the San Andrés government and the National University of Colombia,

Caribbean campus, which have partly funded this study in the Archipelago of San Andrés, Providencia, and Santa Catalina. We also extend our heartfelt appreciation to the fishers who generously shared their knowledge and expertise for this research, as well as the reviewers, whose comments strengthened the manuscript. Thanks to the Corporation Center of Excellence in Marine Sciences, CEMarin, for disseminating our research findings. Finally, we thank Minciencias and Colfuturo (Beca para Estudios de Doctorado-646).

References

Acosta A, Fadilah A, DieiOuaudi Y et al (2021) Perceptions of the western tropical Atlantic and Caribbean stakeholders regarding their role in achieving sustainable fisheries. Ocean Coast Res 69(1):e21047. https://doi.org/10.1590/2675-2824069.21030aa

Aldana DA, Manzano N (2017) Effects of near-future-predicted ocean temperatures on early development and calcification of the queen conch Strombus gigas. Aquacult Int 25:1869–1881. https://doi.org/10.1007/s10499-017-0153-y

Barange M, Merino G, Blanchard JL et al (2014) Impacts of climate change on marine ecosystem production in societies dependent on fisheries. Nat Clim Change 4:211–216. https://doi.org/10.1038/nclimate2119

Bejarano J (2016) Análisis socioeconómico del fallo de la Corte Internacional de Justicia de 2012, sobre los pescadores raizales del departamento de San Andrés y Providencia. Paper presented at the IV Congreso Internacional de Investigación en Gestión Pública, Bogotá, Colombia, 8–9 September 2016

Boavida-Portugal J, Rosa R, Calado R et al (2018) Climate change impacts on the distribution of coastal lobsters. Mar Biol 165:186. https://doi.org/10.1007/s00227-018-3441-9

Bonebrake TC, Brown CJ, Bell JD et al (2018) Managing consequences of climate-driven species redistribution requires integration of ecology, conservation and social science. Biol Rev 93:284–305. https://doi.org/10.1111/brv.12344

Briones-Fourzán P, Lozano-Álvarez E (2015) Lobsters: ocean icons in changing times. ICES J Mar Sci 72:i1–i6. https://doi.org/10.1093/icesjms/fsv111

Buddemeier R, Kleypas J, Aronson R (2004) Coral reefs and global climate change: potential contributions of climate change to stresses on coral reef ecosystems. Pew Center on Global Climate Change, Arlington, VA

Burke L, Reytar K, Spalding M et al (2011) Reefs at risk revisited. World Resources Institute, Washington DC

Caputi N, de Lestang S, Frusher S et al (2013) The impact of climate change on exploited lobster stocks. In: Phillips B (ed) Lobsters: biology, management, aquaculture and fisheries, 2nd ed. Wiley-Blackwell, New Jersey, USA, pp 84–112

Cheal A, McNeil A, Emslie M et al (2017) The threat to coral reefs from more intense cyclones under climate change. Glob Chang Biol 23(4):1511–1524. https://doi.org/10.1111/gcb.13593

Cheung W (2018) The future of fishes and fisheries in the changing oceans. J Fish Biol 92:790–803. https://doi.org/10.1111/jfb.13558

Cheung WWL, Lam VWY, Sarmiento JL et al (2010) Large-scale redistribution of maximum fisheries catch potential in the global ocean under climate change. Glob Chang Biol 16:24–35. https://doi.org/10.1111/j.1365-2486.2009.01995.x

CORALINA (2022) Alerta en San Andrés por agresiva enfermedad de los Corales. Región Caribe. https://regioncaribe.com.co/alerta-en-san-andres-por-agresiva-enfermedad-de-los-corales/

CORALINA-INVEMAR (2017) Informe técnico final. Actualización del conocimiento sobre los ecosistemas sumergidos de San Andrés isla para la gestión ambiental del departamento archipiélago de San Andrés, Providencia y Santa Catalina convenio especial de cooperación No. 007-2017 CORALINA–INVEMAR. Santa Marta, Colombia

Correa S (2012) Procesos culturales y adaptación al cambio climático: la experiencia en dos islas del Caribe colombiano. Boletín De Antropología Universidad De Antioquia 27(44):204–222

Daw T, Adger WN, Brown K et al (2009) El Cambio Climático y la pesca de captura: repercusiones potenciales, adaptación y mitigación. In: Cochrane K, De Young C, Soto D et al (eds) Consecuencias del Cambio Climático para la pesca y la acuicultura: visión de conjunto del estado actual de los conocimientos científicos. FAO Documento Técnico de Pesca y Acuicultura, 530, Rome, pp 119–168

Diouf NS, Ouedraogo I, Zougmoré RB et al (2020) Fishers' perceptions and attitudes toward weather and climate information services for climate change adaptation in Senegal. Sustainability 12:9465. https://doi.org/10.3390/su12229465

Doney SC, Fabry VJ, Feely RA et al (2009) Ocean acidification: the other CO_2 problem. Annu Rev Mar Sci 1:169–192. https://doi.org/10.1146/annurev.marine.010908.163834

Ferrario F, Beck M, Storlazzi C et al (2014) The effectiveness of coral reefs for coastal hazard risk reduction and adaptation. Nat Commun 5:3794. https://doi.org/10.1038/ncomms4794

Food and agriculture organization of the United nations (FAO) (2009) Climate change implications for fisheries and aquaculture, overview of current scientific knowledge. FAO Fisheries and Aquaculture Technical Paper 530, Rome. https://www.ipcinfo.org/fileadmin/user_upload/newsroom/docs/FTP530.pdf

Food and Agriculture Organization of the United Nations (FAO) (2014) The state of world fisheries and aquaculture. Opportunities and challenges. FAO, Rome. https://www.fao.org/fishery/en/publications/66711

Food and Agriculture Organization of the United Nations (FAO) (2016) Brief on fisheries, aquaculture, and climate change in the intergovernmental panel on climate change, fifth assessment report (IPCC AR5). FAO, Rome. https://www.fao.org/3/i5871e/i5871e.pdf

Food and Agriculture Organization of the United Nations (FAO) (2018) Impacts of Climate Change on Fisheries and aquaculture, Synthesis of current knowledge, adaptation, and mitigation options. https://www.fao.org/3/i9705en/I9705EN.pdf

Gobernación Departamental de San Andrés, Providencia y Santa Catalina (2019) Insumos al Plan de Ordenamiento Territorial. Technical report

Gómez-Campo K, Lopez-Londoño T, Gil-Agudelo D et al (2011) Blanqueamiento coralino, amenaza para el futuro de los arrecifes de coral de los archipiélagos Nuestra Señora del Rosario y San Bernardo. In: Zarza-Gonzalez E (ed) El Entorno Ambiental del PNN Corales del Rosario y de San Bernardo. Parques Nacionales Naturales de Colombia, Cartagena de Indias, pp 319–328

Gordon TAC, Harding HR, Clever FK et al (2018) Fishes in a changing world: learning from the past to promote sustainability of fish populations. J Fish Biol 92:804–827. https://doi.org/10.1111/jfb.13546

Grimsditch GD, Salm RV (2005) Coral reef resilience and resistance to bleaching. IUCN Resilience Science Group Working Paper Series, 1, Gland, Switzerland

Guzman L (2021) Respuesta requerimientos Federación de Pescadores de Providencia y Santa Catalina. Ministerio de Agricultura. Radicado Minagricultura No. 20213130119912

Guzmán-Amaya P, Rojas-Carrillo PM, Morales-García GZ et al (2010) Retos para el sector pesquero y acuícola ante el Cambio Climático. In: Botello AV, Villanueva S, Gutiérrez J et al (eds) Vulnerabilidad de las zonas costeras mexicanas ante el Cambio Climático. Semarnat-INE, UNAM-ICMyL, Universidad Autónoma de Campeche, Gobierno del Estado de Tabasco, pp 113–164

Hoegh-Guldberg O, Jacob D, Taylor M et al (2018) Impacts of 1.5 °C global warming on natural and human systems. In: Masson-Delmotte V, Zhai P, Pörtner H-O et al (eds) Global Warming of 1.5 °C. An IPCC special report on the impacts of global warming of 1.5 °C above pre-industrial levels and related global greenhouse gas emission pathways, in the context of strengthening the global response to the threat of climate change, sustainable development, and efforts to eradicate poverty. Cambridge University Press, Cambridge, UK and New York, NY, USA, pp 175–312

Hughes TP, Kerry JT, Álvarez-Noriega M et al (2017) Global warming and recurrent mass bleaching of corals. Nature 543:373–377. https://doi.org/10.1038/nature21707

Hughes TP, Kerry JT, Baird AH et al (2018) Global warming transforms coral reef assemblages. Nature 556:492–496. https://doi.org/10.1038/s41586-018-0041-2

International Court of Justice [ICJ] (2012) Territorial and maritime dispute, Nicaragua v Colombia, Judgment, ICJ GL No 124, ICGJ 436, 19th November 2012

INVEMAR Instituto de investigaciones Marinas y Costeras (2017) Análisis de vulnerabilidad marino-costera e insular ante el Cambio Climático como insumo para la Tercera Comunicación Nacional de Cambio Climático, documento de investigación. Santa Marta, Colombia

INVEMAR-ANH (2012) Estudio línea base ambiental y pesquera en la Reserva de Biósfera Seaflower (Archipiélago de San Andrés, Providencia y Santa Catalina) como aporte al conocimiento y aprovechamiento sostenible de los recursos para la región – fase 1. Informe final pry-var-010-11-itf. Santa Marta, Colombia

IPCC (2018) Summary for policymakers. In: Masson-Delmotte V, Zhai P, Pörtner H-O et al (eds) Global warming of 1.5 °C. An IPCC Special Report on the impacts of global warming of 1.5 °C above pre-industrial levels and related global greenhouse gas emission pathways, in the context of strengthening the global response to the threat of climate change, sustainable development, and efforts to eradicate poverty. Cambridge University Press, Cambridge, UK and New York, NY, USA, pp 3–24

IPCC (2021) Summary for policymakers. In: Masson-Delmotte V, Zhai P, Pirani A et al (eds) Climate change 2021: the physical science basis. Contribution of working Group I to the sixth assessment report of the intergovernmental panel on climate change

Jackson JBC, Alexander K, Sala E (2011) Shifting baselines: the past and the future of ocean fisheries. Island Press, Washington DC

Jyun-Long C (2020) Fishers' perceptions and adaptation on climate change in northeastern Taiwan. Environ Dev Sust 23:611–634. https://doi.org/10.1007/s10668-020-00598-0

Kettle N, Dowb K, Tuler S et al (2014) Integrating scientific and local knowledge to inform risk-based management approaches for climate adaptation. Clim Risk Manag 4–5:17–31. https://doi.org/10.1016/j.crm.2014.07.001

Kim K, Kim PD, Alker A, Harvell C (2000) Chemical resistance of gorgonian corals against fungal infections. Mar Biol 137:393–401

Kough A, Paris C, Behringer D et al (2014) Modelling the spread and connectivity of waterborne marine pathogens: the case of PaV1 in the Caribbean. ICES J Mar Sci 72:i139–i146. https://doi.org/10.1093/icesjms/fsu209

Lestang S, Caputi N, How J et al (2012) Stock assessment for the west coast rock lobster fishery. Fisheries research report, 217. Department of Fisheries, Western Australia

Llanos C (2015) Ayer y Hoy en la pesca artesanal de San Andrés Isla. Observatorio Seaflower, CORALINA. https://observatorio.coralina.gov.co/index.php/es/component/k2/item/407-ayer-y-hoy-en-la-pesca-artesanal-de-san-andres-isla

Lotze HK (2021) Marine biodiversity conservation. Curr Biol 31(19):R1190–R1195. https://doi.org/10.1016/j.cub.2021.06.084

Mallela J, Perry CT (2007) Calcium carbonate budgets for two coral reefs affected by different terrestrial runoff regimes, Rio Bueno, Jamaica. Coral Reefs 26:129–145. https://doi.org/10.1007/s00338-006-0169-7

Mastrandrea MD, Heller NE, Root TL et al (2010) Bridging the gap: linking climate-impacts research with adaptation planning and management. Clim Change 100:87–101. https://doi.org/10.1007/s10584-010-9827-4

McAllister DE (1988) Shiraho coral reef and the proposed new Ishigaki Island airport, Japan, with a review of the status of coral reefs of the Ryukyu Archipelago. International Marinelife Alliance, Ottawa, Canada

McField M. (2017). Impacts of climate change on coral in the coastal and marine environments of Caribbean Small Island Developing States (SIDS). Caribbean Mar Climate Change Rep Card Sci Rev 52–59

Monirul A, Khorshed A, Shahbaz M (2017) Climate change perceptions and local adaptation strategies of hazard-prone rural households in Bangladesh. Clim Risk Manag 17:52–63. https://doi.org/10.1016/j.crm.2017.06.006

Monnereau I, Mahon R, McConney P et al (2021) Fisheries sector vulnerabilities to climate change in small Island developing states. In: Moncada S, Briguglio L, Bambrick H et al (eds) Small island developing states: vulnerability and resistance under climate change. Springer Nature, Switzerland, pp 233–255

Monnereau I, Oxenford H (2017) Impacts of climate change on fisheries in the coastal and marine. Environments of Caribbean small island developing states (SIDS). Centre for Resource Management and Environmental Science (CERMES), University of the West Indies, Cave Hill, Barbados

Mulyasari IG, Waluyati LR, Suryantini A (2018) Perceptions and local adaptation strategies to climate change of marine capture fishermen in Bengkulu Province, Indonesia. IOP Conf Ser: Earth Environ Sci 200:012037. https://doi.org/10.1088/1755-1315/200/1/012037

Mumby PJ, van Woesik R (2014) Consequences of ecological, evolutionary and biogeochemical uncertainty for coral reef responses to climatic stress. Curr Biol 24(10):PR413–R423. https://doi.org/10.1016/j.cub.2014.04.029

Mycoo M, Wairiu M, Campbell D, Duvat V, Golbuu Y, Maharaj S, Nalau J, Nunn P, Pinnegar J, Warrick O (2022) Small islands. In: Pörtner H-O, Roberts DC, Tignor M, Poloczanska ES, Mintenbeck K, Alegría A, Craig M, Langsdorf S, Löschke S, Möller V, Okem A, Rama B (eds) Climate change 2022: impacts, adaptation and vulnerability. Contribution of working group II to the sixth assessment report of the intergovernmental panel on climate change. Cambridge University Press, Cambridge, UK and New York, NY, USA, pp 2043–2121. https://doi.org/10.1017/9781009325844.017

Navas-Camacho R, Acosta-Chaparro A, González-Corredor JD et al (2019) 20 años (1998–2017) de monitoreo de las formaciones coralinas en San Andrés y Providencia. Serie de Publicaciones Generales, 106. INVEMAR-CORALINA, Santa Marta, Colombia

Nurse L (2011) The implications of global climate change for fisheries management in the Caribbean. Clim Dev 3(3):228–241. https://doi.org/10.1080/17565529.2011.603195

OECD (2021) Managing climate risks, facing up to losses and damages. OECD Publishing, Paris. https://doi.org/10.1787/55ea1cc9-en

Ortiz F (2023) Diferendos limítrofes en el Caribe occidental: Reserva de Biosfera Seaflower y el pueblo Raizal. Revista Cuadernos del Caribe. Vol 27. Universidad Nacional de Colombia Sede Caribe, San Andrés Isla, Colombia

Ortiz Roca F (2016) Los Derechos del pueblo Raizal del Archipiélago de San Andrés, Providencia y Santa Catalina, Colombia, Caribe Occidental en el marco del Convenio 169 de la OIT. In: Puyana AM (comp) Consulta Previa y Modelos de Desarrollo: Juego de espejos. Reflexiones a propósito de los 25 años del Convenio 169 de la OIT. Deutsche Gesellschaft für Internationale Zusammenarbeit (GIZ) GmbH, Bogotá Colombia, pp 227–245

Oxenford, HA, Monnereau I (2017) Impacts of climate change on fish and shellfish in the coastal and marine environments of Caribbean small island developing states (SIDS). In: Caribbean marine climate change report card: science review 2017, pp 83–114

Oxenford HA, Monnereau I (2018) Climate change impacts, vulnerabilities, and adaptations: Western Central Atlantic Marine In: Barange M, Bahri T, Beveridge M et al (eds) Impacts of climate change on fisheries and aquaculture, synthesis of current knowledge, adaptation and mitigation options. FAO Fisheries and Aquaculture Technical Paper 627, Rome, pp 185–207. https://www.fao.org/3/i9705en/i9705en.pdf

Pauly D, Cheung WWL (2017) Sound physiological knowledge and principles in modeling shrinking of fishes under climate change. Glob Change Biol 24:e15–e26. https://doi.org/10.1111/gcb.13831

Pidgeon N, Fischhoff B (2011) The role of social and decision sciences in communicating uncertain climate risks. Nat Clim Change 1:35–41. https://doi.org/10.1038/nclimate1080

Prato J, Newball R (2015) Aproximación a la valoración económica ambiental del departamento Archipiélago de San Andrés, Providencia y Santa Catalina – Reserva de la Biósfera Seaflower. Secretaría Ejecutiva de la Comisión Colombiana del Océano- SECCO, Corporación para el desarrollo sostenible del Archipiélago de San Andrés, Providencia y Santa Catalina -CORALINA. Bogotá, Colombia

Rodríguez A (2015) Cambio Climático y acidificación del océano: su efecto sobre las comunidades coralinas. Insumos de Política Pública. Centro de Pensamiento Caribe. Universidad Nacional de Colombia. San Andrés island, Colombia

Rönnback P (1999) The ecological basis for economic value of seafood production supported by mangrove ecosystems. Ecol Econ 29(2):235–252

Ruiz De La Cruz VP (2016) Análisis de las percepciones sobre el deterioro del paisaje y la presión del sector turístico en la Isla de San Andrés. Master's thesis, Universidad Nacional de Colombia

Santos-Martínez A, Mancera JE, Castro E et al (2012) Propuesta para el plan de manejo pesquero de la zona sur del área marina protegida en la reserva de biósfera Seaflower. Instituto de Estudios Caribeños – Jardín Botánico. Universidad Nacional de Colombia. Facultad de Ciencias. Departamento de San Andrés, Providencia y Santa Catalina, Secretaría de Agricultura y Pesca

Santos-Martínez A, Rojas A, Garcia M et al (2019) Dynamics of artisanal fisheries and proposals for sustainable management, providencia and santa catalina, Colombian Caribbean Seaflower biosphere reserve. Proceedings of the 72nd Gulf and Caribbean Fisheries Institute. 2–8 November, 2019, Punta Cana, Dominican Republic

Sheridan J, Bickford D (2011) Shrinking body size as an ecological response to climate change. Nat Clim Change 1:401–406. https://doi.org/10.1038/nclimate1259

Smikle SG, Christensen V, Aiken KA (2010) A review of Caribbean ecosystems and fishery resources using ECOPATH models. Études caribéennes. https://doi.org/10.4000/etudescaribeennes.4529

United Nations Development Program (UNDP) (2007) Fisheries: Fisheries sector, chapter 7. https://www.undp.org/sites/g/files/zskgke326/files/publications/Fisheries_(chapter_7)_ENG.pdf

Universidad Nacional de Colombia, sede Caribe (2019) Insumos para la política pública de ordenamiento pesquero del Archipiélago de San Andrés, Providencia y Santa Catalina

van Hooidonk R, Maynard JA, Liu Y et al (2015) Downscaled projections of Caribbean coral bleaching that can inform conservation planning. Glob Chang Biol 21(9):3389–3401. https://doi.org/10.1111/gcb.12901

Velásquez C (2011) La Percepción del riesgo de los agricultores de la isla de Providencia y Santa Catalina. In: Roman R (ed) Cultura, Sociedad, Desarrollo e Historia en el Caribe colombiano. Universidad Nacional de Colombia, pp 135–167

Velásquez C (2019) Análisis sobre afectaciones del cambio climático en los recursos pesqueros del departamento archipiélago de San Andrés, Providencia y Santa Catalina, reserva de Biosfera Seaflower, en un contexto Caribe. Documento de trabajo. Gobernación departamental del Archipiélago de San Andrés, Providencia y Santa Catalina y la Universidad Nacional de Colombia, sede Caribe. San Andrés isla, Colombia

Wilson R (2017) Impacts of climate change on Mangrove ecosystems in the coastal and marine environments of Caribbean small island developing states (SIDS). Caribbean Mar Climate Change Sci Rev 2017:61–82

Open Access This chapter is licensed under the terms of the Creative Commons Attribution 4.0 International License (http://creativecommons.org/licenses/by/4.0/), which permits use, sharing, adaptation, distribution and reproduction in any medium or format, as long as you give appropriate credit to the original author(s) and the source, provide a link to the Creative Commons license and indicate if changes were made.

The images or other third party material in this chapter are included in the chapter's Creative Commons license, unless indicated otherwise in a credit line to the material. If material is not included in the chapter's Creative Commons license and your intended use is not permitted by statutory regulation or exceeds the permitted use, you will need to obtain permission directly from the copyright holder.

Overcoming Iota: A Reflection on Old Providence and Santa Catalina Cultural Resilience In the Face of Disaster and Climate Change

Ana Isabel Márquez-Pérez

Abstract In November 2020, category 4 Hurricane Iota devastated Old Providence and Santa Catalina (OPSC), small islands located in the Colombian Western Caribbean and home of the Raizal people, an African-descendent ethnic group bearing their own culture and language. Despite the chaotic governmental response, the local community has responded to the situation by adapting and reorganizing their ways of life. In this chapter, I present a reflection on how OPSC people have used culture in different ways to prepare, adapt, and resist during the disaster and post-disaster periods, in the context of increasing climate change, creating new ways to relate to their islands and community that play an important role in their future. To do this, I use some concepts drawn from cultural perspectives on disasters, climate change, and resilience. As I will try to show, the OPSC community has demonstrated a strong cultural resilience through its capacity to recover and its ability to learn from experience and adapt to new situations. This is not an easy process, as the current context exacerbates vulnerabilities. However, cultural resilience and community processes are a source of hope for islanders to continue to inhabit their land and seascapes with well-being and autonomy.

Keywords Culture · Resilience · Hurricane · Caribbean · Raizal people

A. I. Márquez-Pérez (✉)
Assistant Professor, Universidad Nacional de Colombia, Caribbean Campus, San Andres Island, Colombia
e-mail: aimarquezpe@unal.edu.co

Di neks die afta di harikien Ai seh, Ai fi bil bak mai chradishonal huom, and Ai disaid se Ai gwain bil bak mai chradishonal hous ahn Ai staat bil ih, bikaaz wi hafi gat somting fi aidentifai wi. Ahn Ai work haad, bil bak mai hous wozn somting iizi. Bot Ai fiil gud. Ahn wai Ai fiil gud, bikaaz Ai bil bak mai huom miself. Ai had di schrent fi bil bak mai chradishonal hous.[1]

Raizal carpenter from Old Providence Island

1 Introduction

On November 15 and 16, 2020, Category 4^2 Hurricane Iota devastated Old Providence and Santa Catalina (hereafter, OPSC), small islands located in the Colombian Western Caribbean. The impact of Iota was tremendous and disastrous. Devastation was general. Almost 50% of the houses and infrastructure collapsed, while the other 50% was damaged to different degrees (UNGRD 2020). Many trees fell, vegetation was burned and lost its foliage and numerous mangroves died. Although the number of deaths was low, with four people losing their lives as a direct consequence of the hurricane, everybody in the community was affected, in one way or another.

Very few places remained for people to safeguard from the climate conditions, and the population was exposed for weeks, and even months, to rain, cold, and high temperatures, and was without electricity and endured precarious access to water and food.[3] Strong winds also continued, knocking down tents and provisional refuges, and frightening people who had just experienced the trauma of the hurricane. The devastation was so deep that many local people, especially children and the elderly, abandoned the islands in the following days (Valoyes 2020). Meanwhile, those who remained faced great difficulties to survive, overcoming the disaster, and starting the recovery process.

It should be noted that OPSC is home to the Raizal people, an ethnic group of African descent that results from the British colonization process that brought Europeans (mainly British) and Africans of different origins to the Archipelago of San Andrés, Old Providence, and Santa Catalina (hereafter, the archipelago) from the sixteenth century onward, as well as other Caribbean migrants during the eighteenth and nineteenth centuries (mainly Jamaicans, Caymanians, and Central Americans)

---

[1] *"The day after the hurricane I said, I have to build my traditional house again, and I decided that I was going to do it and I did it, because we have to have something that identifies us. And I worked hard, building my house again was not easy, but I feel good, because I built my house myself, I had the strength to build my traditional house again."* Phrase written in Kriol from the Archipelago of San Andrés, Providencia and Santa Catalina.

[2] As noted in the introduction to this book, although Hurricane Iota was initially categorized as category 5, it was later downgraded to category 4. However, in both the archipelago and in general, it is still commonly referred to as category 5.

[3] In June 2022, more than one year and seven months after the hurricane, there were still families living in tents.

and Asians by the last part of the nineteenth century and the beginning of the twentieth (mainly Chinese). Despite their British colonization background, the islands passed to Spain and then to Nueva Granada and Colombia in the nineteenth century, but they retained cultural traits similar to other former British Afro-Caribbean societies, including their language, an English Creole (Parsons 1985; Sandner 2003).

During the twentieth century, Colombia promoted cultural and economic assimilation policies that generated deep changes in the local society and created strong tensions with the state. These led to local movements to reclaim Raizal's rights to culture, identity, and territory, as well as autonomy and self-determination, and also to struggle for the declaration of the islands as a Biosphere Reserve, a recognition that was finally given by UNESCO in 2000. Although the Raizal people were recognized as an ethnic group by the Colombian constitution in 1991, the implementation of cultural and territorial rights has been very difficult (Guevara 2007; Padilla 2010; Márquez-Pérez 2014; Valencia 2015).

Raizal people constitute 90% of the islands' population, and were the main group affected by the disaster of Hurricane Iota. Culture—understood here as ways of being and relating to the world, knowledge, practices, and any other aspect of human life learned from being part of a social group, but also "unevenly distributed and utilized as a resource by individual actors as they construct strategies of action for everyday life" (Swidler cited in Clarke and Mayer 2016: 4) --plays a fundamental role in how Raizal people relate to each other and inhabit their land and seascapes, and includes detailed environmental knowledge and particular ways of relating to marine, coastal, and terrestrial ecosystems (Correa 2012; Márquez-Pérez 2014). As I will try to show, this knowledge and these cultural practices allow us to understand, at least in part, the responses of islanders to attempt to adapt to this new context of devastation.

Nevertheless, the reconstruction process has only superficially included the local community, and paid less attention to the inclusion of social and cultural issues. Governmental actions focused on the recovery of basic infrastructure, such as roads, the airport, and the dock, but seemed to forget to provide substantial aid for local people who, for many days and even months after the disaster, continued to live in precarious conditions (El Espectador 2021). Despite the slow and chaotic governmental response, the community responded to the situation in its own way, overcoming many of the difficulties, adapting to the new situation, and reorganizing their ways of life, in what could be seen as cultural resilience: the capacity of a cultural system to overcome and adapt to extreme perturbations, and to adapt to new circumstances.

Although culture has not been central to the discussions arising in the OPSC emergency context and reconstruction process, mainly managed by outsiders, it is worth reflecting on the role that it has played so far, and the one that it could still play, in shaping how islanders deal with the post-disaster situation, the recovery and reconstruction process, and in the face of the impacts of climate change. Additionally, it is important to ask whether a reconstruction process that does not incorporate a deep cultural approach may constitute a threat to the cultural survival of the Raizal people. Culture is not only important in adapting to the new post-disaster context where the whole material world of the people of OPSC people has changed, and in adapting to the increasing impacts of climate change that intensify these small islands' vulnerability, but also in dealing with the symbolic and emotional process that the victims of such an event must go through. The material impacts are only one dimension of the disaster, while their symbolic ones are numerous, although less visible and analyzed.

On the other hand, as many community members have shown in many ways in the period since the hurricane, Raizal culture does concern local people in the post-disaster context. As many recognize, knowledge, practices, and ways of being and living in the world are severely threatened by the new processes that the islands are subject to within the framework of reconstruction. The community is experiencing a high level of social vulnerability, which makes it extremely sensitive to surrounding events, affecting its resilience. There is apprehension about what the reconstruction process might imply for the cultural survival of the Raizal people and the relationship with their sea and terrestrial living places, and there are expressions of cultural resistance to what has been happening. So, even if culture is not central to those who manage the reconstruction, it is key for the people who lived through the hurricane and who continue to face its consequences.

This chapter is the result of my personal experience of Hurricane Iota and the reconstruction process that I have lived and experienced as both a community member and an anthropologist. In this sense, this account relates to autoethnography, a methodological approach that "seeks to systematically describe and analyze personal experience in order to understand cultural experience" (Ellis et al. 2015: 250). I spent part of my childhood and youth on Old Providence Island, and was living there with my family in November 2020 when the hurricane struck, and I remained there during 2021 which allowed me to participate in different local processes that took place in the first year after the disaster.

Thus, I had the opportunity to develop participant observation of the different stages of the situation, from the day of the hurricane to the weeks and months that followed, and even until today, as I continue to visit the islands every month. During that time, I participated not only in what became day-to-day activities, such as picking up rubbish, helping friends and neighbors, and answering institutional surveys, but I also assumed other roles inside the community, including my participation as an advisor to the OPSC Fishermen's Federation, which emerged and strengthened in the midst of the disaster, a role that I still have today, and as a community overseer of household and health issues in the context of reconstruction.

Based on this, here I present a reflection grounded in participant observation and the analysis I have been able to make through my experience on the role that Raizal culture and its resilience capacity have and should play in the process of rebuilding OPSC. I consider this contribution to be not only an academic exercise, but also a "political, socially fair and socially conscious performance" (Ellis et al. 2015: 250), related to the postulates of autoethnography. In this way, I hope to contribute to a better understanding of the experience OPSC people have gone through trying to overcome Iota, analyzing one of many possible aspects of this process. As Camargo (2020) has pointed out, despite being a disaster-prone country, Colombian anthropology has not paid much attention to these "rather evident and transformative events", so this reflection also hopes to contribute to this area.

In order to do this, I use some concepts drawn from cultural perspectives on disasters, climate change, and resilience, to reflect on what OPSC people have lived. I try to understand and highlight how islanders have used culture in different ways to prepare, adapt, change, and resist the current situation, creating new ways to relate to their islands and community that play an important role in their future. Although cultural dimensions have been underemphasized in discussions on climate change perceptions, adaptation, and mitigation, as well as on those concerning disaster and post-disaster scenarios, the need for perspectives that include these dimensions is clearer every day if we really want to create effective responses to current climate change challenges, as well as reconstruction processes (Adger et al. 2013; Companion 2015; Clarke and Mayer 2016).

2 Some Key Concepts and Discussions

Resilience is a concept that has gained importance in recent decades, while its growing use has given way to different interpretations and meanings, not always consistent (Kendra et al. 2018). The concept has its origins in nineteenth-century natural sciences and was consolidated in the twentieth century to study ecosystem changes. Holling (1973: 14) defines it as "a measure of the persistence of systems and of their ability to absorb change and disturbance and still maintain the same relationships between populations or state variables".

Since the 1950s, the resilience concept has gained importance in the social sciences, especially in ecological anthropology, related to social and cultural change discussions, a question that has been central to this field. For a long time, many in the social sciences assumed that human societies tended towards equilibrium while resisting change. However, this view was questioned from different perspectives, giving resilience more importance, understood as the social and cultural capacity to deal with changes and disruptions (Moran 2022).

In recent decades, social and cultural resilience discussions have broadened and complexified across diverse knowledge fields, so there is no single approximation to it. In this sense, it is important to review some definitions that contribute to a better and more complex understanding of the concept that I consider useful for my discussion. Holtorf (2018: 639) defines it as "the capability of a cultural system (consisting of cultural processes in relevant communities) to absorb adversity, deal with change and continue to develop", where different aspects such as a society's cultural identity, values, norms, knowledge, and practices play a key role in how a social group deals with dramatic changes.

Crane (2010: 2–3) defines it as "the ability to maintain livelihoods that satisfy both material and moral (normative) needs in the face of major stresses and shocks", emphasizing cultural resilience not only on the material but also on the symbolic level. In this sense, the definition "respects the integrity of subjective normative experience, recognizing that people's lives mean something to them, while also accommodating changes in behaviors, values, and social institutions that are inherent in cultural dynamism". Meanwhile, Holfort (2018: 639) also shows how "recent conceptions of resilience de-emphasize notions of 'bouncing back' to a previous state and place more emphasis on processes of 'bouncing forward' involving absorption, learning, adaptation and transformation than on specific outcomes in relation to a previous status quo".

It is important to highlight that Crane's (2010) definition evidences the way symbolic and emotional dimensions also play a role in how societies face disruptive experiences, even though this is usually set aside in the face of the forcefulness of material issues. The definition of Holtorf (2018) emphasizes not only recovery capacity but also change capacity, complexifying the idea of coming back to a previous condition to foreground the capacity to absorb impacts, learn from them, adapt, and transform, as strategies to overcome extreme changes. This last point is significant because it questions the idea that changes only produce negative impacts, by showing that resilience deals with all kinds of changes and not only the negative ones. In this way, resilience could be seen as a conservation mechanism for cultural patterns, which tends to delay and slow down changes but also creates innovative tools that allow societies to adapt even to the worst contexts.

Here, it is worth adding the definition given by Clarke and Mayer (2016: 2) who define cultural resilience as "the capacity of communities to mobilize cultural resources in response to external crises and threats, which in turn shapes individual and community actions related to the recovery process". This definition is important because it gives space to agency, the capacity for individuals to make use of cultural resources in different ways, recognizing the complexity of society. As these authors put it, following Swidler (1986), "these cultural resources exist in conjunction with a social–ecological system, but they reside in the knowledge of individual actors, and they are used to construct "strategies of action" for everyday life" (Clarke and Mayer 2016: 2).

This last definition of cultural resilience comes from disaster studies, a knowledge field that has strengthened in recent decades, as global environmental and technological change increases risk and vulnerability. Here, it is important to highlight that I understand a disaster from an anthropological perspective, as "a process/event involving a combination of a potentially destructive agent(s) from the natural and/or technological environment and a population in a socially and technologically produced condition of environmental vulnerability" (Oliver-Smith 1996: 305). This implies that disasters are not natural but socially constructed, as they are not limited to natural phenomena and result from the interaction between these and society. Barbosa and Zanella (2019) suggest the need to denaturalize the concept of natural disaster, proposing the "socio-environmental disaster" concept, considering that "the social living conditions of a population determine the disaster impact level" (Barbosa and Zanella 2019: 49). From a complementary position, Cupples (2017) highlights the need to take responsibility away from nature, focusing on human-ecological relations and considering how social, cultural, political, and economic factors have a direct relation to the configuration of disasters.

Given the role of disasters as creators of extreme change for societies, resilience has also played an important role in disaster studies, where it has diverse uses and approximations. It is worth noting that there has been an important transformation from a material and technologically centered perspective to the recognition of the role that society and culture play in disasters. However, as different authors point out, culture continues to be underestimated, despite the growing evidence of the key role it plays (Companion 2015; Clarke and Mayer 2016; Kendra et al. 2018; Webb 2018).

Kendra et al. (2018: 87), show some of the ways in which resilience has been used in disaster research:

> For researchers interested in the topic, it is an explanatory framework for systems functioning under stress. For policymakers and officials charged with managing disaster, resilience is an aspirational state to which they might target capacity-building initiatives. Resilience, too, is a positive expression, as opposed to vulnerability, which suggests incapacity or lack of agency. And resilience and the closely associated idea of vulnerability have seemed to provide, either alone or together, unifying frameworks for drawing together streams of scientific findings on what makes people more or less able to deal with risk, or the manifestation of risk as disaster.

Despite the diversity of approaches, it is important to show how the cultural approach is useful to think about different aspects of disasters. In the case of disasters related to natural phenomena, it is important to highlight that all human responses and perceptions of climate change are mediated by culture (Adger et al. 2013). As these authors put it, knowledge, practices, beliefs, and values, all of which belong to culture, can be effective—or not—in adapting to climate change. However, "research and policy on adaptation and mitigation have largely focused on material aspects of climate change" (Adger et al. 2012: 112), while leaving culture aside. Recent research on the topic shows how overlooking culture can lead to maladaptive outcomes (Adger et al. 2012), including the limitation of cultural resilience and thus an increase in vulnerability (Clarke and Mayer 2016).

From a resilience perspective, culture can create short-, medium-, and long-term mechanisms for communities and individuals to deal with these situations, on both the material and spiritual levels, but it can also be a source of vulnerability. In fact, culture can create conditions for society or individuals to ignore or underestimate threats, leading to situations where people expose and put themselves at risk, worsening the scenario (Webb 2018). In any case, as Webb (2018: 109) affirms, "it has become abundantly clear to researchers, policymakers, and practitioners that the key to achieving future societal resilience is gaining a deeper understanding of the role of culture in both producing and preventing disasters".

Based on a deep literature review on resilience and disasters, Kendra et al. (2018: 94) identify nine resilience elements that "are repeatedly discussed in the theoretical literature". These are "capitals or capacities" which include "social, economic, human, institutional, political, and community capital, improvisation, natural resources, and physical resources" (see Fig. 5.1 in Kendra et al. 2018). Although this approach can be questioned for its focus on capital and resources, as well as the almost complete absence of culture, which clearly contradicts the cultural turn in disaster studies proposed by Webb (2018), I consider it useful as a base for the analysis and reflection on cultural resilience in OPSC of this chapter.

In the following pages, I will try to apply these theoretical insights in relation to cultural resilience to reflect on the specific case of OPSC in the context of the disaster and post-disaster events caused by Hurricane Iota, analyzing and reflecting on some of the diverse roles that culture has played in the context of the disaster the islands experienced, and trying to link them to discussions on climate change adaptation, risk, and vulnerability that are currently central to the future of these small islands.

3 Cultural Resilience in Post-Iota Old Providence and Santa Catalina: Islandness, Local Knowledge and Experience

OPSC culture is expressed in particular ways to relate to the environment, knowledge, practices, and livelihoods, as well as specific cultural expressions such as music, dance, cooking, architecture, and activities related to the land and the sea, such as farming, fishing, and navigation (Márquez-Pérez 2014). Although local culture has experienced abrupt changes in recent decades, particularly as the result of Colombian assimilation policies, promoted from the beginning of the twentieth century (Valencia 2015), it should be noted how the Raizal people consider culture to be a stronghold of islanders' identity and a source of pride.

A key cultural aspect to consider here is islandness, understood not as isolation but as particular ways to live in a limited and remote space (Diegues 1998), where capacities and arrangements to deal with limited access to goods and services, such as water, food, and building materials, can be included. In the case of OPSC, some examples of the above are the islanders' custom of building houses with cisterns, where rainwater is captured in order to guarantee water supplies during dry seasons (Aguado 2010; Correa 2012), as pipe water is recent and has never been stable; as well as important knowledge related to food preservation methods, such as food drying, corning, and smoking, as well as sugar and cane syrup recipes, and coal and wood cooking (Ministerio de Cultura 2016). Notwithstanding, the most relevant example are the community and family links and networks that still support reciprocal economies, where common good and solidarity prevail (Márquez-Pérez 2014). Although these practices have weakened over time, they have not totally disappeared and they have been key in facing disaster.

Another cultural aspect relates directly to hurricanes. Although OPSCs have not historically suffered frequent or severe hurricanes, these have been present in people's memories and experience. The most recent was Beta, a category 1 hurricane that struck the islands in October 2005, which many of the current local population experienced and remember. Even though Beta was much less destructive than Iota, it caused various types of damage and showed the usefulness of certain measures, such as sealed cement roofs that can be used as hurricane shelters. This influenced the fact that, after Beta, many people continued to build bathrooms and other spaces in their houses this way, which ended up being vital during Iota, allowing many people to shelter in them and survive.

The experience of the people of OPSC with hurricanes is not limited to those that have affected the archipelago. As frequent migrants across the Caribbean region (Márquez-Pérez 2013), many Raizal islanders have experienced hurricanes in places such as Nicaragua, Honduras, the United States, Jamaica, and the Cayman Islands. Particularly, many islanders lived through Category 5 Hurricane Ivan, which caused severe damage in Grand Cayman in 2004, as well as experienced the reconstruction process that followed, where many even actively participated as construction workers. As a result, many people had experience implementing techniques used by Caymanians before and after Ivan, which also explains the resistance of some roofs and structures to Iota. These two last examples show how resilience is not only a matter of coming back to previous conditions after a disruptive event but is also an experience of learning and adaptation (Holtorf 2018).

The existence of specific climate knowledge (Correa 2012) and abilities acquired fundamentally, although not exclusively, from sea life (Márquez-Pérez 2014) must also be added to this. Even if it was not possible for islanders to predict in detail what was coming with Iota, nor the dimension of the hurricane's impacts, many people were attentive to the situation days before the unfortunate date, and even before institutions alerted the community. Meanwhile, sea and island life abilities relate to a well-known capacity to react rapidly to unexpected events, as well as manual skills that allowed many people to deal with the complex situations experienced during and after the hurricane.

That night, in the middle of winds over 250 km/h, OPSC islanders used these abilities and skills to survive. There were people who tied themselves and their relatives to trees and cisterns, standing all through the night with the sea up to their necks; some emptied their cisterns and sheltered inside them; others hid in sealed rooms (mainly bathrooms), as well as closets and under beds and mattresses. In general, people recurred to anything on hand to secure themselves and to try to stand the wind and the rain, nailing down windows, drilling ceilings, and bailing out water. This can be compared to what Clarke and Mayer (2016: 7) point out in relation to the community they studied in the Gulf of Mexico, where "experience with regular climatic weather is said to lead to a stoic preparedness on their part, both ready and resigned at the same time", as well as to the historical accounts that Crawford (2020: 56) has rescued of Caymanian turtlers, such as the press note of the *Caymanian* newspaper that described a 1909 hurricane offshore event in the following words: "But like brave mariners they stood their ground. There was no excitement among them even at the most trying hour... So the idea struck them that their ropes would be of use to them; and each man securely lashed himself to a tree".

4 A Reflection Based on Resilience Elements

To deepen this reflection, in the following pages, I present an analysis based on the identification by Kendra et al. (2018) of the main resilience elements, which include at least nine different types of capacities that I examine in the light of the OPSC experience. As I will try to show, although culture is not explicitly included by the authors, many of these elements are directly related to it. This is clear in elements such as social capital, considered one of the most important concepts in resilience and disaster discussions (Kendra et al. 2018: 94), where a sense of community and belonging play an important role, as well as in human and community capital, where issues such as temperament, optimism, solidarity, and local understanding of risk are central and directly related to cultural traits. However, it is also very interesting to notice how culture actually influences all the identified elements, including those that are apparently unrelated, such as improvisation (as in technical knowledge), physical and infrastructural resources (as in housing) or natural resources (as in the efficient and traditional use of water). Here I present a revision of each of the nine elements, highlighting some of the different roles culture has played in the post-Iota scenario in OPSC.

4.1 Element 1: Improvisation

Disaster studies research identifies improvisation as a fundamental component of resilience, referring to the capacity to find or create solutions to unexpected situations or problems that characterize disasters (Kendra et al. 2018). As I briefly showed above, improvisation is part of islander culture, particularly for fishers and sailors who are accustomed to dealing with complex and unpredictable problems in contexts of limited resources, such as the ones posed by sea emergencies. It is also a key element in solving minor daily issues that happen in small and isolated environments such as these small islands, where it is difficult to access certain types of expert knowledge and adequate technical resources.

In the OPSC context, improvisation comes together with specific abilities that imply a high degree of training and technical knowledge, usually acquired as part of daily life and enculturation, which includes capacities and abilities related to machine and tool management, mechanic and electrical knowledge, or manual and motor skills such as knot-making, an important expertise developed by seafarers. Islanders usually feel proud and value themselves for these aptitudes which are also elements that contribute to their social reputation, a feature that plays an important role in the Raizal and other Caribbean cultures (Wilson 1973).

Improvisation and creativity played a role in the low number of deaths and injuries during Hurricane Iota. Survivors' memories are full of examples of how people improvised to save their lives, as buildings and other shelters collapsed, finding ways to protect themselves or, at least, to guarantee their survival. Improvisation continued to play a role in the aftermath and even until today, as people have had to improvise refuges and temporary houses, find ways to guarantee food and water, and rebuild their livelihoods.

4.2 Element 2: Natural Resources

OPSC has an important natural base with terrestrial, freshwater, coastal, and marine ecosystems. They are surrounded by an extensive coral reef complex with mangroves, seagrass beds, and rocky and sandy bottoms with a diversity of species. During Hurricane Iota, the barrier reef and coastal mangroves contributed to reducing wave strength, especially on the east coast. Similarly, the dry forest that still covers a good part of the mountains helped to reduce the wind strength. It is important to highlight that these ecosystems' good condition has a direct relation with traditional practices of sustainable use of nature that are part of Raizal culture (Correa 2012; Márquez-Pérez 2014). In fact, the archipelago was declared a UNESCO Seaflower Biosphere Reserve in 2000, as a recognition of the islands' valuable ecological and cultural heritage, and as a result of the local people's struggle to defend their environment and culture (Padilla 2010; Márquez-Pérez 2014).

Social relations with nature have resulted in different knowledge and practices, such as farming and agriculture, fishing, and navigation. There are also important uses of water—a seasonally scarce resource—that is stored in cisterns and other deposits, usually built under or to the side of houses. Iota destroyed the majority of crops, and the large amount of fallen trees limited access to farms. It also destroyed the majority of the artisanal fishing fleet, thus affecting the local food supply. However, fishers and farmers quickly reactivated to a subsistence level. Even with the limitations mentioned, fishers who did not lose their boats and engines started to fish a few days after Iota, as a way to guarantee food access for many and to contribute to the community.

Although the fishing productive chain has been only partially reestablished, as a result of the many obstacles created by the chaotic government response, a year after Iota the local supply was already working. In this sense, the fishers' struggle to recover their productive chain is also one of the best examples of cultural resistance and resilience. This has been led by the Federation of Old Providence and Santa Catalina Fishermen, an organization that joins four cooperatives and associations that existed on the islands prior to the hurricane, and which have been proposing and enacting solutions to the delicate economic, social, and cultural situation that OPSC people have faced since Iota (Jay 2021). A similar process, although less visible, has taken place in relation to agriculture, led by *Agroprovidencia*, a local farmers' organization that links the majority of the people dedicated to this activity.

Here it is worth mentioning the concept of cultural resistance which also plays a role in disaster and post-disaster scenarios, even preceding cultural resilience. According to Marchezini (2015: 294), "before exercising resilience, populations often perform acts of resistance. The concept of resistance includes a sense of action and of opposition to someone or something, which can be visible or invisible, recognized or unrecognized". This author bases his analysis on the disaster caused by a flood in a small Brazilian village in 2010, where "the federal, state, and municipal agencies directed their actions to the process of material reconstruction of the city. This left the luizenses to seek references in the body of their cultural life to reaffirm their identities and recover themselves in the social plane. In this process of recovery, the experiences became meaningful again. This allows the absorption of the event into its history and the body of its culture, rather than negating its existence".

The experience of OPSC fishers and farmers exemplifies cultural resistance as well as agency in the disaster response process. This struggle has allowed people in the community to create tools that help them deal with traumatic experiences, whose effects on the symbolic and emotional plane are frequently ignored, while they might also be crucial for cultural resilience. It has also given these actors a key role in the institutional reconstruction process which has systematically tried to avoid real community involvement.

4.3 Element 3: Physical and Infrastructural Resources

This refers to issues such as housing type and characteristics, as well as institutional infrastructure such as hospitals, schools, churches, and refuges. Although not immediately evident, culture is also related to this element. In the case of OPSC, local infrastructure has some adaptive characteristics, particularly those built in line with local traditions, for example, buildings on stilts that are designed to adapt to swamp areas and/or to resist floods and sea level increases, and roofs designed to favor wind circulation, which guarantee better climate conditions inside, as well as some resistance to strong winds. However, they were not designed to resist the enormous strength of Hurricane Iota, which ended up damaging the majority of the infrastructure on OPSC.

After Iota, infrastructure and housing reconstruction has been subject to many discussions between the local organized community and the central government. Local leaders and many other community members were opposed to constructing only one type of prefabricated house and pressed to guarantee a rebuilding process which respected local and traditional architectonic styles, as well as one which would incorporate anti-hurricane measures. Although not completely successful, given the unbalanced power relations, this has been another strong example of cultural resilience, and resistance, as many people in the community have chosen to fight for their right to houses adapted to their traditional lifestyles and customs. This was even used as an excuse by the central government to justify delays in the reconstruction process which, in reality, resulted from the chaos and corruption previously mentioned.

Despite the difficulties, the outcome is that the reconstruction process in OPSC has been readapting to follow patterns that fit, in some way, with their architectonic traditions, which can be seen as a success of cultural resistance in which the community valued their culture and traditions as fundamental to their ways of being in the world,[4] as well as their possibilities of economic reactivation through tourism. One example of this is the late inclusion in the reconstruction process of a house model built on the remaining foundations of houses destroyed by Iota. This was proposed by the community from the beginning, but only incorporated by external reconstruction managers after delays and problems evidenced its practicality. This shows how the valuation of local knowledge proves fundamental to reconstruction, as well as to climate change adaptation and mitigation, just as its non-incorporation also brings consequences (Adger et al. 2013; Companion 2015).

4.4 Element 4: Institutional Capital

This element includes important aspects such as emergency infrastructure, critical equipment, alternative sources of water and energy, risk mitigation plans, catastrophic events insurance, and disaster management plans and training, amongst others (Kendra et al. 2018). Here it is worth mentioning how disaster preparation at the institutional level was and continues to be very low, linked to the lack of efficiency on different governmental stages and high levels of bureaucracy and corruption, similar to other experiences in the region such as in Honduras, Puerto Rico, and Barbuda (Barrios 2014; García-López 2017; Boger et al. 2019), which corroborates the idea of disasters as not natural but socially constructed phenomena (Oliver-Smith 1996).

Notwithstanding the existence of a National System of Disaster Risk Management and abundant legislation on the matter (but not including hurricanes), as well as previous and permanent disaster experiences across the country, neither the local nor the national government responded adequately. The institutional management of the disaster has been chaotic, with many examples of problems and delays that have permanently affected it. Indeed, the government has not been able to successfully reestablish housing and public infrastructure, and many failures have arisen along the way, with serious delays and dubious priorities in the process. Thus, more than a year after Hurricane Iota, the hospital remained in tents, risking both patients and medical personnel, and there were still families around the islands, including children, living in precarious conditions (Oquendo 2021).[5] This situation worsens with the weakening of environmental, planning, and other regulations, resulting from the government's disaster declaration.[6]

---

[4] There are even examples of people who chose to rebuild their typical houses how they were before Hurricane Iota, as in the example of the local carpenter quoted at the beginning of this chapter.

[5] As it continued to be by mid-2022.

[6] The Colombian government declared by decree a "Disaster Situation" in the archipelago on November 18, 2020, allowing the application of a special normative regime contemplated by Law 1523 of 2012. This situation was extended on November 18, 2021 and continued at the time of writing.

Something that aggravates this is local political culture, where clientelism plays an important role, leading people to choose to participate in it and to tolerate corruption and bad government. In this sense, many people in the community tend to turn a blind eye and do not demand things from their politicians, either because they have lost their faith or prefer not to enter into a conflict with political authorities in order to benefit from them. In the context of the reconstruction, clientelism has also played a role, with the use of gifts and favors as a way to co-opt people from the community, reduce resistance, and gain political power, similar to what Barrios (2014) described for post-Hurricane Mitch Honduras.

Here we can see a scenario in which institutions reproduce inequalities and contribute to several social and environmental injustices, corroborating the unnaturalness of disasters and their link to capitalism and colonialism (García-López 2017; Rivera 2022), an issue that merits further research. From this perspective, the islands' low institutional capital is at the base of the disaster and the reconstruction chaos and conflicts, also marked by inefficiency, authoritarianism, corruption, and other features that make up Colombia's political panorama of permanent institutional crisis. This, of course, limits resilience and reproduces vulnerability.

4.5 Element 5: Community Capital

This refers to community action capacity, as well as to collective flexibility, creativity, efficiency, and empowerment (Kendra et al. 2018), amongst other features that can be understood as culturally based. However, as Barrios (2014: 330) points out, communities are "never static nor bounded", they are in a "constant state of emergence over time" and "are shaped by dynamic, politically and epistemically charged relationships among assisting governments, aid agencies and disaster-affected populations". This helps us to complexify our view of community, showing how resilience and vulnerability are constructed over time, as well as to highlight agency.

In the case of OPSC, community capital has been evident in the strong response to an inefficient reconstruction process maintained by political and economic governmental power. The already mentioned opposition to the standardized house rebuilding, the struggle to reactivate fishing and farming, as well as the strong opposition to the opportunistic attempt to build a coast guard base in an inadequate place, are examples of community capacity and resilience. It is also important to highlight the support given by the Raizal diaspora, a group of organized migrant islanders who have been backing up local processes after Hurricane Iota.

It is worth deepening here on the process triggered by the intention of the Army to build a coast guard base at the Bowden Gully mouth in Old Town Bay, beside one of the main local fishers' organizations (Fish and Farm Coop), which prompted a strong community response and resistance. Here, it is important to consider previous events in 2015 and 2016, when a prior consultation process took place in relation to a coast guard station. The consultation prompted complete opposition from the local community and the result was no agreement, taking into account environmental, social, and security issues. However, a prior consultation process does not imply a veto in Colombian legislation, and higher authorities passed over community interests to approve the base. Nevertheless, the Army never advanced on the project nor even informed the community. Until Iota struck.

A couple weeks after the disaster, the coast guard not only occupied the land acquired by the Ministry of Defense, but also built a dock using sea space where the local fishermen's organization had previously had one, which was destroyed by Iota. In the midst of the humanitarian crisis that islanders were facing, and considering the social discontent that the reconstruction process had already created, this was the straw that broke the camel's back. On February 10, 2021, a group of fishermen started a peaceful protest, occupying the ruins of the Fish and Farm Association and blocking car access to the coast guard base; the news soon spread and many people from the community answered the call and joined what turned into the Dignity Camp (Bent 2021). With the support of the Raizal diaspora and other organizations, the movement grew and consolidated as a strong political actor in the context of reconstruction.

Despite facing several difficulties, including internal crisis, collective empowerment has achieved important victories, even taking into account the unbalanced power relations. Today, the Dignity Camp continues to lead the resistance of Raizal islanders, not only to the coast guard base that was the original trigger, but to all the injustices and cases of corruption that have taken place during the reconstruction. Here, the fight for a culturally-oriented reconstruction has played a central role in the movement's discourse and demands, putting in discussion fishing and farming, not only as key economic activities but also as cultural manifestations central to Raizal people's lives (Bent 2021; Jay 2021).

In this way, OPSC community capital may have been the most important cultural resilience element, which is not surprising if we consider the relevance that community life still has for Raizal people (Márquez-Pérez 2014). In addition, the islands' community also responded with what some authors have denominated therapeutic communities, whose key elements involve "agreement on the nature of the problem, consensus on what to do about it, and an overwhelming outpouring of sympathy and support from others" (Webb 2018 citing Fritz 1961).

Moreover, in community capital, it is possible to include the sum of the individual and collective knowledge and practices that conform to islandness and are key for survival on small isolated islands, as well as boats and ships, which were significantly expressed during the Iota. Other remarkable expressions of this are monitoring and preparedness before the hurricane, and solidarity during and after it, which continues to be evident until now. Given the "relational and emergent nature of social groups" (Barrios 2014: 347), internal social conflicts have always been present in the post-disaster, aggravated by the delicate living conditions people have been experiencing and institutional chaos. Notwithstanding, it is worth highlighting the existence of significant agreements amongst many people in the community, which have permitted a certain degree of union and mutual support, guaranteeing the continuity and cohesion of the struggle, as well as the possibility for many to have certain minimum standards of wellbeing.

Religion also can be included in this analysis showing both the positive and negative roles of culture. The people of OPSC are a strongly religious community, where Christian religions[7] play a key role in everyday life, with religious leaders (pastors and priests) having an important role in what happens on the islands. Although not as visible as other actors, churches helped their members as well as other people in the community after Iota, through the channeling of donations and other help. On the other hand, although the majority of the churches were themselves highly damaged or even destroyed, hindering the possibility to hold services for some time, religion was key for people's emotional containment in the midst of such an extreme and painful situation. Religious beliefs helped people to face reality with stoicism, always thanking God for the opportunity to survive.

However, it should also be mentioned that religious beliefs might also have played a negative role. For example, in the days and hours prior to Iota, many people believed and publicly assured (for example, on social media) that God would protect the islands, believing that nothing extreme was going to happen, relying on collective prayers and leaving everything in God's hands. As religious faith is so strong for many, this may have fostered a situation in which some people did not take enough measures to face the situation, including some who jeopardized their own lives by remaining in high-risk areas, refusing to abandon their homes. Besides religion, the idea that the islands were not a place that is vulnerable to strong hurricanes, reinforced by historical facts (no record of a previous category 5 hurricane, and only 3 hurricanes before Iota in the last one hundred years), might also connect to this belief.

Below, I elaborate on some specific aspects related to community capital:

Monitoring: Weather monitoring is part of local culture, as fishers and seafarers depend on weather conditions to develop their activities safely. Before Iota, many fishers and seafarers manifested their worry about what was forming in the Eastern Caribbean, at least two weeks before Iota struck. This was strengthened by the recent but highly important practice of consulting weather websites and apps (Correa 2012), such as the Hurricane National Center of the USA and Windy App, which is part of current innovations used to reduce risk and uncertainty on sea activities, key for islanders' livelihoods. These websites and apps were checked by many people before the disaster, which allowed for at least part of the community to be aware of the situation days before it happened, preceding official reports that arrived very late. Thus, many were better informed than local authorities, whose response was indeed belated and insufficient, and never included serious warnings of the imminent danger. Based on this individual and collective monitoring, many people had the chance to take precautions, including cutting and pruning trees, tying roofs, and buying provisions, even if these preparations were hardly enough for the eventual strength of the Category 4 Iota.

---

[7] Three major religions peacefully coexist on the islands: Baptists, Catholics and Adventists. Other smaller Christian religions also have presence in the territory.

Preparedness: As Hurricane Iota turned into a real menace, many people started to prepare for it. A general measure was to tie and reinforce roofs, doors, and windows, which probably made a positive contribution even if the impact of Iota overcame many of these efforts. Similarly, some tried to stock up their homes with food and water, while many others trusted in their cisterns and tanks. Cisterns played an important role in guaranteeing the water supply after Iota although, in many cases, the water was spoiled or salinized. However, the traditional practice of disconnecting gutters before this kind of event, as a measure to avoid water being spoiled with organic matter accumulated through the rain, allowed many people to continue to have access to water after the dramatic event. Tanks were less efficient as many fell from high places or their covers blew away with the strong winds, allowing the water to spoil. Some people stocked their homes with food and other tools needed during and after the hurricane, such as lights, batteries, candles, radios, emergency aid kits, hammers, and nails. However, many did not and even many who did, lost them during the disaster. Cellphones were particularly important to communicate, while there was still connection, and as lights and cameras, after and while electric sources were available.

Solidarity: Solidarity and mutual support were and still are basic components of the cultural response that allowed many people to survive during Hurricane Iota, and are also key elements of Raizal cultural traits generally. There were always people helping others who were at risk and, sadly, two of the hurricane victims actually died because of the risks they took to help others. In fact, during the emergency, many people risked their lives to help family and neighbors and, in the aftermath, many of those whose houses resisted the impacts gave shelter to those whose houses did not. Similarly, support from family and neighbors was remarkable after Iota, with many people exchanging water, food, tents, work, and other goods and services needed to guarantee survival. This solidarity extended from several networks conformed after the hurricane, including people from the neighboring island of San Andrés, who were the first to arrive with help one day after Iota hit, and OPSC islanders' relatives and friends living in different places around the world, who have consolidated what is today recognized as the Raizal diaspora.

4.6 Element 6: Political Capital

This element is composed of "capable governance, fair distribution of resources, ability to vote, access to people in leadership or distributing resources" (Abramson et al. 2015 cited in Kendra et al. 2018: 95). As previously mentioned, institutional political capital in OPSC is low, and was overwhelmed by central government interventions which practically substituted the local government by subordinating it and taking its responsibilities and decision-making capability. However, at the community level, political capacities appear greater, despite also being part of what has been described in terms of community capital.

This political capital relates to the struggle people have undertaken to defend their rights and positions in the face of the authorities and, in particular, central government authoritarianism. It is worth mentioning the increase and diversification of the community and ethnic leaderships, which have mobilized different sectors of the community, including fishers, farmers, and young people, as a way to respond to the complex scenario created by both the disaster and the reconstruction. Here, culture has also played a central role in local discourses which have emphasized the defense of particular worldviews and ways of being in the world. As one of the emblematic leaders' communications expressed "Thanks to the Navy, because the transgression of the environmental rights of our people made us gather at the Dignity Camp… Thanks to the mistakes of the reconstruction managers because they caused the most important mobilization in our history, generating an awakening of our ethnic and territorial consciousness" (Raizal Dignity Camp 2022). Thus, Iota may have contributed to a rearrangement of power relations and social mobilization, and a transformation in political consciousness, in ways that we have yet to understand, an issue that has been central in anthropological discussions on political factors in disasters (Oliver-Smith 1996).

4.7 Element 7: *Human Capital*

Human capital includes traits such as education, training, expert knowledge, efficiency, and optimisms, which can also be part of community capital and have been included in some way within this point. In OPSC, formal education is precarious and difficult to access, but there are still many professionals in different areas, many of whom have supported local processes in the post-Iota context. From a different point of view, islanders, including men and women, have an exceptional education for life on the islands, which is part of what was already mentioned as islandness (Diegues 1998): many of them are outstanding swimmers and divers; they drive cars, boats, and motorcycles with great ability, as well as managing diverse tools and equipment such as GPS, cellphones, computers, machetes, drills, and saws; they climb trees and know how to fish, raise animals, and farm. This capital has played and still plays an important function as it has served the victims to be able to deal with many of the situations faced, such as recovering and rebuilding provisional or new houses, as well as working in different areas and creating new businesses, even in activities they had never performed before, as a way to economically reactivate.

4.8 Element 8: *Social Capital*

Many researchers agree that "social capital is at the forefront of thinking about resilience, and many disaster scholars have pointed to social capital as a vital, perhaps even decisive attribute of social systems in places that influence a community's ability to respond and recover from an event" (Kendra et al. 2018: 94). Their components are diverse and range from alternative energy, food and water sources, and volunteers, plans, and resources to attend community needs, to the community organization level, reflected in NGOs and other kinds of organizations (civil, religious, sports), citizen participation and sense of community, with a certain emphasis on formal structures. In the case of OPSC, although social capital is important, in this reflection I have emphasized community capital, as formalization is not very strict, and it is not analytically useful to establish a deep differentiation between the two.

4.9 Element 9: Economic Capital

This economic capital includes a variety of components including income, employment, level and diversity of economic resources, and many others. From the economic point of view, OPSC is a municipality with a relatively high budget in the Colombian context, considering its small population. It is one of the few places in the country where there are no statistics of absolute poverty, although many people face difficult economic conditions with a high level of unmet needs and were already affected by the global pandemic (Márquez-Pérez 2020). Besides that, institutional infrastructure is modest, as a consequence of inefficiency and corruption, as was already signaled. This means that, for example, even before Hurricane Iota, the local hospital only attended to basic health issues and schools were precarious, either in terms of infrastructure or quality. At the community level, houses in OPSC are predominantly modest and basic, many in bad condition, which also contributed to the high level of destruction during Iota. There are few big or luxurious constructions, which can be attributed to cultural patterns, since even those who have the economic means do not usually exaggerate.

Notwithstanding the low economic capital and limited access to many goods and services, a majority of people have good living conditions, something reinforced by the islands' beauty—it is a place that is often qualified by the same locals as a paradise—as well as by the access to certain valuable goods such as fresh seafood and local fruits and crops, all of which were otherwise highly affected by Iota. But it is precisely these limitations imposed by insularity, which also result in moderate lifestyles and traditions that can be linked to islandness (Diegues 1998), as has been described here and also linked to cultural resilience in other isolated contexts such as the case described by Clarke and Mayer (2016). This economic soberness is also linked to cultural resilience and to the high capacity to deal with extreme situations, such as those that islanders have experienced after Iota.

5 Final Considerations

In this chapter, I have tried to present a reflection on the OPSC experience of Hurricane Iota from both my own experience of the hurricane and the perspective of cultural resilience, a concept that helps to understand the social and cultural processes that occur in societies which experience extreme disruptions because of disasters. For these reasons, I have presented a description of different aspects considered important in order to understand what happened before, during, and after Iota, and subsequently offered an analysis, based on the identification by Kendra et al. (2018) of key elements concerning disaster resilience. Although culture is not even considered to be one of the above elements, the framework remains useful as the analysis shows how culture is actually transversal to everything, and is also one of the most important components of many of the elements, including those apparently with no connections.

The analysis allows us to better understand many aspects of what has happened before, during, and after Iota. This includes some negative aspects, such as the resistance of some OPSC inhabitants to believe that something severe could happen, but it mainly contributes to understanding positive aspects, such as the surprisingly low number of deaths, people's capacity to deal with the hard situation created in both the disaster and post-disaster scenarios, and the strong community response to the authoritarian and exclusionary process imposed by the central government that has resulted in many inefficiencies and irregularities during the reconstruction. This last issue is the most remarkable in what has been analyzed regarding the islands' cultural resilience, and resistance, which belongs to what has been termed community capital and is based on community cohesion and cultural identity. This community capital plays an additional role by superseding the weakness of other capitals, such as economic, political, or institutional.

As I have tried to show, Raizal culture has emerged as a stronghold of the struggle that islanders have undertaken in order to guarantee better living conditions and a reconstruction process aligned with their particular livelihoods and worldviews. This has been the reaction to a disaster response and reconstruction that have not really taken into account local views and cultural dimensions, showing not only the importance of the inclusion of these into this kind of process, but also how resilience and vulnerability emerge from the relations between different actors (Barrios 2014).

The OPSC community has shown its flexibility and capacity to recover, as well as the resistance to certain changes considered negative, and the ability to learn from experience and adapt to new situations. This is not an easy process, and it is full of risks, as the current context exacerbates social, cultural, and economic vulnerabilities, ultimately threatening cultural survival. However, local community processes like those described here are a source of hope for those of us who wish that OPSC people may continue to inhabit their land and seascapes with autonomy, as has been done until now. In this sense, the analysis also opens new questions on different social and cultural issues related to the OPSC disaster experience, including social and environmental injustices, disaster capitalism, and other situations that currently threaten the cultural survival of the OPSC Raizal people. These should be addressed in order to deepen our understanding and ensure better actions in the future.

This chapter presents clear arguments to show that culture is not only important, but fundamental, to understand the disaster experience of OPSC islanders and that this experience should be included in the reconstruction process to contribute to its success and reduce risks in terms of the cultural survival of the Raizal people. Additionally, it must be incorporated into the local processes related to climate change adaptation, an issue that gains urgency and relevance given the experience of Iota and the complex regional scenario where hurricanes are projected to increase and worsen. This signifies the valuation and recognition of local knowledge and practices, and the need to promote knowledge exchanges with other types of knowledge. As the complex experience of the OPSC disaster, post-disaster, and reconstruction processes shows, technical and scientific knowledge falls short when it ignores the cultural and territorial realities of the areas it intends to impact.

Acknowledgements I would like to thank my family and friends, as well as all the people from Old Providence and Santa Catalina Islands with whom I have lived, learned, shared, and discussed our experience of Hurricane Iota. I would also like to thank my father, Professor Germán Márquez Calle, for his help, reviews, and suggestions; CEMarin for the interest and economic support; and the anonymous reviewers who made suggestions and adjustments in order to improve this chapter.

References

Adger N, Barnett J, Brown K et al (2013) Cultural dimensions of climate change impacts and adaptation. Nat Clim Change 3:112–117. https://doi.org/10.1038/nclimate1666

Aguado J (2010) El agua en el territorio, la cultura y la política de San Andrés Isla: una historia ambiental del siglo XX para el siglo XXI. Unpublished Master's dissertation, Universidad Nacional de Colombia

Barbosa J, Zanella ME (2019) Desnaturalizando o Desastre: as diferentes concepções teóricas que envolvem o conceito de desastre natural. Revista Da Casa Da Geografia Do Sobral 21(1):40–54

Barrios R (2014) 'Here, I'm not at ease': anthropological perspectives on community resilience. Disasters 38(2):329–350. https://doi.org/10.1111/disa.12044

Bent E (2021) Visión, causa y lucha del 'Campamento por la Dignidad'. El Isleño (30 May 2021). http://xn--elisleo-9za.com/index.php?option=com_content&view=article&id=21973:2021-05-30-12-35-33&catid=36:politica&Itemid=79. Accessed 30 June 2022

Boger R, Perdikaris S, Rivera-Collazo I (2019) Cultural heritage and local ecological knowledge under threat: two Caribbean examples from Barbuda and Puerto Rico. J Anthropol Archaeol 7(2):1–14. https://doi.org/10.15640/jaa.v7n2p1

Camargo A (2020) Thinking through disaster: ethnographers and disastrous landscapes in Colombia. In: García-Acosta V (ed) The anthropology of disaster in Latin America. Routledge, State of the Art, pp 82–101

Clarke H, Mayer B (2016) Community recovery following the deepwater horizon oil spill: toward a theory of cultural resilience. Soc Nat Resour 30(2):1–16. https://doi.org/10.1080/08941920.2016.1185556

Correa S (2012) Procesos culturales y adaptación al cambio climático: la experiencia en dos islas del Caribe colombiano. Boletín de Antropología de la Universidad de Antioquia 27(44):204–222. https://doi.org/10.17533/udea.boan.15621

Companion M (2015) Introduction. In: Companion M (ed) disaster's impact on livelihood and cultural survival: losses, opportunities, and mitigation. CRC Press, pp xxi–xxxi

Crane TA (2010) Of models and meanings: cultural resilience in social-ecological systems. Ecol Soc 15(4):19

Crawford S (2020) The last turtlemen of the Caribbean: waterscapes of labor, conservation, and boundary making. University of North Carolina Press

Cupples J (2017) Comunicación de riesgos en Centroamérica: La naturaleza, la colonialidad y la convergencia mediática [Conference presentation]. "HazMap" Workshop, Antigua, Guatemala, 20 March 2017. https://juliecupples.wordpress.com/2017/05/17/comunicacion-de-riesgos-en-centroamerica-la-naturaleza-la-colonialidad-y-la-convergencia-mediatica/

Diegues AC (1998) Ilhas e mares. Simbolismo e imaginário. NUPAUB – USP

El Espectador (2021) Providencia: 100 días de la tragedia del huracán Iota. El Espectador (24 February 2021). https://www.elespectador.com/colombia/mas-regiones/providencia-100-dias-de-la-tragedia-del-huracan-iota-article/. Accessed 3 February 2022

Ellis C, Adams TE, Bochner AP (2015) Autoetnografía: un panorama. Astrolabio 14:249–273. https://doi.org/10.55441/1668.7515.n14.11626

García-López G (2017) The multiple layers of environmental injustice in contexts of (un)natural disasters: the case of Puerto Rico post-hurricane Maria. Environ Justice 11(3):101–108. https://doi.org/10.1089/env.2017.0045

Guevara N (2007) San Andrés Isla, memorias de la colombianización y reparaciones. In: Mosquera C, Barcelos L (eds) Afro-reparaciones: memorias de la esclavitud y justicia reparativa para negros, afrocolombianos y raizales. Universidad Nacional de Colombia, Bogotá, pp 295–317

Holling CS (1973) Resilience and stability of ecological systems. Annu Rev Ecol Evol Syst 4:1–23. https://doi.org/10.1146/annurev.es.04.110173.000245

Holtorf C (2018) Embracing change: how cultural resilience is increased through cultural heritage. World Archaeol 50(4):639–650. https://doi.org/10.1080/00438243.2018.1510340

Jay E (2021) Providencia: La lucha por la dignidad. El Espectador (17 November 2021). https://www.elespectador.com/colombia/mas-regiones/providencia-la-lucha-por-la-dignidad/ Accessed 3 February 2022

Kendra JM, Clay LA, Gill KB (2018) Resilience and disasters. In: Rodríguez H et al (eds) Handbook of disaster research. Handbooks of sociology and social research. Springer, Cham, pp 87–107. https://doi.org/10.1007/978-3-319-63254-4_5

Marchezini V (2015) Social recovery in disasters: cultural resistance of Luizenses. In: Companion M (ed) Disaster's impact on livelihood and cultural survival: losses, opportunities, and mitigation. CRC Press, pp 293–303

Márquez-Pérez AI (2013) Culturas migratorias en el Caribe colombiano: el caso de los isleños raizales de Old Providence y Santa Catalina. Memorias, Revista Digital De Historia y Arqueología Del Caribe 19:204–229

Márquez-Pérez AI (2014) Povos dos recifes: reconfigurações na apropriação social de ecossistemas marinhos e litorâneos em duas comunidades do Caribe. Unpublished Doctoral dissertation, Universidade Federal Rural de Rio de Janeiro

Márquez-Pérez AI (2020) Crisis económica y alimentaria en el medio del mar Caribe. In: Wagner A, Acevedo R, Aleixo E (eds) Pandemia e Território. Universidade Estadual do Maranhão, pp 885–902

Ministerio de Cultura (2016) Between Land & Sea. Las cocinas tradicionales del Archipiélago de San Andrés, Providencia y Santa Catalina. Ministerio de Cultura

Moran E (2022) Human Adaptability: An Introduction to Ecological Anthropology. Routledge, New York

Oliver-Smith A (1996) Anthropological research on hazards and disasters. Annu Rev Anthropol 25:303–328. https://doi.org/10.1146/annurev.anthro.25.1.303

Oquendo C (2021) Un hospital en carpas y familias sin hogar: Providencia sigue a medio reconstruir un año después de Iota. El País (8 November 2021). https://elpais.com/internacional/2021-11-08/un-hospital-en-carpas-y-familias-sin-hogar-providencia-sigue-a-medio-reconstruir-un-ano-despues-de-iota.html. Accessed 3 February 2022

Padilla K (2010) Entre lo local y lo global: el caso del Movimiento de Veeduría Cívica de Providencia y Santa Catalina islas. Unpublished Master's dissertation, Universidad Nacional de Colombia, Sede Caribe

Parsons J (1985) San Andrés y Providencia, una geografía histórica de las islas colombianas del mar Caribe. Banco de la República, Bogotá

Raizal Dignity Camp (2022) Informe Campamento por la Dignidad 298 días. Facebook (2 January 2022) https://www.facebook.com/Raizal-Dignity-Camp-PVA-696826521127795. Accessed 3 February 2022

Rivera DZ (2022) Disaster colonialism: a commentary on disasters beyond singular events to structural violence. Int J Urban Reg Res 46:126–135. https://doi.org/10.1111/1468-2427.12950

Sandner G (2003) Centroamérica & el Caribe Occidental. Coyunturas, crisis y conflictos 1503 – 1984. Universidad Nacional de Colombia, Sede Caribe

Swidler A (1986) Culture in action: Symbols and strategies. American Sociological Review 51(2):273–86. https://doi.org/10.2307/2095521

UNGRD (Unidad Nacional para la Gestión del Riesgo de Desastres) (2020) Boletín Informativo 214. Oficina Asesora de Comunicaciones

Valencia I (2015) Conflictos interétnicos en el Caribe Insular Colombiano. Controversia 205:171–217. https://doi.org/10.54118/controver.vi205.395

Valoyes K (2020) Damnificados de Iota en Colombia, entre el éxodo y las noches a la intemperie. Agencia Efe (22 November 2020). https://www.efe.com/efe/america/sociedad/damnificados-de-iota-en-colombia-entre-el-exodo-y-las-noches-a-la-intemperie/20000013-4400904. Accessed 3 February 2022

Webb G (2018) The cultural turn in disaster research: understanding resilience and vulnerability through the lens of culture. In: Rodríguez H et al (eds) Handbook of disaster research. Handbooks of sociology and social research. Springer, Cham, pp 109–207. https://doi.org/10.1007/978-3-319-63254-4_6

Wilson P (1973) Crab Antics. A caribbean case study of the conflict between reputation and respectability. Waveland

Open Access This chapter is licensed under the terms of the Creative Commons Attribution 4.0 International License (http://creativecommons.org/licenses/by/4.0/), which permits use, sharing, adaptation, distribution and reproduction in any medium or format, as long as you give appropriate credit to the original author(s) and the source, provide a link to the Creative Commons license and indicate if changes were made.

The images or other third party material in this chapter are included in the chapter's Creative Commons license, unless indicated otherwise in a credit line to the material. If material is not included in the chapter's Creative Commons license and your intended use is not permitted by statutory regulation or exceeds the permitted use, you will need to obtain permission directly from the copyright holder.

Climate Change: A Business Perspective of the Tourism Industry in the Seaflower Biosphere Reserve

Lorena Aldana Pedrozo and Rixcie Newball Stephens

Abstract Mass tourism in the Archipelago of San Andrés, Providencia, and Santa Catalina is depleting natural ecosystems, which are precisely the basis that supports the local economy. Since the establishment of the free port in the territory, the region's development has been promoted without considering the loss of its environmental resources. However, after the islands were declared the Seaflower Biosphere Reserve in 2000, actions have been taken to counteract the situation, envisioning a more environmentally friendly productive performance that provides the income necessary to guarantee the well-being and quality of life of islanders. Despite this, intensive economic activities prevail in the department, showing that the path towards sustainability under these conditions is difficult to achieve. Climate change has highlighted the importance of finding sustainable alternatives for islands. In this case, it is necessary to discourage conventional means that threaten the biodiversity of the territory and to promote business schemes that are more environmentally respectful. This will be possible through the strengthening of governance in the Biosphere Reserve and its coordination with different interest groups. In this article we estimate the relevance of the topic for the most important industry in the territory—tourism—to provide key elements for this discussion.

Keywords Climate change · Mass tourism · Small islands · Sustainable development

L. A. Pedrozo (✉)
Master in Business Innovation Management, San Andrés, San Andres and Providencia, Colombia
e-mail: lorena.aldanap@outlook.com

R. N. Stephens
Corporación Para El Desarrollo Sostenible del Archipiélago de San Andrés, Providencia y Santa Catalina, CORALINA, San Andrés, San Andres and Providencia, Colombia
e-mail: rixcien@gmail.com

© The Author(s) 2025
J. E. Mancera Pineda et al. (eds.), *Climate Change Adaptation and Mitigation in the Seaflower Biosphere Reserve*, Disaster Risk Reduction,
https://doi.org/10.1007/978-981-97-6663-5_11

1 Introduction

Climate change can be explained by natural or external causes. Three decades ago, the United Nations Framework Convention on Climate Change (1992) defined the phenomenon as "a change in climate attributed directly or indirectly to human activity that alters the composition of the global atmosphere and adds to the natural climate variability observed over comparable periods of time" (p. 3). More recently, the Intergovernmental Panel on Climate Change (2023) stated that "human activities, mainly through greenhouse gas emissions, have indisputably caused global warming" (p. 4).

In 2021, the earth's temperature was 0.85 °C above the average for the period 1951–1980 and 1.1 °C warmer than the average at the end of the nineteenth century. Collectively, the past eight years have been the warmest since records began in 1880 (NASA 2022). In the particular case of Latin America and the Caribbean, the World Meteorological Organization (2022) revealed that "the average rate of increase in temperatures in the region was approximately 0.2 °C per decade between 1991 and 2021, compared to the 0.1 °C per decade recorded between 1961 and 1990" (p. 7). Some territories in this area face the Atlantic hurricane season each year, with 2021 being the third most active season recorded, made up of 21 storms including seven hurricanes (NOAA 2022).

The occurrence of anomalies and extreme climate events that are affecting ecosystems also have an impact on economic development. Depending on their magnitude, they constitute a threat that will exacerbate inequality gaps: increasing poverty, food insecurity, deprivation in the provision of basic services and, in general, the habitability of the affected countries.

The effects of climate change are recorded and projected in a particular way for each region. In small islands, to which a specific chapter has been dedicated in the IPCC reports, in consideration of the high level of vulnerability to which they are exposed because of their geophysical characteristics, the main risk has been identified as the loss of livelihoods, settlements, infrastructure, ecosystem services, economic stability, and low-lying coastal areas (IPCC 2014). Furthermore, the IPCC noted that:

> the dependence of many small islands on a limited number of economic sectors such as tourism, fisheries, and agricultural crops, all of which are climate sensitive, means that on the one hand climate change adaptation is integral to social stability and economic vitality but that government adaptation efforts are constrained because of the high cost on the other (p. 1626).

The Archipelago of San Andrés, Providencia, and Santa Catalina (hereafter, the archipelago), a Colombian department (administrative region) declared a Biosphere Reserve by UNESCO in 2000 and the object of study in this article, is not far from the global stage. According to the national results of the multidimensional analysis of vulnerability and risk due to climate change recorded in the Third National Communication to the United Nations Framework Convention on Climate Change, the island territory occupies first place in the risk index (very high), the second place in sensitivity (high) and the last place in adaptive capacity (very low) (IDEAM et al. 2017).

Climate exposure adds to the challenges of the islands in terms of sustainable management and socioeconomic conditions: scarcity of water resources; low coverage of water and sewage services; insufficient capacity for the management and disposal of solid waste; dependence on external trade for the provision of consumer and capital goods; a small domestic market; energy production from fossil resources; high cost of living; high population density; massive influx of visitors.

All these factors affect regional development and its different interest groups. This chapter focuses on local productive dynamics, particularly those mobilized by the tourism industry and the businesses that comprise it, from whom the situation demands skills of analysis and interpretation of the environment to anticipate situations, project future scenarios, and make convenient decisions that allow the adoption of a strategic position in the market and which articulate with the governance purposes of the Biosphere Reserve in particular.

We will carry out a contextualization of the territory; its economic trajectory, a characterization of the tourism sector, and the presentation of other indicators of interest for this purpose. We will evaluate the knowledge, perception of vulnerability, and capacity of the sector regarding climate change. The objective of the research is to determine the relevance of climate change for tourism service providers in the archipelago. The study provides the business perspective as a key element for territorial reflection on the issue of climate change, and the implementation of mitigation and adaptation actions.

2 Literature Review

Scientific literature has made it clear that small island developing states (SIDS) will be particularly affected by climate change, that the dangers will intensify due to the increase in global warming and their socio-economic characteristics will contribute to exacerbating their levels of vulnerability (Thomas et al. 2020). Specifically, it has been pointed out that the Caribbean islands are on the front line of the severe impacts of climate change, more than other parts of the world, due to their geographical location, and because they are home to people who live nearby and depend on them of the sea to guarantee its survival (Mycoo 2018).

The region's beaches will be affected by the rise in sea level and the intensity and direction of the waves. On average, it is estimated that, annually, the beaches will experience a retreat of around 0.16 m, generating a loss of biodiversity and "a negative impact on economic activities and the well-being of the population: less tourism, destruction of coastal infrastructure, population movements and indebtedness" (Bárcena et al. 2020, p. 130).

"The business model currently adopted by many SIDS is mass tourism" (Hampton and Jeyacheya 2020, p. 9). Referring to most small islands in the Caribbean, Cannonier and Burke (2019) found that, "in no other region of the world is the industry as vital to economic sustainability." However, the competitiveness of the tourism sector in the Caribbean is threatened by climate change due to its high dependence on natural resources (Mackay and Spencer 2017). The concentration of the tourism industry on the use of natural resources, and the expansion and profitability of the activity, have not favored economic diversification in tourist destinations, nor the incorporation of regenerative practices that the future of the sector demands (Cave and Dredge 2020).

Uncontrolled tourism growth generates adverse economic, social, and environmental impacts that must be addressed with priority in the small island economies of the Caribbean (Peterson and DiPietro 2021). Peterson (2020) warned that "one of the biggest risks to Caribbean SITEs [small island tourism economies] is underestimating the adverse effects of over-tourism and downplaying the risks of climate change" (p. 31), and that it is currently unsustainable to maintain this position of inertia; it is urgent to transcend the illusion of the extraordinary growth of the tourism industry, especially due to the overflow of activity and the exploitation of environments. In her work, Sheller (2021) pointed out that it is time to think about tourism beyond a profitable business opportunity, and to consider it an expensive social and natural option. Indeed, "determining the carrying capacity of ecosystems is a prerequisite for the success of tourism, social stability, economic prosperity and human well-being (Segrado Pavón et al. 2017, p. 8).

Although tourism has been projected as an opportunity to promote the economic growth of regions, research findings challenge this idea, and instead highlight the importance of the quality of political institutions and the tourism offer in this relationship. Countries with low government efficiency also have low tourism specialization and competitiveness, and must even face the additional costs generated by mass tourism due to pressure on local resources and infrastructure, such as water scarcity and the management of waste (Antonakakis et al. 2019).

Peterson (2020) highlighted that in the sustainable transformation process it is essential to strengthen institutions, overcome the prevailing political and market failures, and involve the private sector and communities to strengthen their resilience capacities. For their part, Santos-Lacueva et al. (2017) found that there is a "close relationship between public policies and vulnerability of destinations" (p. 14). They added that the weakness of regulatory frameworks, the disconnection between tourism policies and climate change, and the disarticulation of the public and private sectors, constitute obstacles to confronting climate change.

At the regional level, the Economic Commission for Latin America and the Caribbean pointed out that resilience and adaptive capacity must be reinforced and aligned with development objectives and practices, also considering that, when viewed as an investment, they have a multiplying influence on the economy, reducing the negative consequences of climate change and contributing to closing gaps in their countries (Bárcena et al. 2020).

Nunn and Kumar (2017) warned that "adaptation failure is a key factor in sustaining, even amplifying, perceptions of island vulnerability" (p. 16). However, "there are numerous barriers to adaptation that affect SIDS, including the need for institutional good governance, climate information, finance, and effective monitoring and evaluation of adaptation efforts" (Thomas et al. 2020, p. 20). "Adaptation action should be guided by each country's unique characteristics and informed by citizen science or run the risk of maladaptation" (Mycoo 2018, p. 2351). Therefore, "fostering 21st century resilience of Caribbean SITEs requires building innovation and institutional capabilities to anticipate shocks, to adapt and learn, and bounce forward toward new paths of development" (Peterson and DiPietro 2021, p. 18).

In this scenario, valuing the knowledge and perception of individuals in a society about climate change is decisive (Corona 2018). For decision-makers, this type of study supports the orientation of strategies and political actions, with there currently being a need for work on adaptation and mitigation processes, in order to favor the involvement of actors (Miranda Vera et al. 2019). It is especially important to recognize that, among the inhabitants of the SIDS, there are those who have knowledge about climate change but not with great precision, while there are others for whom the issue is distant and out of their control, waiting for governments to assume responsibility for addressing the issue (Thomas et al. 2020). Likewise, in the implementation of adaptation measures, it is necessary to understand the risks that climate change represents for companies and destinations from their own perspective (Pandy and Rogerson 2018).

Among the results of the research by González et al. (2019), it draws attention that there are planning instruments that continue to project tourism as an activity with the capacity for growth, and even more so than companies in the sector, far from considering this situation, and the high consumption of natural resources associated with it, as one of the most important issues on which adaptation measures must be established, with strategies to reduce energy consumption among the most important solutions.

As part of sustainable tourism policy and local governance schemes, Caribbean authorities must collect and analyze socioeconomic and environmental data of their territories, in order to monitor the risks and negative externalities of uncontrolled tourism growth and to adopt measures balanced on such impacts (Peterson and DiPietro 2021).

3 Contextual Framework

This study is carried out in the archipelago, declared the Seaflower Biosphere Reserve (BR) by UNESCO in 2000. The archipelago is located at the jurisdictional end of Colombia in the Central—Western Caribbean Sea. It has an area of approximately 180,000 km^2, of which only 57 km^2 correspond to the emerged surface. 99.97% of the territory is eminently oceanic. It has 78% of the country's corals and is the second reef system in the Western Hemisphere. The Seaflower BR has important ecosystems of tropical dry forest, mangroves, seagrasses, and coral sand beaches, in addition to a great diversity of fauna and flora species, offering extensive ecosystem services (CORALINA-INVEMAR 2012). According to the political-administrative division of the country, the archipelago is made up of two municipalities; the municipality of San Andrés and the municipality of Providencia and Santa Catalina.

3.1 Economic Trajectory of the Archipelago of San Andrés, Providencia, and Santa Catalina

In 1953, the national government of General Rojas Pinilla declared the archipelago a free port, while the rest of the Colombian territory moved towards the implementation of an import substitution industrialization model. This exclusive concession for the islands, which allowed the receipt of foreign goods without paying tariffs, set up a new dynamic for the local economy.

Quickly, in addition to the insular agricultural activity that for centuries had in its first lines with export products derived mainly from the cultivation of cotton and coconut, commerce made its way, revolutionizing not only the business context, but also, in its entirety, regional development in a geographic space inhabited and managed by natives.

According to the census population records of the National Administrative Department of Statistics (DANE) shown in Fig. 1, in 1951, only on the island of San Andrés, capital of the archipelago, there were 3,705 people. By 1964, this number increased by 289% with 14,414 citizens.

The demographic explosion of San Andrés was influenced by the opportunity represented by the nascent economic project. Colombians and foreigners from different regions settled on the island to open warehouses, engage in the trade of goods or find a promising job in the territory. The deployment of this commitment required the acquisition of real estate, land, and infrastructure, such as the airport and a ring road.

Private investment efforts were stimulated by Law 127 of 1959, which in its Article 12 decreed "to exempt hotels, restaurants, buildings from the payment of income and complementary taxes, for a period of ten years of apartments and industries that are established or will be established in the territory of San Andrés and Providencia".

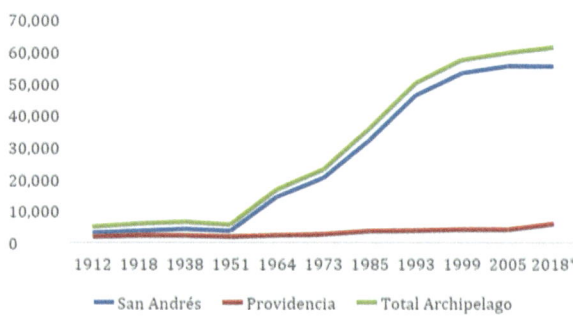

Fig. 1 Census population. Archipelago Department of San Andrés, Providencia, and Santa Catalina 1912–2018. *Source* National Administrative Department of Statistics, DANE. *Note* * Coverage-adjusted population

In the adaptation of the island region for this purpose, construction works and urbanization processes were carried out for housing and productive purposes without sustainable planning, care was not taken in at risk or conservation areas, causing alterations in the natural heritage and consequences that still constitute a challenge for the development of the archipelago today.

At the beginning of the 1990s, Colombia rethought its protectionist scheme and made the transition towards economic openness. The preferential commercial conditions that the islands boasted were no longer attractive in the captive markets and the sector began to lose share in the departmental Gross Domestic Product (GDP). However, thanks to the fact that trade encouraged the mobilization of visitors to the islands and tourism capacities were established, the productive transformation of the archipelago was projected on this other industry, declared by Law 300 of 1996 as "essential for the development of the country and especially for the different territorial entities". Since 1999, the line of hotels, restaurants, bars, and the like has managed to position itself as the main economic activity of the island territory, as shown in Fig. 2.

Otherwise, as presented in Table 1, the local economy is not very diverse and has low profitability in other sectors. Historically, the production of goods and services in the department has been determined by one or certain activities that are used until they are exhausted, taking into account that, from a business perspective, there is no commitment, in a more ambitious way, to the addition of value, nor the development of competitive advantages.

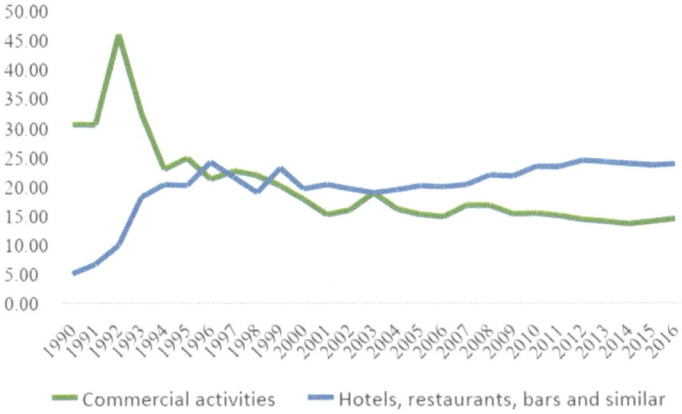

Fig. 2 Percentage share of commercial activities and hotels, restaurants, bars, and the like within the departmental GDP 1990–2016. *Source* National Administrative Department of Statistics, DANE. National accounts

Table 1 Gross domestic product—base 2015, according to economic activity. Archipelago of San Andrés, Providencia and Santa Catalina. Stake (%)

Economic activities	2021[pr]
Agriculture, livestock, hunting, forestry, and fishing	1.4
Exploitation of mines and quarries	0.1
Manufacturing industries	1.2
Supply of electricity, gas, steam, and air conditioning; water distribution; evacuation and treatment of wastewater, waste management and environmental sanitation activities	2.1
Building	2.2
Wholesale and retail; repair of motor vehicles and motorcycles; transportation and storage; accommodation and food services	57.9
Information and communications	1.1
Financial and insurance activities	2.9
Real estate activities	4.3
Professional, scientific and technical activities; administrative and support service activities	3.4
Public administration and defense; compulsory social security plans; education; activities of human health care and social services	15.5
Artistic, entertainment, and recreation activities and other service activities; activities of individual households as employers; undifferentiated activities of individual households as producers of goods and services for their own use	0.7
Gross value added	**92.8**
Taxes	7.2
Departmental GDP	**100.0**

Source National Administrative Department of Statistics, DANE. National accounts
Note [pr] Preliminary

3.2 Characterization of the Tourism Sector in the Archipelago Department of San Andrés, Providencia and Santa Catalina

The tourist potential of the islands has been developed due to their comparative advantages. The natural resources and landscapes of the island territory have been capitalized to boost the productive sector. The image of the destination in the national and international markets was forged around the sun and beach product, and has been complemented with emerging offers in adventure tourism, ecotourism, and cultural tourism. Since 2006, with a couple of exceptions—among them, the COVID-19 pandemic—, the number of tourists who visit San Andrés has registered exponential growth, breaking the barrier of one million travelers three times in recent years, according to statistical information from James Cruz (2013) and the department's Secretary of Tourism, as shown in Fig. 3.

Fig. 3 Total number of tourists arriving annually on the island of San Andrés 1960–2021. *Source* James Cruz (2013) (Data 1960–2011)—Departmental Tourism Secretariat (Data 2012–2021)

This dynamic has been influenced by the purposes of sectoral planning and policy defined at the national level that, combined with territorial action, focused especially on tourism promotion, management for the increase in air routes to the archipelago and the expansion of local accommodation capacity.

Likewise, towards the last decade, the business models of low-cost airlines and tourist inns contributed to the increase in passenger mobilization, providing greater accessibility to market segments than in other fare conditions or travel expense preferences, for people who would possibly not visit the territory otherwise.

Tourism ceased to be the exclusive domain of the businessmen who until then belonged to the sector. Entrepreneurs, self-employed workers, and citizens in general from the department began to get involved in the activity, attracted by the economic benefit of the booming industry. As presented in Table 2, the National Tourism Registry (RNT with its Spanish initials) includes 1,689 tourism service providers in the archipelago; 89% of them are located in San Andrés, and the remaining 11% in Providencia and Santa Catalina.

72% of the region's total service offering is made up of accommodation of different denominations, currently led by the category of tourist housing (69%) where apartments, rural accommodation, and other types of housing are found. In order, the traditional segment of accommodation establishments follows (27%) with options of hostels, apart-hotels, camps, holiday centers, hostels, and hotels. With a smaller participation, there is another type of non-permanent accommodation (4%).

As shown in Fig. 4, the occupancy of the accommodations in the archipelago has been characterized by being in first place at the national level. In 2019, the national monthly average percentage was 48.8% while in the island territory, it was 73.2%. In 2021, the recovery figures due to the pandemic were 40.6% and 62.8%, respectively. Regarding the origin of the hosted visitors, in the national case, in 2019 domestic tourism accounted for 69.4% and foreign tourists accounted for 30.6%. For the archipelago, the distribution was 70.2% and 29.8% respectively. In 2021, the

Table 2 Tourist service providers, aggregated by RNT category. Archipelago of San Andrés, Providencia and Santa Catalina 2022

RNT categories aggregates	San Andrés	Providencia and Santa Catalina	Total Archipelago
Responsive operators	260	32	292
Car rental companies	64	6	70
Bars and gastronomy esta-blishments	79	28	107
Accommodation	1108	112	1220
Total	1.511	178	1.689

Source Own elaboration based on information from the Chamber of Commerce of San Andrés, Providencia and Santa Catalina (2022)

regional performance (85.5% and 14.2%) was similar to the national trend (84.5% and 15.5%).

For the departmental government, tourism activity has also acquired relevance in budgetary matters. Through Decree 2762 of 1991 and Law 47 of 1993, the collection of a tax called a tourism card was authorized for national and foreign citizens who enter the islands as visitors. As shown in Fig. 5, starting in 2002, the money generated became the main source of income for the island territory. In 2021, the money raised by this tourist card reached a 41% share of current income.

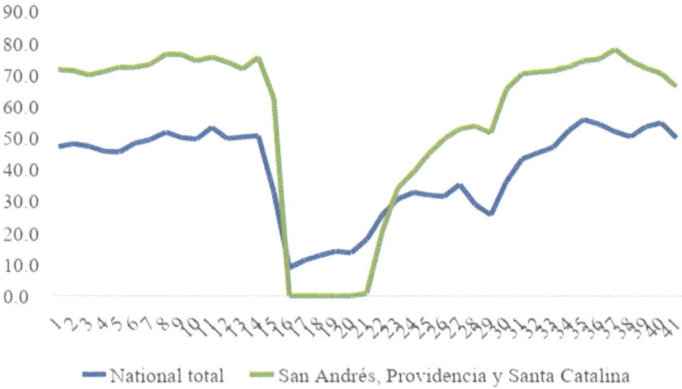

Fig. 4 Monthly accommodation occupancy. National total and total for the Archipelago of San Andrés, Providencia, and Santa Catalina, January 2019–May 2022 (%). *Source* National Administrative Department of Statistics, DANE. Monthly accommodation survey

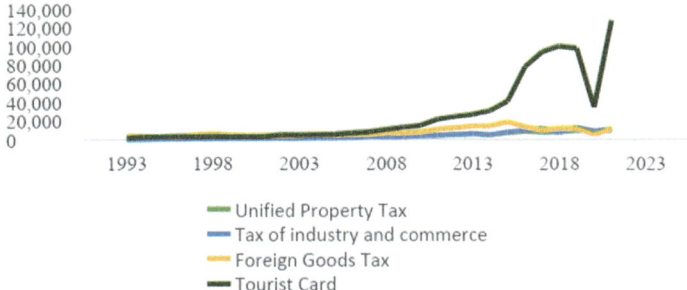

Fig. 5 Main income of the Government of the Archipelago Department of San Andrés, Providencia, and Santa Catalina 1993–2021. *Source* Government of the Archipelago Department of San Andrés, Providencia, and Santa Catalina (2022) *Note* Millions of pesos

3.3 Other Vulnerability and Risk Factors in the Archipelago Department of San Andrés, Providencia, and Santa Catalina

Economic development has favored the high population density in the archipelago, especially in San Andrés. For 2022, according to DANE population projection figures, 2,178.41 inhabitants/km^2 were calculated, in contrast to Providencia and Santa Catalina where the index is located at 356.17 inhabitants/km^2. Between its own citizens and the tourists received annually—which represent 19 times the population of the territory—considerable pressure is exerted on the region's resources.

"Water scarcity is one of the main environmental challenges facing the Seaflower Biosphere Reserve" (Guerrero Jiménez 2020, p. 127). In the past, the community has solved this problem through the collection of rainwater that is stored in cisterns or the exploitation of aquifers through a domestic pumping system to extract the liquid. Both practices are dependent on climatic variability (Guerrero Jiménez 2020).

According to the Quality of Life Survey (ECV) 2021, carried out by the National Administrative Department of Statistics (DANE), 47% of island households live in rented accommodation, they have low coverage of aqueduct services (37.4%) and sewage (27.5%) and assume a higher average monthly expense than the rest of the country, especially in the groups of accommodation, water, electricity, gas and other fuels, and food and non-alcoholic beverages. Since 2018, unemployment on the islands has been increasing, registering the highest rate in 2021 with 13.95%. For its part, labor informality is part of the deprivations faced by 60.3% of families.

4 Methodology

This research is approached from a quantitative approach, considering the intention of this type of study, which seeks to explain, verify, or confirm and predict phenomena, as well as accurately establish behavioral patterns of a population or situation.

The scope of the work is descriptive, adopted to specify the properties, characteristics, and profiles of people, groups, communities, processes, objects, or any other phenomenon that is subjected to analysis, through measurement or data collection that report information on various concepts, variables, aspects, dimensions or components of the phenomenon or problem to be investigated (Hernández and Mendoza 2018).

To develop the objectives of this work, a survey was designed and applied, facilitating access to the required information. The instrument was organized into five components; identification of the business, knowledge about climate change, perception of vulnerability to climate change, capacities of tourism service providers to face climate change, and a free space to write down comments, observations, and reflections.

The population under analysis is made up of tourism service providers, defined as natural or legal persons that usually provide, mediate, or contract directly or indirectly with the tourist, the provision of the services referred to in Article 2.2.4.1.1.13 of Decree 1836 of 2021. These have the obligation to be registered in the National Tourism Registry to carry out operations in Colombia and must update it annually (Law 300 of 1996). The platform operated by the country's Chambers of Commerce is a mechanism to identify and regulate tourism service providers, as well as to maintain an updated information system for the industry. In the case of the archipelago, according to the information provided by the Chamber of Commerce of San Andrés, Providencia, and Santa Catalina, a total of 1,689 active tourist service providers were found as of June 8, 2022.

The sample size calculation was performed using the formula for finite populations. In the operation, a confidence level of 95% and a margin of error of 5.58% was determined. Subsequently, using a simple random technique in Excel, the 262 establishments that should be included in the sample were selected. An online form was used to collect data. The survey link was sent by email to the legal representatives of the businesses that made up the sample.

To analyze the information collected, the data was imported into an Excel file, where the review and preparation of results was carried out.

5 Results

According to the calculated sample size, 262 tourism service providers from the island territory responded to the survey. As a result of probabilistic sampling, 63% of the businesses consulted are located on the island of San Andrés and the remaining

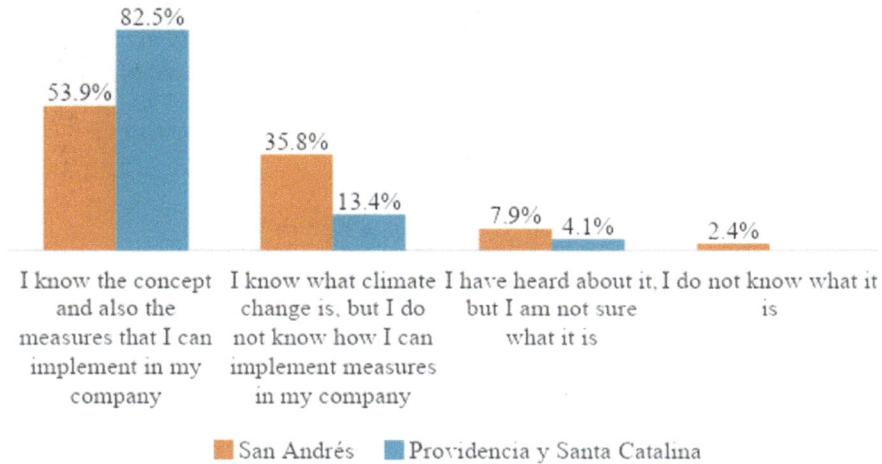

Fig. 6 Knowledge of the tourism service providers of Archipelago of San Andrés, Providencia, and Santa Catalina about climate change. *Source* Self-made

37% in Providencia and Santa Catalina. The research findings will be presented in a general way for the archipelago in the event that the responses obtained are not far from the average, or in a particular way, if they differ between the islands.

When consulting the tourism service providers of the archipelago regarding their knowledge about climate change, it was found that, in the municipality of Providencia and Santa Catalina, they are better informed about the concept and the actions that can be implemented, as shown in Fig. 6. For its part, in San Andrés, a large number of subjects were identified who do not know the topic, who have heard it but are not sure what it is or know what it is, but are not clear about the means to deal with it.

In relation to the quality of information provided by local authorities in the different media and/or social networks about the impacts of climate change on the tourism industry, the majority of respondents (51.5%) expressed that there is little or not at all suitable. 18.7% classified it as very adequate or adequate, while 29.8% rated it as moderately adequate.

Introduced to the topic, 90.5% of the tourism service providers recognized that climate change is impacting and will impact the archipelago, as shown in Fig. 7. Particularly, when asked if they believed that tourism activities would be affected by this issue, 95% of the responses were affirmative. When asked to indicate which of a set of related consequences they would consider to have the greatest impact on the development of tourist activities in the archipelago, in order, 61.1% indicated that hydro-meteorological phenomena (cold fronts, storms, intense rains, hurricanes, etc.), 24% beach erosion, 7.6% sea level rise, and 7.3% water shortages.

Addressing the issue from the development of their own business activity, the findings vary in contrast to the perception that tourism service providers have of the impacts of climate change on the archipelago and the tourism sector. As shown in Fig. 8, the overall proportion of those who responded that they are not currently being

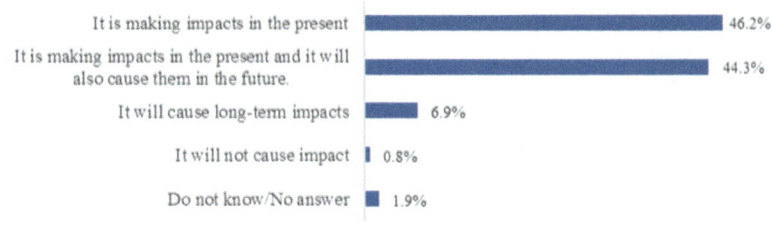

Fig. 7 Perception of the current or future impacts of climate change in the Archipelago Department of San Andrés, Providencia, and Santa Catalina. *Source* Self-made

affected by climate change decreases, especially in Providencia and Santa Catalina, and those who considered that they will be affected in the future increases.

61.1% of tourism service providers stated that they had estimated the risks that their businesses would face due to the effects of climate change in the island territory. Based on their response, this group was specifically asked about the main threat to which they consider their businesses are most exposed. The results were different for the two municipalities. In the case of San Andrés, the first three threats identified were a decrease in demand for the tourist destination (26.6%), floods (18.1%), and vulnerability of its physical infrastructure due to not being adapted to adverse weather events (12.8%). In Providencia and Santa Catalina, flooding (29.2%), coastal erosion (20%), and vulnerability of its physical infrastructure due to not being adapted to adverse weather events (15.4%).

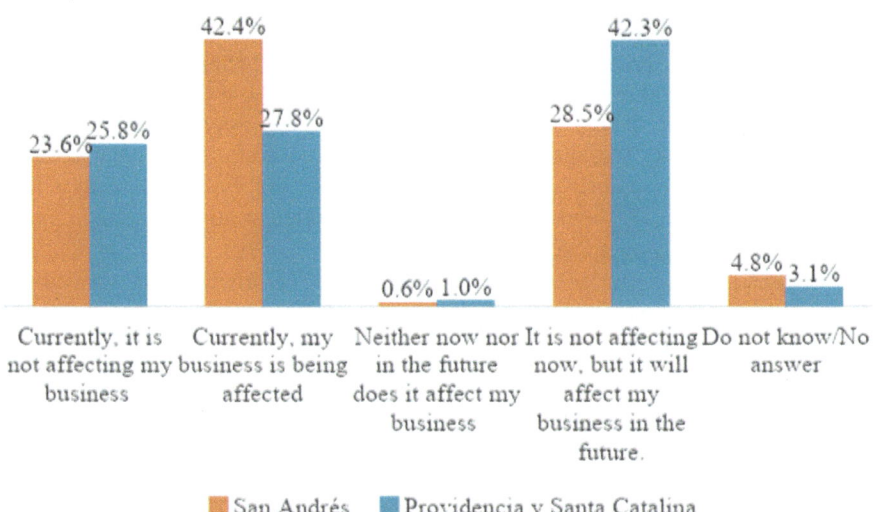

Fig. 8 Perception of the current or future impacts of climate change on the activities of tourism service providers in the Archipelago Department of San Andrés, Providencia, and Santa Catalina. *Source* Self-made

90% of those surveyed expressed that they had considered that their business investment could be affected by the effects of climate change. In San Andrés, 44.2% of tourism service providers mentioned that their businesses are located or carry out activities in an area vulnerable to the risks of climate change, 33.9% said that they did not have this risk and the remaining 21.8% indicated that they did not know. In Providencia and Santa Catalina, on the other hand, 77.3% stated that they do not face this problem, while 20.6% said they do and 2.1% indicated that they did not know. Likewise, when asked if businesses have infrastructure adapted to face adverse weather events, in San Andrés 40% answered yes and 60% said no. In Providencia and Santa Catalina, the trend was reversed, 69.1% said yes and 30.9% said no.

Regarding the main challenge that tourism service providers have identified in their businesses to move towards mitigation and adaptation to climate change, among the most relevant issues was that 26.7% attributed the situation to a lack of knowledge of projects in which the company can participate, 25.6% due to the high costs to implement a measure and 18.7% due to lack of knowledge to implement a measure.

96.9% of the tourism companies studied reported that they have not measured their carbon footprint. Regarding whether they have implemented actions that directly or indirectly reduce the adverse effects of climate change, 52.1% of tourism service providers in San Andrés noted yes, while in Providencia and Santa Catalina those who made this statement represented 33%. In these groups, it is highlighted that, for the first municipality, 59.5% indicated that the main measures adopted were related to good operational practices in environmental and energy management, while 27.4% with actions to optimize water consumption. For the second municipality, in the case of these same measures, the proportions are 37.5% and 40.6% respectively. For their part, only 22.1% of the units analyzed in the archipelago claimed to have environmental quality certification (ISO 14001).

At the end of the survey, a list of five considerations was shared with the tourism service providers so that, according to the importance they deserved, they could select one, taking into account the vulnerability of the island territory due to the adverse effects of climate change. According to what is shown in Fig. 9, it is striking that while in San Andrés, 64.8% of tourism businesses stated that they would implement adaptation/mitigation actions, in Providencia and Santa Catalina 55.7% said that they would continue to develop their business activity as before.

6 Discussion

The present study provides a vision of the knowledge, perception of vulnerability, and capabilities of tourism service providers in the Archipelago Department of San Andrés, Providencia, and Santa Catalina in relation to climate change. The results obtained reveal key challenges and opportunities.

One of the most notable findings of this document is the knowledge gap about climate change between tourism service providers in San Andrés and those in Providencia and Santa Catalina. While in the latter municipality, there is a higher level

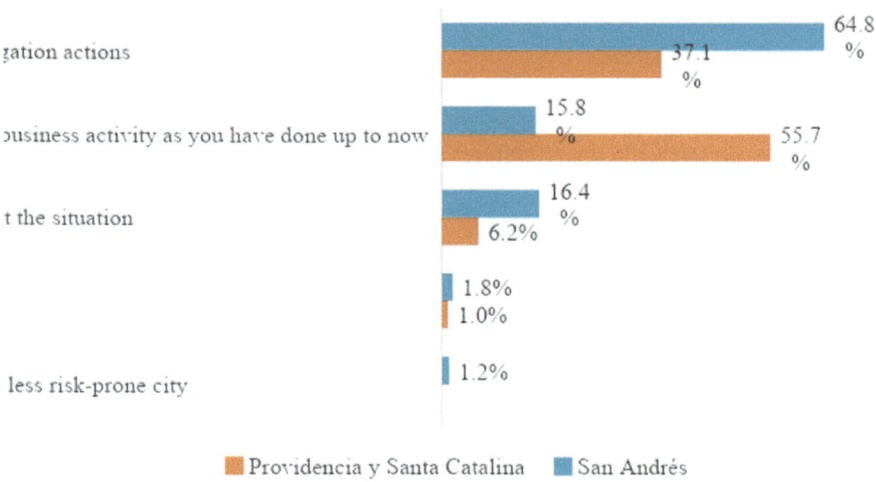

Fig. 9 Business actions against the vulnerability of the Archipelago Department of San Andrés, Providencia, and Santa Catalina due to climate change. *Source* Self-made

of knowledge on the subject, in San Andrés there is a significant segment that has a limited understanding of climate change. This lack of understanding can hinder the implementation of a culture of adaptation and mitigation to climate change in the sector.

It was also found that there is a high perception that climate change is already affecting and will continue to affect the archipelago, particularly due to hydrometeorological phenomena, identified as the main threat to the development of tourist activities. This suggests that the vulnerability of businesses to extreme climate events is recognized and highlights the importance of deploying specific strategies for tourism in this area.

In terms of infrastructure adaptation, the results show notable differences between the islands. In San Andrés, the majority of tourism service providers expressed that they lack such infrastructure. In Providencia and Santa Catalina, a significant group reported that they had it. This may be related to the lessons learned and the interventions that were carried out to rebuild the municipality after Hurricane Iota. However, it is essential that the measures established on this issue are supported by a solid regulatory framework and tourism planning that integrates sustainability and climate resilience.

The lack of concrete action on climate change mitigation and adaptation is another concern that requires attention. A gap was detected between the perception of risk and the effective action of tourism service providers. The lack of knowledge of projects where companies can participate and the costs associated with the implementation of some measures are obstacles that must be addressed to encourage greater mobilization of the industry.

The absence of environmental quality certification (ISO 14001) in the majority of the companies surveyed is an aspect of considerable improvement. Environmental certification can help companies manage their environmental impacts and implement more sustainable practices.

Considering the vulnerability of the island territory due to the adverse effects of climate change, it is important that tourism service providers in the archipelago are prepared to take measures to improve the resilience of their operations. The willingness of some businesses in Providencia and Santa Catalina to continue developing their activities as of now could be a sign of confidence in their ability to face climate challenges, which requires continuous monitoring and support to guarantee that these actions are sustainable in a changing environment.

7 Conclusions

This study highlights the importance of proactively addressing climate change in the tourism sector of the Archipelago Department of San Andrés, Providencia, and Santa Catalina. The results indicate that, although there is widespread awareness of the impacts of climate change, there are variations in knowledge and capabilities across islands.

Developing education programs and investments in climate resilience will be fundamental elements to guarantee that the island territory is consolidated as an attractive and sustainable tourist destination.

Collaboration between the public sector, the private sector, and civil society is essential to address these challenges effectively and guarantee a prosperous future for the tourism industry in the archipelago.

References

Antonakakis N, Dragouni M, Eeckels B et al (2019) The tourism and economic growth enigma: examining an ambiguous relationship through multiple prisms. J Travel Res 58(1):3–24. https://doi.org/10.1177/0047287517744671

Bárcena A, Samaniego J, Peres W et al (2020) La emergencia del cambio climático en América Latina y el Caribe ¿seguimos esperando la catástrofe o pasamos a la acción? Libros de la CEPAL 160. Comisión Económica para América Latina y el Caribe (CEPAL), Santiago, Chile

Cave J, Dredge D (2020) Regenerative tourism needs diverse economic practices. Tour Geogr 22(3):503–513. https://doi.org/10.1080/14616688.2020.1768434

Cannonier C, Burke MG (2019) The economic growth impact of tourism in small island developing states—evidence from the Caribbean. Tour Econ 25(1):85–108. https://doi.org/10.1177/1354816618792792

Congreso de la República de Colombia (21 December 1957). Sobre puerto libre de San Andrés [Ley 127 de 1959]

Congreso de la República de Colombia (26 July 1996) Por la cual se dictan normas especiales para la organización y el funcionamiento del Departamento Archipiélago de San Andrés, Providencia y Santa Catalina [Ley 47 de 1993]

Congreso de la República de Colombia (26 July 1996) Ley General de Turismo. [Ley 300 de 1996]

CORALINA-INVEMAR (2012) Gómez- López DI, Segura-Quintero C, Sierra-Correa PC, y Garay-Tinoco J (eds) Atlas de la Reserva de Biósfera Seaflower. Archipiélago de San Andrés, Providencia y Santa Catalina. Instituto de Investigaciones Marinas y Costeras "José Benito Vives De Andréis" -INVEMAR- y Corporación para el Desarrollo Sostenible del Archipiélago de San Andrés, Providencia y Santa Catalina -CORALINA-. Serie de Publica- ciones Especiales de INVEMAR, 28, Santa Marta, Colombia

Corona Jiménez MÁ (2018) El conocimiento, la percepción y disponibilidad para afrontar el cambio climático en una población emergente, los migrantes de retorno. Estudios Sociales. Revista de Alimentación Contemporánea y Desarrollo Regional 28(52). https://doi.org/10.24836/es.v28 i52.578

González A, Tonazzini D, Klarwein S (2019) Coherencia Política del Turismo de Costa y el Cambio Climático. Calvià, Mallorca

Guerrero Jiménez T (2020) Crisis del agua, turismo y variabilidad climática en la isla de San Andrés. Turismo y Sociedad xxvi:127–154. https://doi.org/10.18601/01207555.n26.06

Hampton MP, Jeyacheya J (2020) Tourism-dependent small islands, inclusive growth, and the blue economy. One Earth 2(1):8–10. https://doi.org/10.1016/j.oneear.2019.12.017

Hernández R, Mendoza C (2018) Metodología de la investigación. Las rutas cuantitativa, cualitativa y mixta. McGraw Hill, México City

IDEAM, PNUD, MADS et al (2017) Tercera Comunicación Nacional De Colombia a La Convención Marco De Las Naciones Unidas Sobre Cambio Climático (CMNUCC). Tercera Comunicación Nacional de Cambio Climático. Bogotá, Colombia

IPCC (2014) Summary for policymakers. In: Field CB, Barros VR, Dokken DJ et al (eds) Climate change 2014: impacts, adaptation, and vulnerability. Part A: global and sectoral aspects. Contribution of Working Group II to the Fifth Assessment Report of the Intergovernmental Panel on Climate Change. Cambridge University Press, Cambridge, UK and New York, USA, pp 1–32

IPCC (2023) Summary for policymakers. In: Core Writing Team, Lee H, Romero J (eds) Climate change 2023: synthesis report. Contribution of Working Groups I, II and III to the Sixth Assessment Report of the Intergovernmental Panel on Climate Change. IPCC, Geneva, Switzerland, pp 1–34. https://doi.org/10.59327/IPCC/AR6-9789291691647.001

James Cruz JL (2013) El Turismo como estrategia de desarrollo económico: El caso de las islas de San Andrés y Providencia. Cuadernos Del Caribe 10(16):37–56. https://revistas.unal.edu.co/index.php/ccaribe/article/view/43409

Mackay EA, Spencer A (2017) The future of Caribbean tourism: competition and climate change implications. Worldw Hosp Tour 9(1):44–59. https://doi.org/10.1108/WHATT-11-2016-0069

Ministerio de Comercio, Industria y Turismo de Colombia (24 December 2021) Artículo 2.2.4.1.1.13. [Sección 1]. Por el cual se modifica y adiciona el Título 4 de la Parle 2 del Libro 2 del Decreto 1074 de 2015, Decreto Único Reglamentario del Sector Comercio, Industria y Turismo, en relación con el Registro Nacional de Turismo y las obligaciones de los operadores de plataformas electrónicas o digitales de servicios turísticos prestados y/o disfrutados en Colombia [Decreto 1836 de 2021]

Ministerio de Gobierno de Colombia (13 December 1991) Por medio del cual se adoptan medidas para controlar la densidad poblacional en el Departamento Archipiélago de San Andrés, Providencia y Santa Catalina [Decreto 2762 de 1991]

Miranda Vera CE, Ramos Palenzuela M, Alomá Oramas RM et al (2019) Percepción social del cambio climático. Estudio en comunidades costeras de la provincia de Cienfuegos. Revista Cubana de Meteorología 25. http://rcm.insmet.cu/index.php/rcm/article/view/479

Mycoo MA (2018) Beyond 1.5 °C: vulnerabilities and adaptation strategies for Caribbean small island developing states. Reg Environ Change 18:2341–2353. https://doi.org/10.1007/s10113-017-1248-8

National Aeronautics and Space Administration (NASA) (13 January 2022) 2021 Tied for 6th Warmest Year in Continued Trend, NASA Analysis Shows. https://www.nasa.gov/press-release/2021-tied-for-6th-warmest-year-in-continued-trend-nasa-analysis-shows

National Oceanic and Atmospheric Administration (NOAA) (13 January 2022) 2021 was the sixth-warmest year on record for the globe. https://www.ncei.noaa.gov/news/global-climate-202112

Nunn P, Kumar R (2017) Understanding climate-human interactions in small island developing states (SIDS): implications for future livelihood sustainability. Int J Clim Chang Strateg Manag 10(2):245–271. https://doi.org/10.1108/IJCCSM-01-2017-0012

Pandy WR, Rogerson CM (2018) Tourism and climate change: stakeholder perceptions of at risk tourism segments in South Africa. EuroEconomica 37(2):104–118

Peterson RR (2020) Over the Caribbean top: community well-being and over-tourism in small island tourism economies. Int Journal of Com WB 6:89–126. https://doi.org/10.1007/s42413-020-00094-3

Peterson R, DiPietro RB (2021) Is Caribbean tourism in overdrive? Investigating the antecedents and effects of overtourism in sovereign and nonsovereign small island tourism economies (SITEs). Int Hosp Rev 35(1):19–40. https://doi.org/10.1108/IHR-07-2020-0022

Santos-Lacueva R, Anton Clavé S, Saladié Ò (2017) The vulnerability of coastal tourism destinations to climate change: the usefulness of policy analysis. Sustainability 9(11):2062. https://doi.org/10.3390/su9112062

Segrado Pavón RG, González Baca CA, Arroyo Arcos L et al (2017) Capacidad de carga turística y aprovechamiento sustentable de Áreas Naturales Protegidas. CIENCIA Ergo-Sum 24(2):164–172

Sheller M (2021) Mobility justice and the return of tourism after the pandemic. Mondes Du Tourisme 19. https://doi.org/10.4000/tourisme.3463

Thomas A, Baptiste A, Martyr-Koller R et al (2020) Climate change and small island developing states. Annu Rev Environ Resour 45(6):1–6. https://doi.org/10.1146/annurev-environ-012320-083355

United Nations (9 May 1992) United Nations framework convention on climate change. https://unfccc.int/sites/default/files/convention_text_with_annexes_english_for_posting.pdf

World Meteorological Organization (WMO) (2022) State of the climate in Latin America & the Caribbean 2021. WMO 1295, Geneva, Switzerland. https://library.wmo.int/idurl/4/58014

Open Access This chapter is licensed under the terms of the Creative Commons Attribution 4.0 International License (http://creativecommons.org/licenses/by/4.0/), which permits use, sharing, adaptation, distribution and reproduction in any medium or format, as long as you give appropriate credit to the original author(s) and the source, provide a link to the Creative Commons license and indicate if changes were made.

The images or other third party material in this chapter are included in the chapter's Creative Commons license, unless indicated otherwise in a credit line to the material. If material is not included in the chapter's Creative Commons license and your intended use is not permitted by statutory regulation or exceeds the permitted use, you will need to obtain permission directly from the copyright holder.

Climate Change Education and Research

Archeology Expanded—a Multidisciplinary Approach for Natural Disaster Response

Long-Term Vulnerability and Climate Change: Analyzing Three Archeological Sites on the Colombian Caribbean Island of Santa Catalina

Víctor Andrés Pérez Bermúdez and Daniela Vargas Ariza

Abstract This article evaluates the vulnerability index of three elements associated with the historical and fortified heritage of Santa Catalina Island, a volcanic promontory located in the northwest of the Colombian Caribbean. The extremely active 2020 Atlantic hurricane season intensified the loss of valuable heritage assets of the Raizal community. Taking into account their intangible and historical values, the analysis focuses on the impacts of climate change on these elements over time, with the aim of contributing to scientific debate on the awareness and protection of cultural heritage for future generations.

Keywords Climate change · Hurricane · Vulnerability · Defense and fortified systems · Santa Catalina

V. A. P. Bermúdez (✉) · D. V. Ariza
Fundación Apalaanchi, Bogotá, Colombia
e-mail: victorabperezarq@gmail.com; gerencia@fundacionapalaanchi.com

1 Introduction

The island of Santa Catalina is located in the Archipelago of San Andrés, Providencia, and Santa Catalina[1] (hereafter, archipelago), a group of several islands in the northwest of the Colombian Caribbean, around which a 400 km^2 group of atolls and coral banks was created.

In November 2020, the category 4 Hurricane Iota devastated the island. The disaster revealed the high degree of vulnerability to which the island's community is exposed, the lack of response measures in this type of event, and the latent dangers that surround the archipelago. The reconstruction process after the hurricane has been led by a centralized discourse that disregards the local knowledge of the territory.

Currently, this lack of knowledge is reflected in a state narrative where the instrumentalization of the landscape from a commercial logic predominates, overlooking traditional and historical elements that could offer key elements to achieve a transversal view during the reconstruction of the islands. This contrasts with works using multicomponent approaches in different sectors of the Caribbean, in which a series of experiences, decisions, and transformations have left their mark on the landscape and have offered elements to the discussion surrounding the impact of climate change on archeological heritage (Cooper and Peros 2010; Fitzpatrick 2010; Gomez and Jeong 2022; García-Herrera et al. 2007; Hetzinger et al. 2008; Higuera-Gundy et al. 1990; Lul and Liu 2005; McCloskey and Keller 2009; Malaizé et al. 2011; Mora, Miller and Grissino-Mayer 2006; Rivera-Collazo 2019; Rivera-Collazo and Declet-Pérez 2017; Rivera-Collazo et al. 2018; Trouet et al. 2016).

In examining all the different perspectives from which this problem can be faced, this article seeks to discuss, from the study of three archeological sites on Santa Catalina, the relationship between natural disasters and the transformation of the fortified context of the island, in order to understand how human beings have interacted with their landscape and how they have modified the coastal areas. This will, in turn, help grasp how societies have or have not developed adaptation strategies in response to natural hazards and natural dynamics between coasts and ports, as well as to the vulnerabilities to which these societies are exposed.

---

[1] In the sixteenth century, these islands were a single geographic unit known as Santa Catalina. However, during the 1930s, British occupants opened a narrow strip between an isthmus to strengthen the island's defenses against Spanish attacks. Blanco Barros (2009) confirms this fact, yet the archived documentation that describes the island in 1635 mentions that the segregation could already be observed. As time went by, the small island to the north became known as Santa Catalina, while the larger one to the south of the gap was named Providencia.

2 Climate Change, Vulnerability, and Cultural Heritage

The Caribbean islands are considered to be particularly vulnerable to the various effects of climate change. They offer a unique setting due to the multiple responses that past communities used when faced with sudden environmental changes (Cooper 2012). Climate change is not a recent event in the contemporary world. In fact, from the moment humans harnessed fire and all the way to the beginnings of agriculture, humans began to considerably impact their environment. However, it wasn't until the first wave of globalization and the development of the industrial revolution that the planet started manifesting with more frequency a series of phenomena that would impact life on Earth.

Archeology, as the discipline that focuses on the study of the historical development of human groups from material remains, has developed approaches to account for the environmental impact of past societies, studying how they applied pressure on different natural resources (Butzer 2012, 2009; Hudson et al. 2012; Van de Noort 2015). The role humans play in changing the landscape and contributing to climate change has been examined since the publication of Childe's landmark work *The History of Civilization* in 1936.

Since the 1980s, with the latest scientific advances in new techniques to tackle studies of the past, the study of archeological heritage in the face of climate change has been gaining strength[2] since this phenomenon exposes societies, economic sectors, and ecosystems to risks.[3]

If we assume that heritage is a non-renewable asset that, due to its properties and particularities, presents a high degree of exposure to exogenous agents, ranging from brief phenomena such as violent storms to chronic events such as droughts over several decades or the rise in sea level over several centuries, the vulnerability[4] to which an archeological site is exposed will be influenced not only by the same environmental events, but also by multi-scale social, economic, historical, and political factors (Thomas et al. 2019).

Identifying these environmental and human factors that influence or have influenced the sovereignty of a site enables the understanding of *differential vulnerability*, a function of exposure, sensitivity, and adaptive capacities (Thomas et al. 2019, p. 2), and, in this sense, heritage, as a product of this decision-making process, is susceptible to being instrumentalized to analyze possible responses to dynamic and spontaneous phenomena, such as hurricanes.

---

[2] Climate change is the variation of average weather conditions due to natural processes or external factors, such as 'modulation of solar cycles, volcanic eruptions, or persistent anthropogenic changes of the composition of the atmosphere or land use' (IPCC 2014, p. 129).

[3] We understand risk as a potential danger with an uncertain outcome; it is the probability of a dangerous event or trend. Risks result from the 'interaction of vulnerability, exposition, and danger' (IPCC 2014, p. 195).

[4] We understand vulnerability as the propensity of a sensitive system to be negatively affected by an external agent, without the ability to generate an adequate response (Oppenheimer et al. 2014; Rhiney 2015; IPCC 2014).

3 The Sites

Archeological studies on the islands of Providencia and Santa Catalina have been carried out since 2013 and, to this day, there are only eight studies, six of which come from preventive archeology programs (Fajardo and Rodríguez 2017; Londoño 2014; Pérez et al. 2020; Romero 2013, 2016), while the remaining two come from academic research (Mayfield et al. 2019; Moreno and Baez 2021). These investigations identified seven archeological sites, particularly in the 2019–2021 field season (Pérez et al. 2020).

This article seeks to evaluate the vulnerability index from the case study of three archeological sites on Santa Catalina, one of the areas of the archipelago that was most affected by Hurricane Iota and the one with the highest archeological potential. For this purpose, we define an archeological site as a place or space in which evidence of past activity is preserved (Renfrew and Bahn 1991). Understanding the possible risks at the site level is vital to ensure that appropriate adaptation measures are put in place. For this reason, a vulnerability perspective is an advantage because it focuses not only on an evaluation of the site in terms of exposure and sensitivity to hazards, but also on the ability of a site to adapt and recover (Daly 2014, p. 269).

All of these structures were part of the fortification system that was constructed around the seventeenth century. Only one is registered as a *Bien de Interes Cultural*[5] that is, an asset of cultural interest. The three structures are currently known as *Fuerte Warwick*, *Playa del Fuerte*, and *Las Ruinas*, and we decided to preserve the labels with which the sites were registered in the Colombian Institute of Anthropology and History (ICANH, with its Spanish acronym) (see Table 1 and Fig. 1) (Fajardo and Rodríguez 2017). Accordingly, *Fuerte Warwick* is PRV005. The structure is located on a volcanic rock promontory, occupying an area of 7,000 m^2, where it is protected from the northwest and northern winds. Since its construction by the British in around 1634, this site has suffered transformations related to the landscape adaptation caused by the clearing of tropical dry forest, soil leveling, and ditch construction, as well as its toponymy. After the island was recovered by Spain towards the middle of the seventeenth century, the island was renamed *Fuerte de la Cortadura de San Jerónimo*, while *Fuerte la Libertad* was built over its foundation in 1822.

In regard to the second site, PRV013, between the 1630s and 40s, a small battery, called the *Plataforma San Mateo*, was found on an artificially elevated location on a beach, with an area of 1400 m^2. This coastal deposit of unconsolidated fine sand is made up of small and rounded fragments of corals and lithic volcanic rocks (Álvarez-Gutiérrez et al. 2014). Even before Hurricane Iota, November 16, 2020, a natural well was found that would've served as a watering point. After the natural phenomenon,

---

[5] *Bien de Interés Cultural* (BIC), or 'Asset of Cultural Interest' is a category of the heritage register in Colombia. It includes tangible and intangible cultural heritage. It could have national, regional or community recognition. Some BICs in Colombia enjoy international protection as World Heritage Sites. *Fuerte Warwick* was declared a BIC in national law through *Resolución 788 del 31 de julio de 1998 del Ministerio de Cultura* (Resolution 788 of 31 July 1998 of the Ministry of Culture).

Table 1 The different names over time of the sites studied

Sixteenth to eighteenth centuries	Nineteenth century	Twentieth century	ICANH record
Fuerte de la Cortadura de San Jerónimo	Fuerte La Libertad	Fuerte Warwick	PRV005
Plataforma San Mateo	–	Playa del Fuerte	PRV013
Plataforma Santa Cruz	–	Las Ruinas	PRV011

Source Archivo General de Indias, MP-PANAMA, 77 MP-PANAMA, 78 and SANTA FE, 223

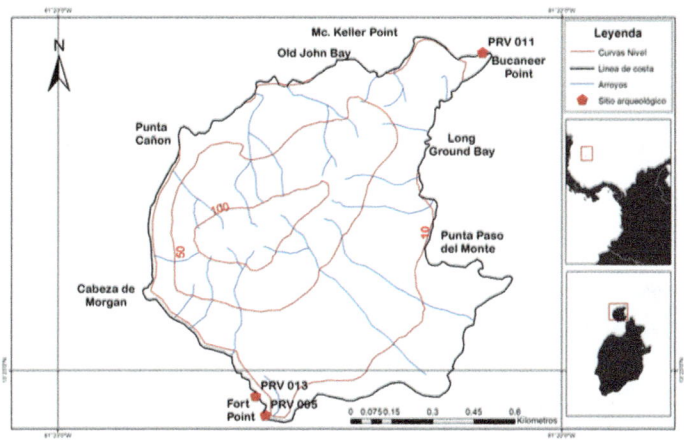

Fig. 1 Location of the sites studied. *Source* Archivo General de Indias, MP-PANAMA, 77 MP-PANAMA, 78 and SANTA FE, 223

the coverage of tropical dry forest disappeared, and the landscape resources were buried beneath a 50 cm layer of pebbles and sandstone.

Finally, the third one (PRV011) was found in the northwestern end of the island, which was known as *Plataforma Santa Cruz* in the seventeenth century, and similarly to the previous element, didn't change its name, although its remains are currently referred to as *Las Ruinas*. This element was placed on a slag foundation occupying an area of 297 m^2. This volcanic material usually has a pink coloring (Álvarez-Gutiérrez et al. 2014), but appears black when exposed to a high degree of weathering, as was our case here.

The three sites are characterized by having been military constructions; PRV005 and PRV011 are located on a slag promontory, while PRV013 is located on a small beach, and PRV005 and PRV011 are on elevations not greater than 10 m in height, while PRV013 is just above sea level. All of them sought to ensure access to the marine channel that led to the different anchorages of the port. Platforms were built from brick and volcanic rock, which differed from those fortifications of the Spanish defense system in the Caribbean. Each unit is composed of ceramic, cannons, glass, and construction materials such as tiles and brick.

The sites are located on a deposit in an abrasion area, which implies that they are exposed to erosive processes due to the swell that causes the detachment and sliding of blocks and the formation of wide scarps and necks in the lower areas of scarps and crevices. These erosive processes contribute to the disappearance of the three sites since they represent a constant rate of land loss (Table 2).

Table 2 PRV005, PRV013, and PRV011 before (column A) and after (column B) Hurricane Iota

4 Materials and Methods

Considering all the above, a multi-factorial approach is needed to evaluate the degree of response of the populations, in the past and present, when facing potential risks from climate change. This can be done by evaluating the Climate Change Vulnerability Index (CCVI), which measures a society's exposure, sensitivity, resilience, and ability to adapt to the impact of climate change (CAF 2014, p. 5). This index is obtained from a formula that was first introduced by the Development Bank of Latin America, in which vulnerability equals the sum of the sensitivity and exposure indices[6] minus the adaptive capacity index:

> The indices are presented in a scale of 0–10, where values closer to 0 represent a greater risk, while values closer to 10 represent a lower risk. To support this interpretation, the indices are divided in four categories: extreme risk (0–2.5), high risk (>2.5–5), moderate risk (>5–7.5), and low risk (>7.5–10) (CAF 2014, p. 5).

In order to present a vulnerability framework adapted to the three archeological sites it was necessary to get to know the three sites over time to know which variables to use for the three categories involved: exposure, sensitivity, and adaptive capacity. To gain an understanding of these aspects, different factors were evaluated using a deep and multiscale temporal perspective to understand the socio-natural characteristics of each site (Rivera-Collazo et al. 2015, p. 2) and the changes that occurred after the 2020 event. The idea of using a multiscale temporal approach implies data categorization into a defined chronology, specifically a major and a minor scale of analysis.

With the use of qualitative primary data from historical sources, three chronological periods that involve the past and the present of the structures were defined. The first period was between the sixteenth and nineteenth centuries, the second was between the twentieth and twenty-first centuries until Hurricane Iota, and the third came after the hurricane. Since we were on the island before, during, and after this natural phenomenon we were able to compare the vulnerability index of each landscape resource, allowing us to contrast the different landscape transformations and observe to what extent responses were created or not, specifically to such an atypical high-energy event that weakened the three archeological sites, increasing the risk of loss and destruction of these memories from the past.

5 Results

The identification of the criteria mentioned above was established from the review of archival documentation housed in different Spanish and English repositories. Santa Catalina and Providencia have a good documentation arsenal composed of different

---

[6] The exposure index evaluates the risk of a site being impacted by extreme related events, while the sensitivity index examines the current relative sensitivity of exposure to extreme related events. Finally, the adaptive capacity evaluates the ability of the site to adjust to climate change.

types of archives, among which the files from Sevilla's *Archivo General de Indias (AGI)* stand out, as well as those from London's National Archives, which were transcribed by Newton (1985) and Kupperman (1993). These archives were the type of information used for the analysis in the major chronological scale because they offer a panoramic view of the relationship between humans and the different elements within the observed landscape, which allowed the identification of a series of actions for the care and maintenance of these particular archeological sites. The islands represented a problem for Spain since the seventeenth century as they were a place from which it was possible to attack the Silver Route between Cartagena and Portobelo; there are innumerable actions and reports from governors outlining a series of guidelines sent from Spain to try to maintain possession of the islands.

For the two minor chronological scales, the perception was quantified by implementing repeated field observations and interviews with stakeholders. The field observations aimed to collect data from the three sites' locations and to contrast their observations before and after the hurricane. Site visits develop a first-hand understanding of the relationship between the heritage values and the surrounding environment, such as topography, patterns, and land use (Daly 2014, p. 273). As a result, different elements and the state of deterioration were visually recorded and georeferenced. Additionally, context descriptions of the sites were made. Regarding the interviews with stakeholders, semi-structured interviews were undertaken with local residents, visitors, and researchers. The interviewees gave important insights about the way in which they perceive the three sites, and how climate has impacted heritage.

5.1 Sensitivity

Based on these results, to establish the Vulnerability Index, we first obtained the Sensitivity Index from the quantification of sensitivity factors, such as biophysical and ecological factors (vegetation, soil, animals, sea), visual factors (size, slope, shape, height), and anthropic factors (human presence and accessibility). The last one also covered some historical aspects like war, and since the structures were part of a defense system, their sensitivity is measured by the stage of conservation of their most significant elements: cannons. Each of these variables were carefully selected and evaluated as follows: 0 (null), 1 (low), 2 (medium), and 3 (high) (Table 3).

5.2 Exposure

To establish the Exposure Index of each of our resources, it is key to understand how the archeological site and its values are exposed to climatic variations. In this case, we considered the most vulnerable factors, such as population, real estate, movable heritage, and tourism. These were chosen by combining the information gathered

Table 3 Sensitivity factor assessment

Site		Sensitivity factors													
		Biophysical and ecological factors				Visual factors				Anthropic factors					
		Plants	Soil	Air	Animals	Sea	Size	Slope	Form	Height	Human presence	War elements	Buildings	Residential structure	Accessibility
Major scale (16th to 19th century)	PRV005	2	1	1	2	0	3	2	1	2	1	1	2	2	1
	PRV013	2	1	1	2	2	2	0	0	3	1	1	2	2	1
	PRV011	2	1	3	2	2	3	2	1	1	1	1	2	2	1
Site															
Minor scale I (20th century–before hurricane Iota)	PRV005	1	2	1	1	2	1	2	1	2	2	2	2	2	2
	PRV013	1	2	1	1	3	1	1	3	3	2	2	2	2	2
	PRV011	1	3	3	1	2	1	2	0	1	3	2	2	2	1
Site															
Minor scale II (20th century–after hurricane Iota)	PRV005	3	3	2	3	2	3	3	3	2	3	1	3	3	3
	PRV013	0	3	3	2	3	1	1	2	3	3	0	0	0	3
	PRV011	3	3	3	3	3	2	3	3	2	3	1	3	3	3

Table 4 Exposure index evaluation

Site		Climate exposure				
		Anthropic factors				
		Population	Immovable heritage	Movable heritage	Tourism (economic attractions)	Sensitivity + exposure
Major scale (16th to 19th century)	PRV005	1	1	1	0	1.3
	PRV013	1	1	0	0	1.2
	PRV011	1	0	1	0	1.4
Site						
Minor scale I (20th century–before hurricane Iota)	PRV005	0	2	1	3	1.6
	PRV013	0	1	1	3	1.7
	PRV011	0	2	1	1	1.6
Site						
Minor scale II (20th century–after hurricane Iota)	PRV005	0	1	0	1	2.2
	PRV013	0	0	0	1	1.4
	PRV011	3	3	3	3	2.8

during the visits and the degree to which the anthropic factors were affecting the context of the sites (Table 4).

5.3 Adaptive Capacity

Regarding the assessment of the Adaptive Capacity Index, we adopted natural, anthropic, and governance factors that would allow the greatest response capacity to climate challenges (Villarreal 2019, p. 464) (Table 5).

Once each index was obtained, we calculated the Vulnerability Index by applying the formula, and we obtained a value for each of the studied resources (Table 6).

6 Discussion

The estimation of the Vulnerability Index provides a useful reference for developing parameters and indicators for future scenarios that must be kept under constant review. In order to understand the results, it is important to compare these values with the historical record of each site and how the Vulnerability Index has changed over time.

Table 5 Adaptive capacity index evaluation

Site		Adaptive capacity					Sum total
		Anthropic factors					Adaptive capacity
		Arborization	Ecological connectivity	Protection and conservation	Formal construction	Governance	Adaptive capacity
Major scale (16th to 19th century)	PRV005	3	2	3	3	3	2.8
	PRV013	3	2	3	3	3	2.8
	PRV011	3	2	3	3	3	2.8
Site							
Minor scale I (20th century–before hurricane Iota)	PRV005	2	2	1	1	1	1.4
	PRV013	2	2	1	1	1	1.4
	PRV011	2	2	1	1	1	1.4
Site							
Minor scale II (20th century–after hurricane Iota)	PRV005	1	0	0	0	1	0.4
	PRV013	1	0	0	0	1	0.4
	PRV011	1	0	0	0	1	0.4

Table 6 Vulnerability index estimation

Site		Vulnerability index
Major scale (16th to 19th century)	Prv005	–1
	Prv013	–2
	Prv011	–1
Site		
Minor scale I (20th century–before hurricane iota)	Prv005	0.2
	Prv013	0.3
	Prv011	0.2
Site		
Minor scale II (20th century–after hurricane iota)	Prv005	1.8
	Prv013	1.0
	Prv011	2.4

The Vulnerability Index shows that in both the present and the past, PRV005, PRV013, and PRV011, have been at great risk due to their exposure to natural phenomena. The main difference between them is their adaptive capacity because, in the major chronological scale, the sites had greater governance due to the importance of the island within the geopolitical situation of the Caribbean in the seventeenth century.

6.1 Major Scale (Sixteenth and Seventeenth Centuries)

In the past, PRV005, PRV013, and PRV011 were a part of the fortified landscape of the island of Santa Catalina. At some point between 1632 and 1634, the *Company of Adventurers of Westminster City for the Plantation of the Islands of Providencia or Catalina, Henrietta or Andrea, and the Lands Adjacent to the Coasts of America*, began the fortification of the island. These fortifications constituted platforms with artillery, with some small sashes or terraced ditches with three or four artillery pieces spread around the island; this became a type of fortified defense system at the entrance of the port (Pérez et al. 2021, p. 25).

In 1620, Santa Catalina was already known for its port potential and for the need to include it in the Spanish fortification system of the Caribbean. The text *Descubrimiento de la isla Santa Catalina*, written by the North Sea pilot Simón Zacarías, a possible pirate with flamenco origins, describes the island as a promontory near Portobelo that presents on the windward side, between a possible shallow or atoll, an ideal cove that could serve as a port. In contrast, the natural characteristics of the leeward side made safe navigation impossible due to the high presence of shoals and reefs. In the end, Zacarías concluded that the ideal place to fortify this port would be at its entrance since the leeward side could act as a natural barrier.

In 1634, the engineer Juan Somovilla Tejada made a series of observations on the spatial organization of their defense and the qualities of their ports. The dominance of the landscape by the British reflected their knowledge and strategic usage of the natural and visual resources offered by the island, which is why the engineer did not hesitate to mention how the different defense elements were found in places that defended the entry into the areas where it was possible to disembark, where they could communicate visually with each other, and where the thickness of the mountain played in favor of the settlers by preventing easy access to the island.

In 1819, the majority of the defense elements that were rebuilt by the Spanish were abandoned. In certain cases, only their foundations remained. Since Santa Catalina is a mountainous island covered by jungle, it was necessary to reestablish old communication routes, which is why Agustín Codazzi insisted on manufacturing a series of elements from the ruins of the Spanish system. He was sure that the characteristics of the port offered the best outpost for espionage during the Independence period (Codazzi 1973, p. 334). Around this time, there was news of the first recorded hurricane, because it destroyed all properties. The engineer mentioned that the unfinished fortifications were ruined and that hunger, disease, humidity, and lack of care caused a good part of the population to succumb (Codazzi 1973, p. 336). A year later, new settlements were built using prefabricated houses brought from the USA, some fortifications—such as *Fuerte de La Libertad*—were built, and the islands reestablished communication with the Caribbean.

It is interesting to note that after the Spanish settlers abandoned the island most elements of the defense system were in disuse but, due to their strategic location, the most important ones were rebuilt, even after the hurricane, because maintaining military strategy was more important. The lack of homes in the areas where we work shows that these areas were never intended for residential use, even if there was a small encampment in one of them.

Even if the different elements could present a high Vulnerability Index due to their exposure, it was necessary for each of them to have constant maintenance by creating ditches, building wells to save fresh water, leveling the soil, and controlling pests: in this way, the greatest number of risks that could be controlled by the military contingent was reduced, leaving the island exposed only to natural phenomena such as the hurricane of the nineteenth century.

6.2 Minor Scales (Twentieth to Twenty-First Centuries, Before and After the Hurricane)

The data obtained show a high score in the Vulnerability Index in the contemporary era. The sensitivity variables have increased noticeably. In 2014, the archeologist Wilhelm Londoño recorded the context of PRV005 and described a high exposure to volcanic rock, specifically over its maritime façade (Londoño 2014).

Since *Fuerte Warwick* (PRV005) was declared a *Bien de Interes Cultural*, a series of tourist service adaptations have been established, such as roads and displays of various construction materials corresponding to the fortification. Even though PRV005 is the most visited site for tourism, it has a high rate of degradation. During this major scale period, the element was held within a fortified landscape and expected to participate in the port's defense, meaning it was necessary to keep the area free of weeds and animals. Constant abandonment has been observed from the middle of the nineteenth century until today: on one hand, vegetation invades these structures and, on the other, the lack of maintenance leads to their collapse, as in the case of PRV005.

After Hurricane Iota, PRV005 showed drastic changes, the first of them being the total loss of its coverage, when some materials from the military construction broke apart and scattered around the site. The hurricane has made evident, even today, the lack of a national policy of prevention against disasters in the sites that house cultural heritage, which resulted in the loss and destruction of a good part of the existing cultural heritage on the islands. In the case of PRV005, PRV011 and the remains scattered along the coastline of Santa Catalina of the constituent elements of the old battery that can hardly be reconstructed. The case of PRV011 is worse than PRV005, since the lack of visible structures made it uninteresting to the local government in their efforts to integrate it into a government prevention or promotion plan. This led to its disappearance and it was, in turn, exacerbated by Hurricane Iota.

On the other hand, PRV013 is a touristic beach with well-known elements that could have been part of *Plataforma San Mateo*, such as Morgan's Cave, the old well, and an old cannon. Even though these elements present tourist signage, this site does not have a protection plan and, for many years, the legend of treasures hidden in the cave has attracted treasure hunters. After Hurricane Iota, this beach was covered by a layer of pebbles 50 cm thick, which completely buried the well and cannon. This phenomenon also dispersed cultural materials across the beach, such as traditional European ceramics, construction materials, and the fragmented and disjointed remains of a shipwreck. Although Iota allowed the discovery of new archaeological sites, the high vulnerability of the island's cultural heritage condemns them to disappear due to the low adaptive capacity of these elements.

Even though Law 2134 of 2021 was issued after Hurricane Iota, recognizing the material and intangible cultural heritage of the Raizal people, and proposing studies to generate corresponding declarations, the truth is that today, there has not been a single recognition or technical visit by the state agencies in charge of protecting cultural heritage. As Wilhelm Londoño highlighted in 2015, the abandonment and degradation of the different archeological sites can still be observed.

Finally, PRV011 is very remote from the city and other sites. It is currently only possible to reach the site by sea since it is located at the northwestern end of Santa Catalina. However, PRV011 is not a cultural or heritage reference of the island, many people do not consider this element to be a part of their collective memory. Unlike PRV005, this site does not have a high presence of stubble, presenting only a patch of dry tropical forest behind it. The base, on which *Plataforma Santa Cruz* was founded, was made of volcanic material that turned black over time due to weathering, which led to constant erosion and conglomerate loss when it began yielding to the swell. A small portion of it was found at sea before Hurricane Iota, and after the hurricane, 50% of the platform fell into the sea while the other covered the old beach.

7 Conclusions

The archeological study of climate change offers analytical tools that allow us to note changes in the landscape, as well as actions that have been taken to face natural disasters. Considering the environmental vulnerability of the Caribbean islands, a multidisciplinary approach can offer insights into the way culture interacts with climate risks, as well as tools for designing public policy.

The islands of Providencia and Santa Catalina are part of what is known as Forgotten Islands, Useless Islands, or Hidden Ports (Bassi 2021; Schwartz 2015). The political imbalance of these islands, in addition to the lack of interest in the Spanish economic sector, led its occupants to keep these centers fortified and ready for defense. Military strategy was more important than their high vulnerability, which maintained the three elements we analyzed until the beginning of the period of independence from the Spanish. The arrival of the Europeans drastically transformed, both culturally and ecologically, the numerous islands and archipelagos of the world (Braje et al. 2017).

The landscape resources that were studied here, with a series of patrimonial, economic, and political values, present a constant fragility that has been increasing over time. These non-renewable assets face risks derived from climate change because the Seaflower Biosphere Reserve has been presenting warmer waters which have contributed to high-energy events like Hurricane Iota, creating coastal erosion that has caused elements associated with military constructions from the seventeenth century to be found today at sea or spread out along these new coastlines.

The islands' archeological heritage has exemplified how a natural phenomenon can destroy a source of information regarding the islands' past. The most critical area, from the elements studied, is the northwestern side of the islands, toward Punta Bucanera where PRV011 is located. Here we observed how coastal erosion, and later Hurricane Iota, dismantled the different components of this archeological site, practically destroying it in its entirety. Total destruction was also found in PRV013, since Hurricane Iota not only changed the coastline but also destroyed the water supply and buried the cannon that was previously found on the beach. PRV011, despite presenting greater integration of the previous elements, presents a negative outlook due to constant neglect.

All three elements could disappear both materially and from the Raizal population's collective memory, due to the circumstances previously stated, if the protection policies necessary to face the present and the future of a Caribbean clearly affected by climate change remain unmanaged.

References

Primary Sources

Archive
Archivo General de Indias (AGI)
Panamá, MP, 77 & 78
Indiferente, I, 1528, N.19
Santa Fe, SF, 223

Primary Sources (Print)

Codazzi A (1973) Memorias de Agustín Codazzi. Banco de la República, Bogotá

Secondary Sources

Álvarez-Gutiérrez Y, Amaya-López C, Barbosa-Mejía LN et al (2014) Descripción e Interpretación Geológica de Las Islas de Providencia y Santa Catalina. Boletín De Ciencias De La Tierra 35:67–81

Bassi E (2021) Un territorio acuoso. Universidad del Norte, Barranquilla, Colombia, Geografías marineras y el Gran Caribe transimperial de la Nueva Granada

Blanco Barros J (2009) El Archipiélago de San Andrés y Providencia. Boletín De Historia y Antigüedades 96(845):369–388

Braje T, Dillehay T, Erlandson J et al (2017) Finding the first Americans: the first humans to reach the Americas are likely to have come via a coastal route. Science 358:592–594. https://doi.org/10.1126/science.aao5473

Butzer KW (2011) Geoarchaeology, climate change, sustainability: a Mediterranean perspective. In: Brown AG, Basell LS, Butzer KW (eds) Geoarchaeology, climate change, and sustainability: geological society of America special paper, vol 476, pp 1–14. https://doi.org/10.1130/2011.2476(01)

Butzer K (2012) Collapse, environment, and society. PNAS 109(10):3632–3639. https://doi.org/10.1073/pnas.1114845109

Cooper J (2012) Fail to prepare, then prepare to fail: rethinking threat, vulnerability, and mitigation in the precolumbian Caribbean. In: Cooper J, Sheets P (eds) Surviving sudden environmental change: understanding hazards, mitigating impacts, avoiding disasters. University Press of Colorado, Boulder, USA, pp 91–116

Cooper J, Peros M (2010) The archaeology of climate change in the Caribbean. J Archaeol Sci 37:1226–1232. https://doi.org/10.1016/j.jas.2009.12.022

Corporación Andina de Fomento (CAF) (2014) Índice de vulnerabilidad y adaptación al cambio climático en la región de América Latina y el Caribe. Banco de Desarrollo de América Latina, Argentina

Daly C (2014) A framework for assessing the vulnerability of archeology sites to climate change: theory, development, and application. Conserv Manag Archeol Sites 16:262–282. https://doi.org/10.1179/1350503315Z.00000000086

Fajardo S, Rodríguez AM (2017) Estudios de impacto ambiental para las obras de protección del litoral de la bahía de black sand en la Isla de Providencia y para la recuperación de las playas north end (spratt bight) y los estudios y diseños estructurales para la construcción de un mirador turístico en el sitio donde se localiza el espigón tiuna, Isla de San Andrés. Informe final y planes de manejo arqueológico. FINDETER, Bogotá, Colombia

Fitzpatrick S (2010) On the shoals of giants: natural catastrophes and the overall destruction of the Caribbean's archaeological record. J Coast Conserv 16:173–186. https://doi.org/10.1007/s11852-010-0109-0

García-Herrera R, Gimeno L, Ribera P et al (2007) Identification of Caribbean basin hurricanes from Spanish documentary sources. Clim Change 83:55–85. https://doi.org/10.1007/s10584-006-9124-4

Gomez Pretel W, Jeong MS (2022) Shipwrecks on Roncador Cay, the Caribbean Sea and their relationship with hurricanes, 1492-1920. Int J Hist Archaeol 26(2):498–528.https://doi.org/10.1007/s10761-021-00612-9

Hetzinger S, Pfeiffer M, Dullo W et al (2008) Caribbean coral tracks Atlantic multidecadal oscillation and past hurricane activity. Geology 36(1):11–14. https://doi.org/10.1130/G24321A.1

Higuera-Gundy A, Brenner M, Hodell A et al (1990) A 10,300 14C yr record of climate and vegetation change from Haiti. Quat Res 52:159–170. https://doi.org/10.1006/qres.1999.2062

Hudson MJ, Aoyama M, Hoover KC et al (2012) Prospects and challenges for an archaeology of global climate change. Wiley Interdiscip Rev Clim Change 3(4):313–328. https://doi.org/10.1002/wcc.174

IPCC (2014) Cambio climático 2014 Impactos, adaptación y vulnerabilidad. Contribución del Grupo de trabajo II al Quinto Informe de Evaluación del Grupo Intergubernamental de Expertos sobre el Cambio Climático. World Meteorological Organization, Geneva, Switzerland

Kupperman K (1993) Providence Island, 1630–1641. The Other Puritan Colony. Cambridge University Press, Cambridge, UK. https://doi.org/10.1017/CBO9780511583834

Londoño Díaz W (2014) Informe del Programa de Arqueología Preventiva del proyecto de reconstrucción del fuerte de La Libertad. Santa Catalina, municipio de Providencia, Departamento de San Andrés, Colombia. Augusto Rico S.A.S., Bogotá, Colombia

Lul H-Y, Liu K-B (2005) Phytolith assemblages as indicators of coastal environmental changes and hurricane overwash deposition. Holocene 15(7):965–972. https://doi.org/10.1191/0959683605hl870ra

Malaizé B, Bertran P, Carbonel P et al (2011) Hurricanes and climate in the Caribbean during the past 3700 years BP. Holocene 21(6):1–14. https://doi.org/10.1177/0959683611400198

Mayfield T et al (2019) Reportaje sobre la temporada arqueológica 2019 Islas de Providencia y Santa Catalina, Colombia. Proyecto Arqueológico de las islas de Providencia y Santa Catalina / Old Providence and Santa Catalina Islands Archeological Project, OPSCIAP

McCloskey TA, Keller G (2009) 5000 year sedimentary record of hurricane strikes on the central coast of Belize. Quatern Int: J https://doi.org/10.1016/j.quaint.2008.03.003 Int Union for Quatern Res 195(1–2):53–68.

Mora CI, Miller DL, Grissino-Mayer HD (2006) Tempest in a tree ring: paleotempestology and the record of past hurricanes. Sedim Record 4(3):4–8. https://doi.org/10.2110/sedred.2006.3.4

Moreno Calderón M, Báez Santos LV (2021) Expedición Seaflower: el paisaje cultural marítimo de Providencia y Santa Catalina. Resultados preliminares. Bol Cient CIOH 40(1):83–90. https://doi.org/10.26640/22159045.2021.566

Newton AP (1985) Las actividades colonizadoras de los puritanos ingleses en la isla de Providencia. Banco de la República: Bogotá, Colombia

Oppenheimer M, Campos M, Warren R et al (2014) Emergent risks and key vulnerabilities. In: Field CB, Barros VR, Dokken DJ et al (eds) Climate change 2014: impacts, adaptation and vulnerability. Part A: global and sectoral aspects. Contribution of working group II to the fifth assessment report of the intergovernmental panel on climate change. Cambridge University Press, Cambridge, UK and New York, NY, USA

Pérez V, Gabaldón M, Vargas D, Chía A, Peña W (2020) Informe de avance en la ejecución del Plan de Manejo Arqueológico para el Programa de Arqueología Preventiva para el proyecto de dragado y profundización del Canal de Acceso al puerto de Providencia. Consorcio Dragado de Providencia-INVIAS

Renfrew C, Bahn P (1991) Archeology: theories, methods, and practice. Thames & Hudson, London, UK

Rhiney K (2015) Geographies of Caribbean vulnerability in a changing climate: issues and trends. Geogr Compass 9(3):97–114. https://doi.org/10.1111/gec3.12199

Rivera-Collazo IC (2019) Severe weather and the reliability of desk-based vulnerability assessments: the impact of hurricane Maria to Puerto Rico's coastal archaeology. J Isl Coast Archaeol 15(2):1–20. https://doi.org/10.1080/15564894.2019.1570987

Rivera-Collazo IC (2020) Severe weather and the reliability of desk-based vulnerability assessments: The impact of hurricane Maria to Puerto Rico's coastal archaeology. J I Coast Archaeol 15(2):244–263. https://doi.org/10.1080/15564894.2019.1570987

Rivera-Collazo IC, Declet-Pérez M (2017) Contribuciones de la arqueología a la mitigación de riesgos ante el cambio climático: lecciones recuperadas de Tibes y de Los Bateyes de Viví. Puerto Rico. Cuba Arqueológica 10(2):5–15

Rivera-Collazo IC, Rodríguez-Franco C, Garay-Vázquez J (2018) A deep-time socioecosystem framework to understand social vulnerability on a tropical island. Environ Archaeol 23(1):97–108. https://doi.org/10.1080/14614103.2017.1342397

Rivera-Collazo IC, Winter A, Scholz D et al (2015) Human adaptation strategies to abrupt climate change in Puerto Rico ca. 3.5 ka. Holocene 25(4):627–640. https://doi.org/10.1177/0959683614565951

Romero Pico Y (2013) Reconocimiento de arqueología submarina no intrusiva para la profundización de los canales navegables de acceso a las islas de San Andrés y Providencia. Programa de arqueología preventiva y Plan de Manejo. INVIAS, Bogotá

Romero Pico Y (2016) Programa de arqueología preventiva, fases de reconocimiento submarino no intrusivo y propuesta de manejo para los estudios y diseños de la construcción del área de boyaje y su zona de recepción, municipio de Providencia Archipiélago San Andrés, Providencia y Santa Catalina. INVIAS, Bogotá

Schwartz S (2015) Sea of storms. A history of Hurricanes in the Greater Caribbean from Columbus to Katrina. Princeton University Press, Princeton, NJ, USA

Thomas K, Hardy RD, Lazrus H et al (2019) Explaining differential vulnerability to climate change: a social science review. Wiley Interdiscip Rev Clim Change 10(2):e565. https://doi.org/10.1002/wcc.565

Trouet V, Harley GL, Domínguez-Delmás M (2016) Shipwreck rates reveal Caribbean tropical cyclone response to past radiative forcing. PNAS 113(12):3169–3174. https://doi.org/10.1073/pnas.1519566113

Van de Noort R (2015) Conceptualising climate change archaeology. Antiquity 85(329):1039–1048. https://doi.org/10.1017/S0003598X00068472

Villarreal H (2019) Estrategias de paisaje para la adaptación al cambio climático. Caso Cartagena de Indias. Doctoral Thesis, Universidad de Granada, Spain

Open Access This chapter is licensed under the terms of the Creative Commons Attribution 4.0 International License (http://creativecommons.org/licenses/by/4.0/), which permits use, sharing, adaptation, distribution and reproduction in any medium or format, as long as you give appropriate credit to the original author(s) and the source, provide a link to the Creative Commons license and indicate if changes were made.

The images or other third party material in this chapter are included in the chapter's Creative Commons license, unless indicated otherwise in a credit line to the material. If material is not included in the chapter's Creative Commons license and your intended use is not permitted by statutory regulation or exceeds the permitted use, you will need to obtain permission directly from the copyright holder.

Taking Seaflower to the Classroom: A Proposal to Bring Sustainability Education to High Schools in an Oceanic Archipelago (Western Caribbean, Colombia)

Juan F. Blanco-Libreros, Sara R. López-Rodríguez, Jairo Lasso-Zapata, Beatriz Méndez, Nairo De Armas, and Margareth Mitchell-Bent

> "The concern about progress has no other solution: education"
> Enrique Cortés, Radical Liberal Party Leader (1876)

Abstract Geographic understanding is an important ability to be developed in learning communities in islands, since, as previously established, they have unique interactions with natural and human systems. In the case of the Archipelago of San Andrés, Providencia, and Santa Catalina, the curriculum standards, guidelines, and textbooks are designed mostly considering mainland learning communities of white/mestizo ancestry and overlooking the different backgrounds and needs of ethnic minority students, resulting in a total absence of a culturally responsive approach. In this chapter, we propose the geo-literacy umbrella as a tool to improve teacher training and institutional capabilities, previously proposed by academics as points of intervention to close the educational breach by 2030. The seascapes and landscapes protected by the Seaflower Biosphere Reserve offer a rich spatial template for context-based teaching and learning, particularly in local high schools where Geo-literacy might be also useful for introducing general knowledge about sustainability in islands. The General Law for Education in Colombia, provides guidelines that could be used by teachers as anchor points to integrate concepts and methods about sustainability in islands that challenge students to think about real-world problems.

J. F. Blanco-Libreros (✉) · S. R. López-Rodríguez
Faculty of Exact and Natural Sciences, Institute of Biology, Universidad de Antioquia, Medellín, Colombia
e-mail: juan.blanco@udea.edu.co

J. Lasso-Zapata · B. Méndez · N. De Armas · M. Mitchell-Bent
Institución Educativa Antonia Santos, Phillip Beekman Livingston Campus, Archipelago of San Andrés, Providencia and Santa Catalina, Colombia

Keywords Geo-literacy · Archipelago of San Andrés · Providencia and Santa Catalina · Context-based education · Seaflower biosphere reserve

1 Introduction

Taking the outside world and global sustainability issues into the classroom is a major challenge in high school (secondary or middle) education, particularly in the natural sciences (Church and Skelton 2010; McGee et al. 2018; Zoller 2012), given the pressing need for the involvement and further mobilization of young people in discussions on the global climate crisis (Bentz and O'Brien 2019; Sanson and Bellemo 2021; Han and Ahn 2020). This is particularly critical in small insular states and territories where high schools are in a position of double isolation caused by the sum of geographic distance and technological barriers (see discussion for Trinidad and Tobago: De Lisle 2012). Despite the fact that students in those learning communities experience strong negative impacts derived from the globalized economy and global climate change, they may not be fully aware of global sustainability issues and discussions (Douglas 2006; Crossley and Sprague 2014). In particular, computer rooms and internet connectivity are lacking or, at best, limited in many places (e.g. islands off the coast of Belize: Curry et al. 2018), a situation that has clearly worsened during the COVID-19 pandemic, even at the higher-education level (Seetal et al. 2021). In Colombia, schools located in the insular territories of the Caribbean coast, suffer from the prescribed highly-segmented official curriculum which offers few opportunities for dialogue among disciplines or knowledge areas, for example, between the social and natural sciences (Medina Cobo 2022). In particular, mathematics, arts, native and foreign languages, and other courses, are taught independently from the rest of the curriculum, leaving little room to incorporate the learning process into daily life and social and environmental contexts.

In the case of the Archipelago of San Andrés, Providencia, and Santa Catalina (hereafter, the archipelago), the major islands of Colombia's territory in the Southwestern Caribbean, high school teachers face major difficulties to establish a dialogue between the prescribed official curriculum and the island's reality, as well as with national-level contexts related to nature, society, history, and culture (Sanabria James 2014; Cadena Livingston 2018). The curriculum standards, guidelines, and textbooks are designed by the Ministry of Education, mostly considering mainland learning communities of white/mestizo ancestry and overlooking the different backgrounds and needs of ethnic minority students, which in turn results in a total absence of a culturally responsive approach (Medina Cobo 2022). Unsurprisingly, high school students from these islands perform poorly in national tests, ("*Pruebas Saber 11*", Cárcamo Vergara and Mola Ávila 2012). In addition, developing an optimal learning environment for students in the archipelago represents a challenge given the uniqueness of the geographic setting (see Parsons 1985; CORALINA-INVEMAR 2012; CCO 2015 for a detailed description of the physical, biological, and human geography of the archipelago). On the islands, school communities are located nearly 800 km from the Colombian mainland and 240 km off the coast of Nicaragua in Central America, but despite their condition as Caribbean islands, many dissimilarities are found between them and the Greater and Lesser Antilles (e.g. Heartsill Scalley 2012; López Marrero and Heartsill Scalley 2012; López Marrero et al. 2012). For instance, while the climate is predominantly maritime, the influence of strong cyclones is not as frequent as in the Leeward Islands and the volcanic and karstic geology of this large archipelago is proximate to features of the Nicaraguan continental shelf.

On the historical side, local populations remain strongly influenced by the mixed heritage from the cotton field workers and slaves from Jamaica that occupied the deserted island in the eighteenth century, after the English Puritans that arrived in Providencia on board the Seaflower vessel in 1629 (Parsons 1985; Aguilera Díaz 2016; Meisel-Roca 2016a, b, and references therein). This anglophone heritage poses difficulties for students on the islands because Spanish is not their first language, and thus reading comprehension, grammar, and compositional skills are deficient (Abouchaar Velásquez and Moya 2005; García León and García León 2012). Spanish was first introduced as the official language in 1818 when the islands became part of the Colombian territory following independence from Spain. But it was not until 1953 that the mainland culture was strongly brought to the archipelago, when San Andrés was declared a "Free Trade Port". As a consequence, scholars commonly express that the island culture has been "continentalized" or "colombianized" (Parsons 1985; García 2013 Meisel-Roca 2016b) during the second half of the twentieth century. Moreover, the case of the archipelago can be defined as *"tri-lingüismo"* (three languages), given the predominance of the creole (*"kriol"*) dialect -a legacy of the blending of African and British heritages- among the native Raizal population, posing further tensions with formal English language learning (García León and García León 2012). Finally, performance in mathematics and natural sciences is below the national average, which can be attributed to language barriers between teachers and students, as well as to a lack of a contextualized learning process in complex topics such as chemistry (Cárcamo Vergara and Mola Ávila 2012; De La Rosa 2011). In response to this problem, efforts have been made by the local government to provide complementary guides to teachers to support ethno-education, highlighting the particular conditions derived from living on an island, as well as the creole (*"Raizal"*) ethnicity (CORALINA and ORFA 2016a, b, c).

Finally, the IPCC's AR6 has recently documented an increase in tropical storms and hurricanes, as well as an increase in mean sea surface temperature and a reduction in annual precipitation as the result of global warming (IPCC 2023). These trends are expected to continue over the next 50 years. Recurrent El Niño events are also a natural hazard, as demonstrated by the economic and ecological impacts observed in 2015–2016 (DANE 2016). The recent COVID-19 pandemic also demonstrated the vulnerability of the socioeconomic system of the islands, particularly related to limited health coverage, deficient public services, and strong dependence on tourism as the main source of income (Bonet-Morón et al. 2021). Despite these threats and hazards, the natural system of the archipelago is so unique that it was declared a Biosphere Reserve in 2000 by UNESCO's "Man and Biosphere Program", and it offers multiple ecosystem services that have not been fully integrated into the planning of the socioeconomic system.

In summary, we consider that the main challenge faced by high school teachers in the islands is to connect students with Raizal cultural heritage, the wonders of local nature, and sustainability issues while achieving the specific goals prescribed by the Ministry of Education. Accordingly, the objectives of this chapter are the following: (1) to describe the most salient features of the Seaflower Biosphere Reserve (SBR) that can be used in different areas like educational resources, (2) to summarize the main sustainability issues, (3) to propose geo-literacy as a foundation to connect different areas of the curriculum while allowing both students and teachers to contextualize new information, and (4) to propose a general guideline for incorporating novel geographic approaches—both conceptual and methodological—into secondary education, while providing an opportunity for dialogue among disciplines in the classroom. The final aim is to promote sustainability education as a permanent element of schools' learning goals.

2 The Seaflower Biosphere Reserve as an Opportunity for Context-Based Education

The SBR protects 180,000 km^2 of a prime coral reef barrier (Fig. 1), the third largest worldwide, behind the Australian Great Barrier and the Mesoamerican Barrier Reef System. Only 57 km^2 corresponds to terrestrial areas, including the large islands (San Andrés, Providencia, and Santa Catalina), various cays, and numerous sandbanks (CORALINA-INVEMAR 2012; CCO 2015). The Colombian Government protects 65,000 km^2 as a marine protected area (Seaflower MPA, SMPA; Fig. 1).

The SMPA preserves *ca.* 2000 km^2 of seagrasses, coral reefs, and mangroves, providing habitat to numerous terrestrial and marine species, including fish, sea turtles, and marine and shorebirds. This is a biodiversity hotspot, particularly for marine life. There are also small fragments of tropical dry forests that provide habitat for terrestrial birds, reptiles, and amphibians, as well as large invertebrates. Some of them are charismatic endemic species such as the black crab (*Gecarcinus ruricola*), and the endemic anoles and lizards. Other non-endemic but charismatic species are the queen conch (*Strombus gigas*), the hawksbill turtle (*Eretmochelys imbricata*), the butterflyfish (*Chaetodon striatus*), and the emerald parrot fish (Labridae family: *Scarus* and *Sparisoma* genera). Finally, the conservation goals of the SBR are the following: (a) promoting sustainable human economic development with social, cultural, and ecological goals in mind, and (b) supporting logistics for demonstration, education, training, and research projects.

The seascapes and landscapes protected by the SBR offer a rich spatial template for context-based teaching and learning (Rose 2012), particularly in local high schools.

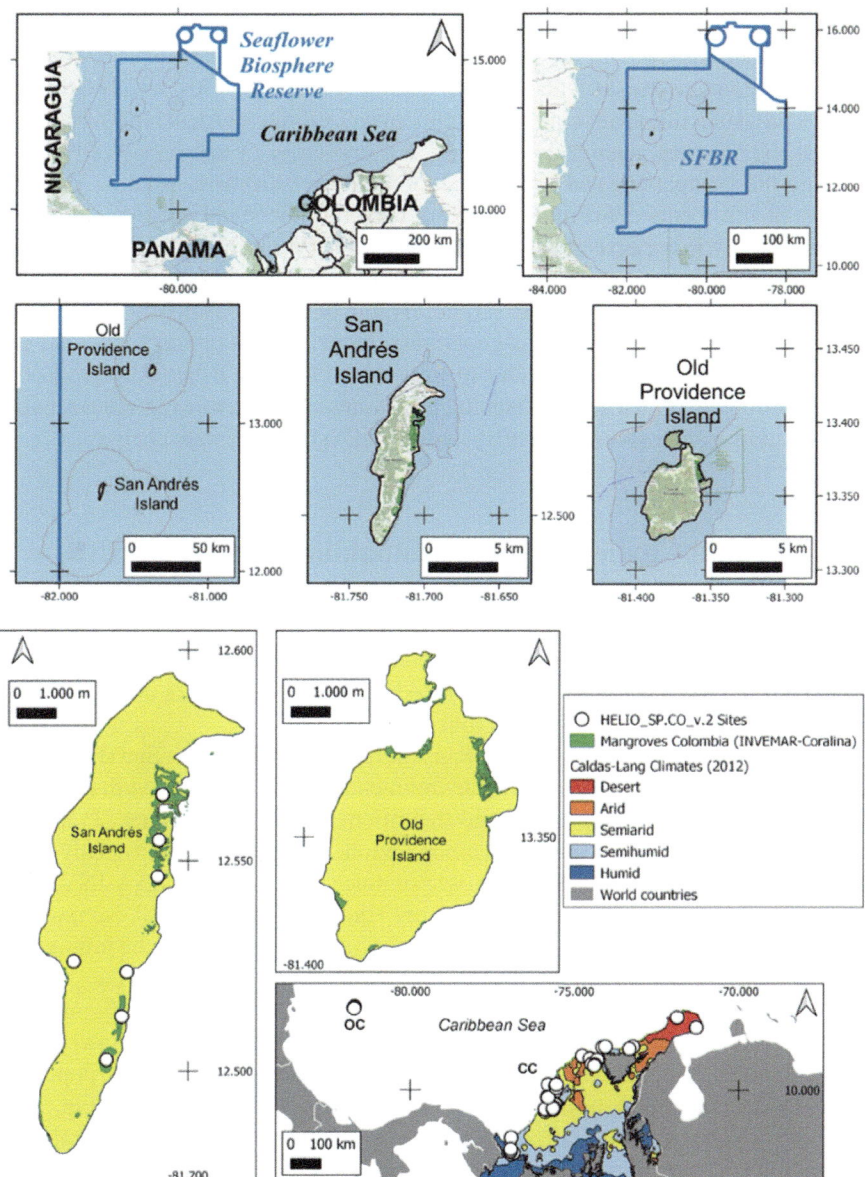

Fig. 1 Location map of the Seaflower Biosphere Reserve (SBR; blue line). Top: Zoom in from the SBR and the archipelago to San Andrés and Providencia using OpenStreetMap as a base-map. Bottom: Maps showing the location of mangroves on both islands (note the semi-arid climate in the islands relative to the Colombian mainland). Cartographic data: IGAC, IDEAM, INVEMAR, SIAC

For instance, the extensive maritime territory offers an opportunity for "virtual navigation and exploration" in the Caribbean Sea, while learning concepts about geography and cartography, as well as about natural sciences and history. The different marine, coastal, and terrestrial ecosystems may facilitate the teaching of concepts such as the physical, chemical, and biological properties of nature, as well as highlighting the complexity and uniqueness of Western Caribbean islands. The great variety of life forms constitutes an opportunity to understand ecological concepts such as biodiversity, ecological interactions, roles within ecosystems, and evolution. The presence of migratory species, both terrestrial (e.g. the black crab) and marine (e.g. sea turtles), as well as shorebirds, can be seen as an opportunity to introduce population concepts such as geographic range, habitat connectivity, and gene flow. Finally, some of these species of importance for folk fisheries might also be useful examples for discussing sustainability issues such as overexploitation, habitat deterioration, conservation efforts (including MPAs), and climate change.

3 The Main Challenges for Sustainability Within the SBR

By observing San Andrés using nightlights (Fig. 2), the high level of urbanization (particularly at North End) and the concomitant overpopulation are evident, in great contrast to Providencia, which is dominated by grassroots culture and ways of life (e.g. traditional wooden houses and folk fisheries).

By 2013, 73% of the inhabitants of San Andrés lived in the North End (INVEMAR 2014). Many studies, as well as the environmental authority, indicate that the main sustainability issues (using traditional definitions) in San Andrés are (1) land use conflicts and urban sprawl, and (2) overpopulation, both of which are partially related to unregulated growth of the tourism sector since 1953 (CORALINA-INVEMAR 2012). The promotion of the island as a "Free Trade Port" and the "3S" (sun, sea, and sand) tourism policy after the construction of the airport speeded up the immigration of investors from the mainland, with the consequent increase of hotel constructions and displacement of Raizal people from their lands, which were usually purchased at very low prices (James Cruz 2009; García 2013). This rapid urban sprawl has also been explained by poor planning (Parra 2009). As a consequence, the total population increased exponentially during the second half of the twentieth century (Fig. 3).

After slow linear growth between 1800 and 1950 (<400 to *ca.* 4,000 inhabitants, respectively, during the agriculture-based economy period), the population grew exponentially reaching *ca.* 56,000 inhabitants by 1993, as the consequence of the economic shift to commerce, industry (processed coconut), and tourism (James Cruz 2009; Meisel-Roca 2016a; DANE 2020). Ethnicity also shifted in dominance from Raizales to immigrants from the mainland. Such rapid population growth and illegal immigration required intervention from the national government in response to local civil groups' advocacy (e.g. Sons of Soil Movement 1984) (García 2013). In 1991, the reformed National Constitution of Colombia included Article 30, which states the need to regulate immigration to San Andrés, while Decree 2762 of 1991 issued

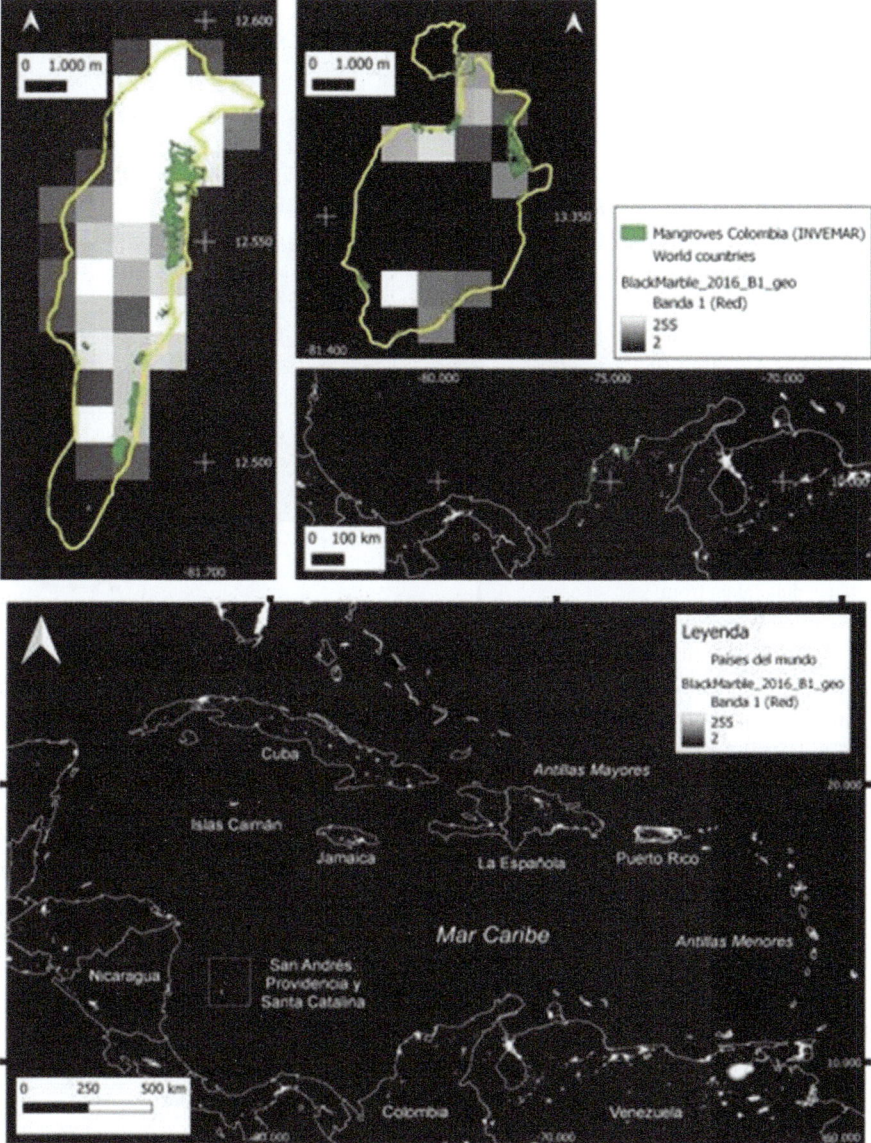

Fig. 2 The Caribbean in the dark. Top: San Andrés and Providencia. Bottom: Northern South America, Central America, and The Antilles. Nightlight emission intensity denotes the degree of urbanization. Note the intensity of North End on San Andrés island. *Source* NASA Black Marble 2016. Cartographic data: IGAC, IDEAM, INVEMAR, SIAC

Fig. 3 San Andrés and Providencia in the Anthropocene. **a** Aerial view of North End, San Andrés, the most densely populated and urbanized area within the Seaflower Biosphere Reserve (photo: Juan F. Blanco-Libreros, September 2021). **b** Two centuries of population growth in San Andrés (note the logarithmic scale in the y-axis, and the exponential increase after 1953). **c** Logistic growth in Providencia (exponential growth between 1953 and 1991 was reverted by the enforcement of immigration regulations and the establishment of a carrying capacity. Open points indicate data taken from different sources, and black points official data taken from the DANE webpage). Data projections are shown until 2035. Data from: Parsons (1985), Meisel-Roca (2016a), and DANE

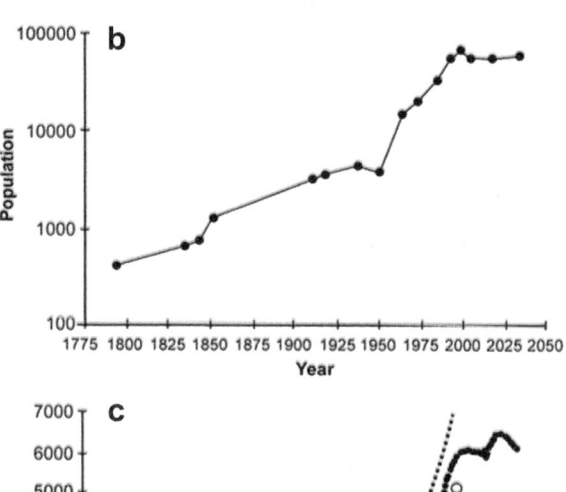
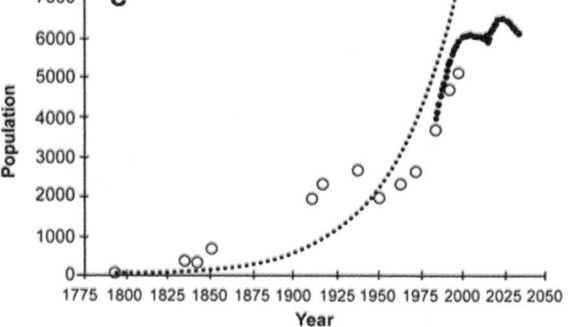

norms to control population density and regulate the rights to circulate and reside on the islands. In addition, the Office for Control of Circulation and Residency was established. Accordingly, a carrying capacity was defined and the population is forecasted to reach *ca.* 59,000 inhabitants by 2035, as a consequence of multiple regulations, law enforcement, and economic shifts (DANE 2016, 2020).

The semi-arid climate of the archipelago combined with the low-elevation topography and the carbonate lithology, limit the permanence of superficial freshwater runoff. Therefore, inhabitants of San Andrés rely on collecting rainfall and underground water for household purposes and drinking. However, the island suffers from water stress, as consumption by the tourism sector is double that of domestic consumption (James-Cruz and Barrios-Torrejano 2020). There is poor sewage and stormwater infrastructure that, in the lowlands, collapses during heavy rainfall events and spring tides, thus polluting the aquifer and wetlands in the North End, as well as coastal waters. As a partial consequence of overpopulation, the enormous production of solid waste has collapsed various dump sites over the past three decades. Solid waste recollection systems and recycling are very limited, particularly in rural areas.

Given the dependence of the archipelago on fishing resources, populations of several species have declined, thus requiring management actions such as ban periods and minimum-size regulations (e.g. Sánchez Jabba 2016a). During the mid-1990s, the decline of the queen conch (*Lobatus gigas*) urged the local authority to issue a ban on its fishing, purchasing, and selling from June 1 to October 31 each year. Other species protected with similar regulations, either permanent or temporary, are the spiny lobster (*Panulirus argus*) and some species of white fish.

Finally, illicit activities such as drug trafficking provide extra income that can worsen these sustainability issues by increasing constructions in the littoral zone or in prime ecosystems (e.g. mangroves and swamps) and by producing more solid wastes with the import of goods or with increased tourism (Sánchez Jabba 2016b). Finally, illegal economies also spill over to marine resources, adding pressure on valuable fish and shellfish stocks by expanding fleets and promoting demand.

4 Geo-literacy as a Platform for Teaching Interdisciplinarity and Sustainability

Various scholars in Colombia have emphasized the need for innovative policies as a means to close the breaches between high schools in the Caribbean region and those in the Andean region (Meisel-Roca 2011). Specifically, the Caribbean region has a deficit in terms of school numbers and education quality in the public system (Bonilla-Mejía and Martínez-Gonzalez 2017). In order to tackle this situation, teacher training and institutional capabilities are the points of intervention that have been proposed to close the educational breach by 2030. We here propose that the geo-literacy umbrella may help to achieve such goals.

Geo-literacy is defined by the National Geographic Society (n.d.) as "the ability to use geographic understanding and geographic reasoning to make far-reaching decisions". This geographic understanding deals with three components of the world we live in: interactions, interconnections, and implications. Geographic understanding is an important ability to be developed in learning communities living on islands, since, as previously established, they have unique interactions with natural and human systems. Nowadays, these communities are interconnected with other proximate islands and distant geographic regions through maritime and air transportation, as well as through the Internet. At this point in time, when students have gained more internet access, due in part to the unusual learning environment imposed by the COVID-19 pandemic, geo-literacy is a timely approach for building on cultural identity and values. The implications of geo-literacy as an educational resource are broader than simply offering new information to students, it also helps in setting innovative goals for teaching and learning, designing activities that best support learning and, ultimately, encouraging students to envision environmental scenarios for their islands.

Geo-literacy activities start by discussing basic cartographic concepts and geographical names. Three important cartographic concepts are: the spatial or graphical scale, the location of geographic north, and the convention or legend. Additional concepts such as projection and coordinate systems can be introduced to more advanced students. Learning about geographical names also offers students the opportunity to further explore the history and geography of continents, countries, and regions, as well as their physical and biotic features. This allows teachers to use geo-literacy as a framework for interdisciplinarity by establishing dialogues with mathematics, natural sciences, and the arts.

Geo-literacy might be also useful for introducing general knowledge about sustainability in islands. Emerging from the Brundtland Report, sustainable development can be defined as the use of natural resources by present generations without compromising their availability for future generations. More recent definitions such as environmental sustainability also stress the right to a healthy environment, including healthy ecosystems, and clean air and water. Finally, climate justice is a crucial contemporary concept behind youth climate activism such as "Fridays For Future" (https://fridaysforfuture.org/).

The complex geographic setting of the archipelago, added to the well-defined historical phases, provides an excellent template for geo-literacy. For instance, with more than 2,000 species (197 of them included on the IUCN's Red List), the SBR offers the possibility to discuss issues around the geographic range of distribution (including the concept of scale) of marine and terrestrial species, and how species with a restricted range are more susceptible to human threats and prone to extinction. As another example, coral reefs and seagrasses, as extensive marine ecosystems in shallow areas, offer an opportunity for teaching about the geography of the archipelago in the context of the wider Caribbean and the mainland, as it exhibits biogeographic affinities with Central America and Jamaica. As a third example, sand beaches, rocky shores, and mangroves are the dominant coastal ecosystems, and hence they might be useful for touring around the islands while discussing local variability of rock types (lithology), interactions with the marine environment (waves and currents), and climate conditions. Finally, as a salient feature, in the largest islands (San Andrés and Providencia), fringing mangroves form small patches within embayments or behind large coral reef barriers protecting them from strong waves on the eastern coasts (Old Point Regional Mangrove Park and McBean Lagoon National Natural Park, respectively). In contrast, with the exception of Cove Bight, the northern, southern, and western coasts of San Andrés are almost deprived of large mangrove areas. It is also noticeable that the tall canopies of inland mangroves in Salt Creek and Smith Channel are shaped by the strong winds, a feature also observed in sand dune vegetation.

Under the geo-literacy umbrella, while discussing the geography of San Andrés, teachers from social sciences can also introduce the influence of climate on the variability of a single ecosystem. Conversely, teachers from natural sciences (particularly biology) can discuss the island's geography and climate while explaining selective topics of botany, zoology, or ecology. For Raizal people, land and landscapes are not only physical spaces but imaginaries (sensu James Cruz and Soler Caicedo 2018), thus geo-literacy can be a useful approach to establish a dialogue between individual preconceptions of the world and the outside world in high school learning communities. Eventually, improving geo-literacy among high school students could be the foundation for a shift from the currently dominant "3S" tourism in San Andrés to a more sustainable, high-value model, such as the "3L" (landscape, leisure, and learning) (see James Cruz 2009).

Over the past two decades, geo-literacy initiatives in high schools around the world have benefited from the proliferation of computational free or open-access platforms. Google Maps is a good example of a user-friendly application that can be installed on most mobile devices (phones and tablets), and it can be easily browsed from desktop and laptop computers with an internet connection. Although many students and teachers are familiar with this software for basic navigation purposes such as address searches and geo-location, this application has many other functions that can be incorporated into classrooms as didactic technologies, for example, different layers (relief, traffic, public transportation routes and bicycle lanes, satellite imagery, and street view) and the image exploration menu. As a complement, Google Earth Pro, a more advanced program, can be installed for free on mobile devices, desktops, and laptops, or it can even be used online. This application offers tools such as distance and area measurements, and allows object drawings that can be used in advanced grades for introducing cartographic concepts and basics of map creation. Google Earth Outreach (www.google.com/earth/outreach) is an educational program to promote geo-literacy as well as to provide networking opportunities for educators using geo-tools.

OpenStreetMap (OSM, www.openstreetmap.org) is a growing open-source crowd-mapping platform that allows contributions to the making of a global map while building a global network of mapping communities (www.youthmappers.org). With basic training (e.g. TeachOSM, teachosm.org) teachers and students can learn to create objects (buildings and roads), and, after signing in, they can contribute to enriching the local map using their knowledge. With further training, they can contribute to humanitarian mapping initiatives elsewhere, either independently or as part of coordinated efforts called "mapathons" (www.hotosm.org). Finally, Mapillary (www.mapillary.com) is an open-source mobile application and computer platform that improves ground-level navigation by capturing street-level imagery (pictures or video), multi-scale integration, automatic mapping, and computer vision. High school communities can use Mapillary only as a navigation platform, or they can be actively involved in imagery capture with smartphones to help improve local maps.

5 A Proposal for High Schools: Geo-literacy for Sustainability Education (GeoSE)

Applying the sustainability concept in the case of the archipelago requires the use of a definition from a systems perspective (Ben-Eli 2018). This definition emphasizes the limit to human population growth imposed by the carrying capacity of the environment—a fundamental issue in clearly-delimited geographic spatial units like islands—based on five core principles: (1) material domain, (2) economic domain, (3) domain of life, (4) social domain, and (5) spiritual domain. Thus, here we apply this definition to the SBR as follows: (a) it is a complex system, hence sustainability depends on the interactions of many islands (elements) (populated, large to small islands, and unpopulated cays and shoals), (b) each element can also be defined as a complex, human-nature coupled system, (c) SBR as a system isolated by distance from Central America, South America, North America (the Florida peninsula), and other Caribbean islands, (d) isolation by distance is offset by sea- and air-borne transportation, further strengthening the dependence on local resources defining the carrying capacity, (e) the complex system can be seen as a hierarchical arrangement of the space, with the entire SBR exhibiting a certain carrying capacity while each element exhibits its own carrying capacity depending on the characteristics of the human dimension, and (f) at different spatial scales, either a single island (e.g. San Andrés) or a group of islands (e.g. San Andrés, Providencia and Santa Catalina), the complex systems consist of five domains (materials, economics, life, society, and cultural values). Finally, we also advocate for including education as part of the sustainability definition (see the discussion on "education as sustainability" versus "education for sustainability").

According to the environmental authorities with jurisdiction in the archipelago, the sustainability concept must involve adaptation to climate change (INVEMAR 2014). As a consequence, the departmental adaptation plan to cope with climate change outlines the following strategic lines, according to which actions and projects were proposed: (a) ecosystem-based adaptation, (b) infrastructure for adaptation, (c) adaptation-oriented planning, (d) empowerment and local development, (e) education and training, and (f) research and monitoring. This plan also includes an extensive glossary that must be understood by decision-makers, stakeholders, and learning communities, in order to move from agendas to actions, as broadly discussed in climate change adaptation literature worldwide. Here we propose improving the use of such terms in the vocabulary of students and teachers in local high schools through geo-literacy.

The national legal framework for the first stage of the present "Geo-literacy for Sustainability Education" (GeoSE) proposal is the Institutional Pedagogical Project (IPP; *Proyecto Educativo Institucional, PEI*, in Spanish), defined by the General Law for Education in Colombia (Law 115 of 1994). The IPP is the framework that allows high schools to adopt a conceptual and methodological approach to ensure full compliance with curricular requirements. This law allows flexibility to cover the minimum topics required for the curriculum to be assessed in national tests (*Pruebas Saber 11*), as long as schools follow national educational standards. In addition, the School's Environmental Project (SEP; *Proyecto Ambiental Escolar*,[1] *PRAE*, in Spanish) is an important dimension of the IPP. The SEP provides an extracurricular platform for students to integrate concepts and methods learned in the curriculum while involved in activities related to the conservation of the environment, thus challenging them to think about real-world problems. Some umbrella topics used by high schools nationwide include: recycling, responsible consumption, sustainable agriculture, urban agriculture, school biodiversity, water protection and conservation, wetland conservation, and urban ecosystems. We encourage local schools in the SBR to redesign their SEP (and even their IPP) using the umbrella concept, which constitutes the goal and main methodological approach of the "Geo-literacy for Sustainability Education" (GeoSE) proposal.

The GeoSE, as a SEP, is built on three pillars: **Sustainability** as a concept, improved **Adaptation** to climate change as a goal, and **Geo-literacy** as a methodological approach (comprising geo-concepts and geo-tools). This triad can be referred to as SAG. In turn, these three pillars are supported in ethno-education (specifically Raizal education). The present proposal for the pilot stage has been developed through collaboration among university researchers (*Universidad de Antioquia*, Medellín) and high school teachers (*Institución Educativa Antonia Santos, Sede Phillip Beekman Livingston, INEDAS*, San Andrés).

The Antonia Santos School (*Institución Educativa Antonia Santos, INEDAS*), is a public institution, consisting of two separate locations in different parts of San Andrés, depending on the age group they serve. The Phillip Beekman location is situated on the eastern coast, in the El Bay sector, south of the downtown area, surrounded by a mosaic of residential and natural patches. The main facilities comprise a three-story building containing classrooms, meeting spaces, and a lunchroom. This school hosts middle and high school levels, but it also offers students the opportunity to obtain a technical degree—Culinary Services or Network Systems—as part of a strategic partnership with the Colombian public system of trade schools (*Servicio Nacional de Aprendizaje, SENA*). The school also features an adult education program, intended to assist older neighborhood residents in meeting their graduation requirements. Traditionally, 90% of the student population comes from the surrounding neighborhoods (San Luis area, Little Hill, Tom Hooker, Pepper

---

[1] https://www.mineducacion.gov.co/1621/article-90893.html.

Hill, Loma Barack, Orange Hill, South End, and Schooner Bay) and belong to the Raizal community, with Creole being their mother tongue. The academic performance of the students has been classified as low according to the analysis of the results obtained in the Colombian national standardized test (ICFES), between 2014 and 2022. The analysis identified four areas of improvement that included Literacy and Social Sciences skills in addition to critical thinking skills applied to Mathematics and Natural Sciences (De Armas-Castañer 2023).

The general approach was discussed by email and a pilot extracurricular activity ("*Experiencia significativa*", or "Meaningful experience") was designed for volunteer teachers and students at the secondary level. The pilot extracurricular activity, named "Taking the Seaflower to the classroom" took place on September 9, 2021, and an assessment was carried out with teachers the next day. The pilot was developed in two stages. During the first stage, which took place in the computer room, a focus group of teachers and advanced students was introduced to the general concepts of geography and cartography and the basics of navigating in Google Maps and Google Earth Pro (including "Street View" mode). The second stage incorporated other students in addition to the focus group. An icebreaker included an outdoor ludic activity, where each participant introduced themselves and "invited" the coordinator of the workshop to visit a place in San Andrés, to introduce a "sense of space" within the island's geography. They then formed a circle and passed out an inflatable world map globe while singing the rhyme (in English) "Taking the Seaflower to the classroom", to reinforce listening comprehension and fine-tune coordination among the participants. Finally, a short game ("Human compass") was played to refresh the "cardinal points" concept. The indoor activity was similar to the first part. The morning finished with an exhibition of drone piloting. On September 10, 2021, the teachers and coordinators, including the school principal and the academic coordinator, had a hands-on workshop to draw a map of the island or a neighborhood, using the concepts learned the previous day, working in small teams. After completing the drawing, each team presented their results in front of the class, followed by feedback from the coordinator (Fig. 4).

The session ended with an evaluation. It must be noted that the activity involving students was run mostly in English, while the activity with teachers was mostly in Spanish. This strategy was used as a way to make the pilot activity more culturally responsive, as Creole, followed by English, are the languages Raizal students use in their daily lives. A graphic abstract of the activity is shown in Fig. 5.

The teachers provided feedback on the activity based on three factors: pertinence to Competency-Based Education (CBE), interdisciplinarity, and alignment with the IPP. In general, they concluded that this kind of activity facilitates effective communication with the students as they are familiar with new technologies and devices like smartphones and apps. Better communication, in this case, translated into increased motivation and collaborative effort. In terms of CBE, the activity helped improve scientific, social, and mathematical competencies. For example, students classified numerical and spatial–temporal elements using established criteria, worked in teams, and contrasted ideas about natural phenomena. The teachers' workshop encouraged the exchange of opinions and perceptions to solve unfamiliar tasks by means of

Fig. 4 Two maps produced by teachers during the team activity. Both representations include labels for features perceived by the participants as the most salient characteristics of the neighborhood (left) and of San Andrés (right). The maps include basic geo-literacy concepts such as geographical names and the location of geographic north

comparing different approaches based on individual expertise. The teamwork also proved effective as it helped to set short-term goals and anticipate results, therefore, the general consensus among teachers was that interdisciplinarity may help in developing common learning objectives. Finally, teachers stated that the learning objectives and results of the pilot align with the PEI and the SEP, which makes "geo-literacy" a concept of ample applicability.

Since this workshop, the first three authors of this chapter have led the drafting of recommendations for a SEP titled "Geo-literacy for Sustainable Education in the SBR" (Box 1 after the conclusions, see also the word cloud in Fig. 5). We followed the guidelines from the Ministry of Education, the departmental Secretary of Education, and the "Raizal Heritage, Nature, Tradition and Culture" guides[2] (CORALINA and ORFA 2016a, b, c), with inputs from the coauthors, high school teachers, and other actors. The ten recommendations can be summarized as follows: (1) to define the umbrella and ancillary sustainability problems, (2) to define immediate and extended geographic areas of interest, action, and application, (3) to declare concepts, activities, methods, and technologies to be covered during each encounter, (4) to select target ecosystems and flagship or charismatic species, representative of the biodiversity present on the islands, (5) to include elements of human society to introduce sustainability issues, (6) to promote dialogue among teachers from different disciplines, (7) to establish an annual calendar of activities (e.g. monthly or bi-monthly), (8) to outline grade-specific themes to be covered and skills to be developed, and geo-concepts and sustainability concepts to be incorporated into the vocabulary, as well as geo-tools and activities.

---

[2] https://observatorio.coralina.gov.co/index.php/es/seaflower-aprende

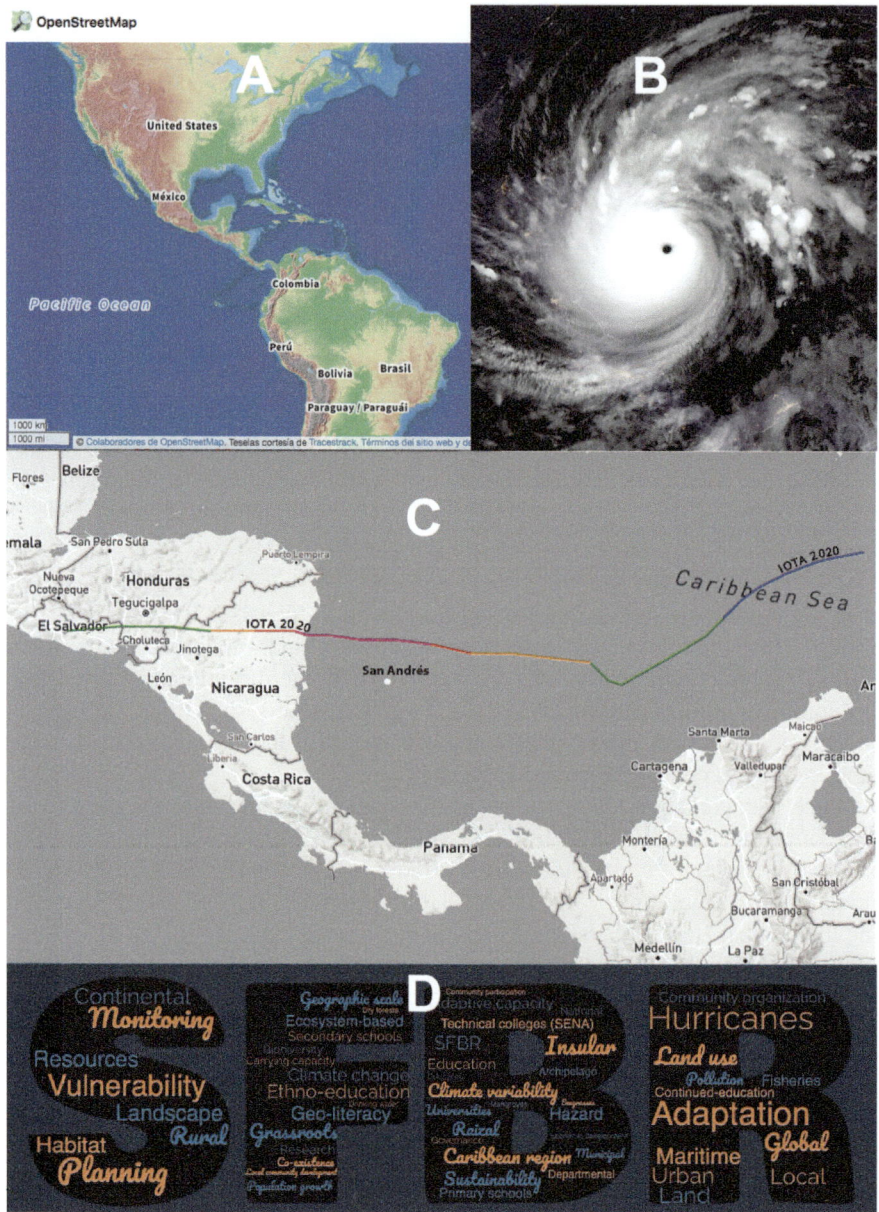

Fig. 5 Online resources employed during a GeoSE 1-day activity in INEDAS. **a** OpenStreetMap for measuring distances between San Andrés and different places on Earth, and to complement Google Earth Pro for "street-viewing" tours around the island. **b** Hurricane Iota information on Wikipedia. **c** Hurricane Iota's passage across the Caribbean Sea on the NOAA website (note the location of San Andrés; available at: https://coast.noaa.gov/hurricanes). **d** Word cloud of selected geographic and sustainability concepts covered during a GeoSE activity with high school teachers. Hurricane Iota photo available on Wikipedia in Spanish: date: 2020-11-16-1500Z; NOAA: https://console.cloud.google.com/storage/browser/gcp-public-data-goes-16. Panel A: Courtesy OpenStreetMap (R) open data license (CC BY-SA 2.0) by OpenStreetMap Foundation

6 Conclusions

By way of conclusion, here we propose a guide for teachers and students for a 3-h activity (*Experiencia significativa* or "Meaningful experience"), conceived by one of the co-authors (J. Lasso-Zapata, *INEDAS*), as a concrete example of geo-literacy for sustainability education. This guide, titled "Mapping mangroves in my island", is an activity aimed at introducing spatial concepts and basic knowledge about a key ecosystem within the SBR to 6th-grade students, bearing in mind the socioeconomic and cultural characteristics of the learning community. Guidelines for teachers are included in Box 2, whereas Box 1 contains recommendations for an activity with 6th-grade students. These include guidelines on how to define a specific theme, expected skills, concepts, geo-tools, and sustainability problems. We suggest carrying out this activity either on the International Day for Mangrove Conservation (July 26th) or the International Wetlands Day (February 2nd). This school is located between Cocoplum and San Luis, thus online tours or field tours can target mangroves nearby (i.e. Old Point Regional Mangrove Park). Tours can also explore the linkages between mangroves and coral reefs in the Rocky Cay area, just outside the school.

As an end result of GeoSE, we anticipate that activities with 10th- or 11th-grade students will provide simple cloud-computing skills to assess the impacts of Hurricane Iota on different natural and human systems by browsing and analyzing open-access satellite imagery online (e.g. Sentinel Hub EO Browser). As a sample output, in Fig. 6, we show the one-year aftermath on mangroves in McBean Lagoon National Natural Park, the area most severely damaged by the storm surge of Hurricane Iota. The extensive tree mortality is evident in both a Sentinel-2 satellite image and the ground panoramic picture taken during field verification. According to the visual information provided by these images, it is possible to confirm that red mangrove dominated areas (*Rhizophora mangle*) are not resilient to strong hurricanes (for comparison with black-mangrove—*Avicennia germinans*—and white mangrove—*Laguncularia racemosa*—dominated areas in South Florida after two major hurricane landfalls in 2017, see Lagomasino et al. 2021). As a final recommendation, we propose that the experiences gained during the implementation phase of the GeoSE program in *INEDAS* and other high schools in San Andrés, Providencia, and Santa Catalina should be shared at specialized meetings such as the "Geo for Good Summit", "Geospatial for Good", and "Free and Open Source Software for Geospatial Conference (FOSS4G)", as well as in specialized conferences on high school education.

Box 1 Basics of the proposal "Geo-literacy for Sustainable Education in the SBR" as a guideline to develop SEP in San Andrés, Providencia, and Santa Catalina Archipelago.

Ten recommendations to design GeoSE-based SEP:

1. Define the umbrella sustainability problem to be tackled by the SEP.

Fig. 6 Advanced geo-literacy skills (e.g. online satellite imagery browsing and interpretation) to be developed in grade 10 and 11 students to learn about ecosystem resiliency. **a** and **b** Sentinel-2 true color imagery before and one year after the passage of Hurricane Iota close to Providencia (see dates in images). Compare the insets to note the pervasive loss of mangrove "greenness" (brown color denoting tree mortality in McBean Lagoon National Natural Park, MBL NNP, to the east of the airport runway). Satellite images downloaded from the Sentinel Hub EO Browser. **c** Panoramic photo showing the low recovery or low resiliency of mangroves in MBL NNP (see the airport runway for reference. Photo: Daniela Valentina Pacheco Brieva, November 2021)

2. Define one or two additional sustainability problems that are linked to the umbrella problem, either as potential causes or potential consequences.
3. Define the geographic extent of immediate action (e.g. the school, the neighborhood).
4. Define the geographic area for application of the lessons learned (e.g. one island, the archipelago, the SBR).
5. Define concepts, activities, methods, and technologies (ICTs, Information and Communication Technologies) to be employed within each encounter (develop a guide for teachers and students).
6. Select one or various ecosystems (i.e. mangroves, coral reefs, seagrasses, dry forests) to provide context about the natural world in the islands. Also, select one or various species of plants or animals to discuss sustainability problems and biodiversity.
7. Include elements of human society as an anchor for introducing sustainability issues.
8. Promote dialogue among teachers from different knowledge areas to increase interdisciplinarity (at least two; e.g. natural and social sciences, natural sciences and arts, natural sciences and mathematics).
9. Establish an annual schedule of meetings for carrying out activities (Experiencias Significativas or "Meaningful activities") to maintain interest among participating students. A way to do so is to organize activities around celebrations within the environmental awareness calendar (e.g. Earth Day, Day of Wetlands, Day of Mangroves, Day of Biodiversity, Day of Soils). There are numerous environmental awareness days in Colombia that would allow monthly or bimonthly encounters.
10. Grade-specific themes, expected skills to develop, concepts, geo-tools, sustainability concepts, and example activities are described as follows:

Grades 6–7: Theme: Maps and navigation around my world. Expected skills to develop: basic mapping. Concepts: Location, distance, travel time, map elements or objects (points, lines, polygons), geographic North, geographic scale, geographic coordinates. Geo-tools: Google Maps, Open Street Maps, Google Earth, and Scribble Maps. Sustainability problem: e.g. solid waste production and disposal. Example activity: "Mapping my school and neighborhood".

Grades 8–9: Theme: Increasing geographic literacy to cope with hurricanes: from island to global geography. Expected skills to develop: Intermediate-level mapping (making maps with online software, map verification in the field). Concepts: Series of polygons, lines and points, thematic maps, legends, export and import files (i.e. kmz). Geo-tools: Open Street Map (Id Editor), Google Earth Pro, Mapillary. Sustainability problem: urbanization, urban sprawl, hurricanes in urbanized islands. Example activity: "Mapping sustainability issues in my island".

Online resources on hurricanes and OpenStreetMap:

https://coast.noaa.gov/hurricanes
https://wiki.openstreetmap.org/wiki/Education
https://learnosm.org/es/.

Grades 10–11: Theme: Using geographic literacy to respond to hurricanes. Expected skills to develop: Advanced-level mapping: Active and crowd-based online and desktop mapping, basic satellite imagery browsing. Concepts: Vector and raster data, sources of open data (official Colombian government repositories: e.g. IGAC, SIAC, SIAM, IAvH), satellite imagery, remote sensing, and maps for communicating ideas. Geo-tools: JOSM, HOTOSM, TeachOSM, QGIS, Sentinel Hub EO Browser. Sustainability problem: Global changes, globalization, unsustainable consumption, carrying capacity, global warming and hurricanes, resilience, adaptive capacity. Example activity: "Mapping the impacts of Hurricane Iota: a comparison between San Andrés and Providencia".

Online resources about Hurricane Iota:

https://appliedsciences.nasa.gov/what-we-do/disasters/disasters-activations/hurricane-iota-2020
https://es.wikipedia.org/wiki/Huracán_Iota.

Box 2 Simple description of a teachers' guide for a GeoSE-based extracurricular activity with 6th graders in San Andrés.

Extracurricular activity: "Mapping mangroves in my island (San Andrés)".

Grade: 6

Theme: Maps and navigation around my world.
Expected skills to develop: Basic mapping on paper.

Concepts: Location, distance, travel time, map elements or objects (points, lines, polygons), geographic north, geographic scale, geographic coordinates.

Geo-tools: Google Maps, OpenStreetMaps, Google Earth, and Scribble Maps.

Sustainability problem: Solid waste disposal in mangroves.

Icebreaker: Tell a story about a fascinating mangrove critter, for instance the gray heron (bird) using mangroves for fishing, a marine catfish using mangroves as a nursery area, or the blue crab using mangroves as a habitat and migration pathway to the sea.

Exploration of previous knowledge: A word search with mangrove-related concepts and names of mangrove sites around the island.

Research questions (to guide the quest or inquiry process): Where are the largest areas of mangroves located on my island? Are there mangroves on the West coast of the island? How do I get to Smith Channel mangroves?

Online reading and explanation: Select a website with basic information about mangroves in general or mangroves in San Andrés (e.g. Seaflower Biosphere Reserve website or Wikipedia). Select a website explaining the basics of cartography and summarize the key points to the class (geographic north, scale, symbols/legends, points, lines, polygons, geographic coordinates). Topics can be reinforced by navigating Google Maps and OpenStreetMap.

Work in break-out groups (small groups): Mapping on paper using the concept learned and online resources: Google Maps and OpenStreetMap. Make each group identify itself with the name of a mangrove species or a mangrove animal.

Sustainability problem discussion: Inquire among the students: Why is there so much solid waste among mangrove roots? What problems does this waste cause to animals and plants? How can this problem be solved? What should we do? (a YouTube video can be used to deliver the main message while eating a snack).

Optional activity: Make a map with the main sources of solid waste on the island.

References

Aguilera Díaz M (2016) Geografía económica del archipiélago de San Andrés, Providencia y Santa Catalina. In: Meisel-Roca A, Aguilera Díaz M (eds) Economía y medio ambiente del archipiélago de San Andrés, Providencia y Santa Catalina. Colección de Economía Regional, Banco de la República, Bogotá, Colombia, pp 47–116

Abouchaar Velásquez A, Moya S (2005) Dominio de la lengua española entre estudiantes de quinto grado en la isla de Providencia. Cuadernos del CES (Centro de Estudios Sociales) 9:8–12. Universidad Nacional de Colombia

Ben-Eli MU (2018) Sustainability: definition and five core principles, a systems perspective. Sustain Sci 13:1337–1343. https://doi.org/10.1007/s11625-018-0564-3

Bentz J, O'Brien K (2019) Art for change: transformative learning and youth empowerment in a changing climate. Elementa Sci Anthrop 7:52. https://doi.org/10.1525/elementa.390

Bonet-Morón J, Ricciulli-Marín D, Peña D (2021) San Andrés y Providencia en el siglo XXI y la pandemia del COVID-19. Documentos de Trabajo sobre Economía Regional y Urbana. Banco de la República, Centro de Estudios Regionales (CEER), Cartagena, Colombia

Bonilla-Mejía L, Martínez-González EF (2017) Educación escolar para la inclusión y la transformación social en el Caribe colombiano. Documentos de Trabajo sobre Economía Regional y Urbana. Banco de la República, Centro de Estudios Regionales (CEER), Cartagena, Colombia

Cadena Livingston D (2018) Pertinencia socio-cultural de los modelos pedagógicos que contextualizan la educación preescolar en San Andrés isla. M.Ed. thesis, Department of Humanities, Universidad de La Costa, San Andrés, Colombia

Cárcamo Vergara C, Mola Ávila JA (2012) Diferencias por sexo en el desempeño académico en Colombia: un análisis regional. Economía y Región 6(1):133–188

Comisión Colombiana del Océano (CCO) (2015) Aportes al conocimiento de la Reserva de Biósfera Seaflower. Comisión Colombiana del Océano. Bogotá. https://cco.gov.co/cco/publicaciones/83-publicaciones/133-aportes-al-conocimiento-de-la-reserva-de-biosfera-seaflower.html

CORALINA-INVEMAR (2012) Atlas de la Reserva de Biósfera Seaflower. Archipiélago de San Andrés, Providencia y Santa Catalina. Instituto de Investigaciones Marinas y Costeras "José Benito Vives De Andréis" -INVEMAR- y Corporación para el Desarrollo Sostenible del Archipiélago de San Andrés, Providencia y Santa Catalina -CORALINA-. Serie de Publicaciones Especiales de INVEMAR, 28, Santa Marta, Colombia http://www.invemar.org.co/redcostera1/invemar/docs/10447AtlasSAISeaflower.pdf

CORALINA-ORFA (2016a) Herencia raizal, naturaleza, tradición y cultura. Guía educativa 3. Grados 6° y 7°. https://observatorio.coralina.gov.co/index.php/es/seaflower-aprende

CORALINA-ORFA (2016b) Herencia raizal, naturaleza, tradición y cultura. Guía educativa 4. Grados 8° y 9°. https://observatorio.coralina.gov.co/index.php/es/seaflower-aprende

CORALINA-ORFA (2016c) Herencia raizal, naturaleza, tradición y cultura. Guía educativa 5. Grados 10° y 11°. https://observatorio.coralina.gov.co/index.php/es/seaflower-aprende

Church W, Skelton L (2010) Sustainability education in K-12 classrooms. Journal of Sustainability Education 1 https://www.susted.com/wordpress/content/sustainability-education-in-k-12-classrooms_2010_05/

Crossley C, Sprague T (2014) Education for sustainable development: implications for small island developing states (SIDS). Int J Educ Dev 35:86–95. https://doi.org/10.1016/j.ijedudev.2013.03.002

Cruz JLJ, Torrejano DJB (2020) Valoración del uso del agua en la isla de San Andrés: turistas, hoteles y viviendas turísticas. PASOS Revista de Turismo y Patrimonio Cultural 18(2):293–308 https://doi.org/10.25145/j.pasos.2020.18.020

Curry KL, Sabina LL, Sabina KL et al (2018) Advancing the promise of educational equity in Belize: a case study. Am J Soc Sci Res 4(2):22–32

De Armas-Castañez NK (2023) Proyecto de Coordinación para el mejoramiento de los desempeños académicos de los estudiantes. Institución Educativa Antonia Santos, San Andrés, Colombia

De La Rosa LR (2011) Problemáticas y alternativas en la enseñanza de la química en la educación media en la isla de San Andrés. Thesis, Faculty of Sciences, Universidad Nacional de Colombia, Bogotá, Colombia, Colombia. MSc

De Lisle J (2012) Explaining whole system reform in small states: the case of Trinidad and Tobago. Secondary education modernization program. Curr Issues Comp Educ 15(1):4–82

Departamento Administrativo Nacional de Estadísticas (DANE) (2016) Informe de coyuntura económica regional 2015. Departamento de San Andrés (Archipiélago de San Andrés, Providencia y Santa Catalina). Bogotá, Colombia

Departamento Administrativo Nacional de Estadísticas (DANE) (2020) Boletín técnico. Encuesta de hábitat y usos socioeconómicos 2019. San Andrés, Providencia y Santa Catalina. Bogotá, Colombia

Douglas CH (2006) Small island states and territories: sustainable development issues and strategies – challenges for changing islands in a changing world. Sustain Dev 14:75–80. https://doi.org/10.1002/sd.297

García MC (2013) "Colombianizar" a toda costa o ser raizal allende los mares. Cien Días (CINEP/PPP) 77:48–55

García León J, García León D (2012) Políticas lingüísticas en Colombia: tensiones entre políticas para lenguas mayoritarias y lenguas minoritarias. Bol Filol XLVII 2:47–70. https://doi.org/10.4067/S0718-93032012000200002

Han H, Ahn SW (2020) Youth mobilization to stop global climate change: narratives and Impact. Sustainability 12(10):4127. https://doi.org/10.3390/su12104127

Heartsill Scalley T (2012) Freshwater resources in the insular Caribbean: an environmental perspective. Carib Stud 40(2):63–93. https://doi.org/10.1353/crb.2012.0030

INVEMAR (2014) Plan de adaptación al cambio climático para el archipiélago de San Andrés, Providencia y Santa Catalina. Informe técnico. Convenio interadministrativo No. 277 de 2014 entre el MADS y el INVEMAR-MADS-CC-277

IPCC (2023) Summary for Policymakers. In: Core Writing Team, Lee H, Romero J (eds) Climate change 2023: synthesis report. contribution of working Groups I, II and III to the sixth assessment report of the intergovernmental panel on climate change. IPCC, Geneva, Switzerland, pp 1–34. https://doi.org/10.59327/IPCC/AR6-9789291691647.001

James Cruz JL (2009) El papel del estado en la construcción de desarrollo sostenible: el caso del turismo en el Caribe insular. Cuad Econ 25(1):265–281

James Cruz JL, Soler Caicedo CSI (2018) San Andrés: cambios en la tierra y transformación del paisaje. Cuad Geogr Rev Colomb Geogr 27(2):372–388. https://doi.org/10.15446/rcdg.v27n2.65356

Lagomasino D, Fatoyinbo T, Castañeda-Moya E et al (2021) Storm surge and ponding explain mangrove dieback in southwest Florida following Hurricane Irma. Nat Commun 12:4003. https://doi.org/10.1038/s41467-021-24253-y

López-Marrero T, Heartsill Scalley T (2012) Get up, stand up: environmental situation, threats, and opportunities in the insular Caribbean. Carib Stud 40(2):3–14

López-Marrero T, Yamane K, Heartsill Scalley T et al (2012) The various shapes of the insular Caribbean: population and environment. Carib Stud 40(2):17–37

McGee S, Durik AM, Zimmerman JK et al (2018) Engaging middle school students in authentic scientific practices can enhance their understanding of ecosystem response to hurricane disturbance. Forests 9(10):658–675. https://doi.org/10.3390/f9100658

Medina Cobo O (2022) El currículo oficial en las dos últimas reformas educativas en Colombia. Rev Educ Polít Soc 7(1):9–30. https://doi.org/10.15366/reps2022.7.1.002

Meisel-Roca A (2011) El sueño de los radicales y las desigualdades regionales en Colombia: La educación de calidad para todos como política de desarrollo territorial. Documentos de trabajo sobre economía regional. In: Sánchez Jabba A, Otero Cortés A (eds), Educación y desarrollo regional en Colombia. Banco de la República de Colombia, Cartagena, Colombia, pp 261–278

Meisel-Roca A (2016a) La estructura económica de San Andrés y Providencia en 1846. In: Meisel-Roca A, Aguilera Díaz M (eds) Economía y medio ambiente del archipiélago de San Andrés, Providencia y Santa Catalina. Colección de Economía Regional, Banco de la República, Bogotá, Colombia, pp 1–14

Meisel-Roca A (2016b) La continentalización de San Andrés Islas, Colombia: panyas, raizales y turismo, 1953–2003. In: Meisel-Roca A, Aguilera Díaz M (eds), Economía y medio ambiente del archipiélago de San Andrés, Providencia y Santa Catalina. Colección de Economía Regional, Banco de la República, Bogotá, Colombia, pp 15–46

National Geographic Society (n.d.) What is Geo-literacy? https://www.nationalgeographic.org/media/what-is-geo-literacy/

Parra E (2009) Ordenamiento territorial costero en el Caribe colombiano. Las directrices del Estado en los casos de estudio de Coveñas y San Andrés. Master's thesis. Faculty of Architecture, Universidad Nacional de Colombia, Medellín, Colombia

Parsons J (1985) San Andrés y Providencia. Una geografía histórica de las islas colombianas en el Caribe. El Ancora Editores, Bogotá, Colombia

Rose DE (2012) Context-based learning. In: Seel NM (ed), Encyclopedia of the sciences of learning. Springer, Boston, MA, pp 799–802. https://doi.org/10.1007/978-1-4419-1428-6_1872

Sanabria James LA (2014) El sistema educativo insular. Cuadernos del Caribe 4(8):11–27. Universidad Nacional de Colombia, Sede Caribe

Sánchez Jabba A (2016a) Manejo ambiental en Seaflower: reserva de biósfera en el archipiélago de San Andrés, Providencia y Santa Catalina. In: Meisel-Roca A, Aguilera Díaz M (eds) Economía y medio ambiente del archipiélago de San Andrés, Providencia y Santa Catalina. Colección de Economía Regional, Banco de la República, Bogotá, Colombia, pp 157–190

Sánchez Jabba A (2016b) Violencia y narcotráfico en San Andrés. In: Meisel-Roca A, Aguilera Díaz M (eds) Economía y medio ambiente del archipiélago de San Andrés, Providencia y Santa Catalina. Colección de Economía Regional, Banco de la República, Bogotá, Colombia, pp 235–250

Sanson A, Bellemo M (2021) Children and youth in the climate crisis. Bjpysch Bull 45(4):205–209. https://doi.org/10.1192/bjb.2021.16

Seetal I, Gunness S, Teeroovengadum V (2021) Educational disruptions during the COVID-19 crisis in small island developing states: preparedness and efficacy of academics for online teaching. Int Rev Educ 67:185–217. https://doi.org/10.1007/s11159-021-09902-0

Zoller U (2012) Science education for global sustainability: what is necessary for teaching, learning, and assessment strategies? J Chem Educ 89(3):297–300. https://doi.org/10.1021/ed300047v

Open Access This chapter is licensed under the terms of the Creative Commons Attribution 4.0 International License (http://creativecommons.org/licenses/by/4.0/), which permits use, sharing, adaptation, distribution and reproduction in any medium or format, as long as you give appropriate credit to the original author(s) and the source, provide a link to the Creative Commons license and indicate if changes were made.

The images or other third party material in this chapter are included in the chapter's Creative Commons license, unless indicated otherwise in a credit line to the material. If material is not included in the chapter's Creative Commons license and your intended use is not permitted by statutory regulation or exceeds the permitted use, you will need to obtain permission directly from the copyright holder.

Advances and Needs in Marine Science Research in the Archipelago of San Andrés, Providencia, and Santa Catalina: A Literature Analysis

Camilo B. García🅓 and Johan Sebastián Villarraga🅓

Abstract A searchable database of marine science bibliographic references relating to the Archipelago of San Andrés, Providencia, and Santa Catalina is presented. A total of 422 documents were located, including scientific articles, books and book chapters, and thesis works, plus 103 internal technical reports. The database is used to formulate a diagnosis of advances and needs in marine science research in the archipelago. Despite having received the most attention, the species inventory is not complete and is biased toward certain groups. The biology of most species is poorly known, as well as their interactions. Static aspects like the co-occurrence of species in a certain time period predominate, while dynamic aspects including responses to climate change have been barely touched upon. Hence, there is a wide scope and need for new and modern initiatives in marine science research for the archipelago.

Keywords Marine science · Seaflower biosphere reserve · Literature review · Archipelago of San Andrés · Providencia and Santa Catalina · Colombian Caribbean

1 Introduction

A problem that contributes to the slow advance in the marine sciences in developing countries is the lack of visibility of local scientific production in international databases. International search engines like ScienceDirect or Scopus present a biased picture of what is produced, as most journals in developing countries, including Colombia, do not reach their standards.

Although there are regional initiatives like Scielo (https://scielo.org/es/), a non-specialized literature search engine intended to alleviate this situation, not all

---

C. B. García (✉) · J. S. Villarraga
Department of Biology, Faculty of Sciences, Universidad Nacional de Colombia, Bogotá, Colombia
e-mail: cbgarciar@unal.edu.co

sources are covered and searchability is low. Additionally, much scientific production in Colombia remains as gray literature, for example, both undergraduate and postgraduate thesis works that are never published, and institutional reports.

This chapter tackles this issue by presenting a searchable database with an easy and user-friendly interface containing the bibliographic references of the scientific literature, gray or not, produced to date concerning the Archipelago of San Andrés, Providencia and Santa Catalina (hereafter, the archipelago). All marine science topics are included in the database, from physical, chemical, and biological oceanography to ecosystem services and fisheries. Although an effort was made to include all scientific documents, no claim is made as to the exhaustibility of the compilation of references. Nevertheless, the texts compiled in the database do represent an accurate picture of what has been produced in the marine sciences regarding the archipelago.

The constructed database is currently hosted and, ideally, will be maintained and regularly updated, in the library of the Universidad Nacional de Colombia, Caribbean Campus, and should be freely accessible to any interested user.

The database serves as an information source for the diagnosis presented in this chapter on topics and themes in marine science that have historically been touched upon in reference to the archipelago. Here we aim to produce an overview of the current status and highlight patterns and tendencies in marine science research, as well as to provide suggestions for future research needs in marine science in the archipelago. We achieve this by means of a bibliographic analysis of the compiled references in the database. Aspects and features of the marine sciences previously unconsidered or only occasionally dealt with are highlighted. Research needs, including climate change, are identified and characterized.

2 The Database

2.1 Data Collection

The bibliographic search was carried out in the following databases (1–6) and online search engines (a–j):

1. Scopus (https://www.scopus.com/)
2. SciELO (https://scielo.org/es/)
3. RedALyC (https://www.redalyc.org/)
4. Latindex (https://www.latindex.org/latindex/inicio)
5. Mendeley (https://www.mendeley.com/)
6. Google Scholar (https://scholar.google.com/)
a. SpringerLink (https://link.springer.com/)
b. JSTOR (https://www.jstor.org/)
c. Colombian Network of Scientific Information (http://redcol.minciencia.gov.co/vufind)

d. Universidad Nacional de Colombia institutional repository and digital library (https://repositorio.unal.edu.co/)
e. Universidad de Bogotá Jorge Tadeo Lozano repository and digital library (http://unicornio.utadeo.edu.co)
f. Universidad de los Andes repository (https://repositorio.uniandes.edu.co)
g. Universidad del Valle repository (https://bibliotecadigital.univalle.edu.co/)
h. Pontificia Universidad Javeriana repository (https://repository.javeriana.edu.co)
i. INVEMAR Bulletin of Marine and Coastal Research (http://boletin.invemar.org.co)
j. The Seaflower Biosphere Reserve Observatory (https://observatorio.coralina.gov.co/index.php/es/publicaciones).

For the search in the different databases, the following keywords were used: "San Andrés", with the following combinations: "Seaflower Biosphere Reserve", "archipelago", "Providencia", "Santa Catalina", or "Colombian Caribbean". Words in English and Spanish were used to generate as many results as possible, without any time restriction on the search. The search was carried out between August 2021 and February 2022.

Bibliographic references were classified into scientific articles, books and chapters, theses, and other documents. Once the search work was done, an Excel file was created with the following fields: author or authors' surname(s), author or authors' name(s), author or authors' affiliation(s). In the case of articles: year of publication, title, journal, volume, number, pages, and DOI. In the case of books and chapters, additional fields were the ISBN, the editorial house, and the APA citation.

The information of the documents stored in the Excel file was transferred to an Access database and a search engine was created, with an interface that allows the user to conduct bibliographic searches on the marine science documents referring to the archipelago. Upon starting the application, instructions on how to conduct searches in the database are given, including a video tutorial.

A total of 422 documents were located, of which 201 were scientific articles, 41 were books and book chapters, and 77 were undergraduate and postgraduate theses. A compilation was also made of internal documents, technical reports, and abstracts, which amounted to 103 entries.

2.2 The Metric

The search for patterns and tendencies in the database is based on the titles of the documents and a cursory review of their content. To that end, a system of keywords was developed so that the number of associations between the document and the keywords, herewith called mentions, was used as a metric to discover patterns in the focus and orientations of the documents. A document may be associated with more than one keyword. Titles of documents in English, German, and French were translated into Spanish before the application of the keyword system.

Two sets of keywords were developed, one dealing with biotic groups i.e., not strictly taxonomic groups, and the other dealing with general themes in biology, ecology, and environmental studies. Biotic groups were as follows: Fishes, Mollusks, Bacteria (Bacteria and Cyanobacteria), Coral (Reefs—"Arrecife", Atolls—"Arrecifal", Coral—"Coralino"), Macroalgae, Mammals, Sponges, Hexacorals (Anemones and Zoanthids), Turtles, Polychaetes, Echinoderms, Plankton (Phytoplankton, Zooplankton, Ichthyoplankton, Dinoflagellate, Foraminifera, Zooxanthellae), Seagrasses, Octocorals (Octocorals and Gorgonacea), Mangroves, Crustaceans, and Birds.

The thematic keyword set included: Physical Sciences (Physical Oceanography, Chemical Oceanography, Geology, Geomorphology, climate events), Genetics, Paleontology, Coraline Affections (diseases and bleaching), Interactions (ontogenic changes according to habitat, predation, synergies on "blue carbon", responses on presence of snappers, allelopathic relations, symbiosis), Distribution, Temporal Change (seasonal or long term), Abundance (density, abundance), Population (individual growth, ontogeny, mortality, reproduction including size at first reproduction and reproductive aggregations, recruitment, demography, physiology, active molecules), Fisheries, Assemblages (meaning structural co-occurrence of species), Management and Environment (ecosystem services including economic valuations, planning, impact of Hurricane Beta, water quality, algae blooms, environmental threats including algae toxicity and contamination, marine protected areas, management, sustainability), and Lists (new species records, new species to science, lists of species).

3 Patterns in Scientific Articles

3.1 Biotic Groups in Scientific Articles

Of the biotic groups defined, Corals, Fishes, Mollusks, Crustaceans, and Macroalgae represent the bulk of mentions with a cumulative percentage of 68.7% (Fig. 1). As expected, Corals is the biotic group that has received the most attention (22.4% of all mentions). One notable finding is that several major groups important in reef landscapes have scarcely been touched upon, like seagrasses, sponges, or echinoderms while important groups like Sipunculids and Bryozoa are absent. Looking within groups, several absences can be noticed, for instance, peracarid crustaceans and nonspecific mention of Bivalves among others.

Not only are several major biotic groups poorly or not represented in the articles. Within the biotic groups, a very small number of species have been the subject of focused studies. This is the case of the queen conch *Aliger gigas* (Linnaeus 1758) among Mollusks, representing 73.9% of mentions of Mollusks, and the spiny lobster *Panulirus argus* (Latreille 1804), representing 68.2% of mentions in the Crustaceans group. In both cases, the other mentions mostly refer to the respective biotic groups

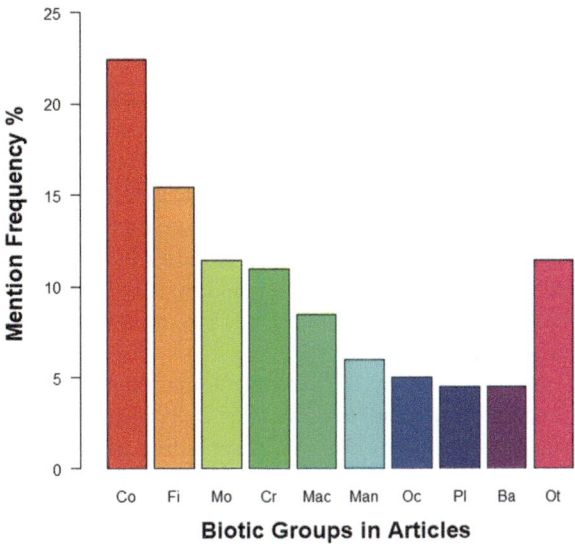

Fig. 1 Percentage distribution of biotic group mentions in scientific articles referring to Marine Science in the Archipelago of San Andrés, Providencia, and Santa Catalina. Co = Coral, Fi = Fishes, Mo = Mollusks, Cr = Crustaceans, Mac = Macroalgae, Man = Mangroves, Oc = Octocorals, Pl = Plankton, Ba = Bacteria, and Ot = Others. The biotic group Others includes Seagrasses, Echinoderms, Polychaeta, Sponges, Turtles, Hexacorals, and Mammals, each with less than 3% of mentions in the database

in general. A similar situation was found for Corals, for which just two mentions were recorded on a particular species *Dendrogyra cylindrus* (Ehrenberg 1834) and only one to *Agaricia undata* (Ellis 1786).

In the case of fishes, the same pattern emerges with most mentions referring to the general group. Thus, two mentions were recorded for snappers, serranids, and pelagic fishes, one for parrotfishes, groupers, Acanthurids, and the invader lionfish (*Pterois volitans*, Linnaeus 1758) while for cartilaginous fishes four mentions were recorded. The case of Macroalgae is no different, with just one mention recorded of the genera *Sargassum*, one mention of *Crouania pumila* (Gavio et al. 2013) a new species, and one mention of *Griffithsia capitata* (Børgesen 1930) as a new record. For octocorals, three mentions were made of *Antillogorgia elisabethae* (Bayer 1961).

These findings are a clear indication that, in the archipelago, the biotic groups as defined here have received uneven attention, with the bulk of mentions associated with certain groups, subgroups, and species. This indicates that the inventory of marine biodiversity of the archipelago is incomplete, and that studies on single species are exceptional.

The focus on the queen conch and the spiny lobster is not surprising given the commercial importance of both of these species, but their overwhelming representation in mentions inside their respective biotic groups reinforces the statements above. The scarcity of mentions recorded for snappers, groupers, parrot fishes, and cartilaginous fishes is surprising, as these fishes are an essential part of the ecological dynamics in coral reefs.

3.2 Thematic Groups in Scientific Articles

The theme with the highest number of mentions was Lists with 16.7% of all mentions (Fig. 2). Lists focus mostly on Macroalgae, Fishes, and Corals, collectively representing 53.3% of mentions inside this theme, with the other biotic groups amounting to less than 6% of mentions each. Clearly, a great effort has been made in characterizing the taxonomic aspect of biodiversity in the archipelago, with this theme being the most frequent in terms of mentions, but this effort has been biased to a limited number of biotic groups. This finding is in line with the findings of mentions by the biotic groups discussed above. Thus, mentions of biotic groups often relate to lists of species.

Management and Environment, and Physical Sciences, share the second position with 13.9% of mentions each (Fig. 2). For Management and Environment, three topics stand out: ecosystem services including economic valuation of services, management planning, and impacts due to anthropogenic activities or natural events (Hurricane Beta). Clearly, the establishment of the Seaflower Biosphere Reserve provoked significant attention to the management of the archipelago and its governance may look well-sustained on theoretical grounds.

In the Physical Sciences theme group, the most common topic is water movement, be it waves or currents (Physical Oceanography). The general current system of the archipelago appears to be well-characterized, as is also the case with geomorphological features. Interestingly, few articles were found that related oceanographic features with biological features: a simulation on the dispersion of queen conch larvae, salinity related to hydroperiod and mangrove roots, and ecosystem responses

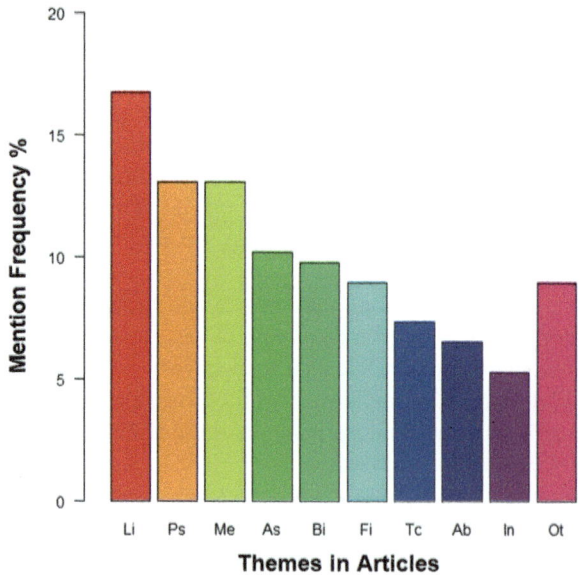

Fig. 2 Percentage distribution of thematic group mentions in scientific articles referring to Marine Science in the Archipelago of San Andrés, Providencia, and Santa Catalina. Li = Lists, Ps = Physical Sciences, Me = Management and Environment, As = Assemblages, Bi = Biology, Fi = Fisheries, Tc = Temporal Change, Ab = Abundance, In = Interactions and Ot = Others. The theme group Others includes Distribution, Paleontology, Genetics, and Coraline Affections, each with less than 4% of mentions in the database

to extreme climatic events. This line of research should be promoted, as it is central to the question of biological connectivity.

Assemblages is the third most frequently mentioned keyword (10.2% of mentions, Fig. 2). The structure of coral reef communities has been the focus of this theme, with detailed descriptions of co-occurrence patterns of coralline species (51.6% of mentions in this theme). This is not surprising, as coral reefs dominate the sea landscape. Assemblages of fishes rank second with 16.1% of mentions in the theme. Thus, the characterization of species co-occurrence patterns is a rather uncommon topic and one that is biased towards certain biotic groups.

Biology comes next in the frequency of mentions, followed by Fisheries (9.8% and 8.9% respectively, Fig. 2). Of the mentions in the Biology theme, 55.2% refer to the queen conch, Mangroves, and Coral. Mentions of the queen conch focus mostly on reproductive aspects. Few other species or groups are mentioned (*Cittarium pica*—Linnaeus 1758—, *Dendrogira cylindrus*—Eherenber 1834—, the spiny lobster, the black crab *Gecarcinus ruricola*—Linnaeus 1758, and Serranids), most of them with just one mention. Interesting is the case of *Antillogorgia elisabethae* (Bayer 1961) which has been investigated in relation to bioactive molecules.

The scarcity of Biology studies on the spiny lobster comes as a surprise, as this species is highly valued in economic terms and thus subject to fisheries. Beyond that, it is quite clear that knowledge of the biology of the species present in the archipelago is very shallow and incomplete. Topics of life history, vital rates, and demography are practically non-existent in the articles found on marine science in the archipelago, with the exceptions mentioned.

In the case of Fisheries, most mentions alluded to the characterization of capture and effort, and some alluded to the management of fishery resources. Fishery mentions included Fishes, spiny lobster, and the queen conch with 40%, 35%, and 20% of mentions, respectively. The black crab received just one mention in this thematic group. This finding clearly highlights the economic importance of the spiny lobster and the queen conch for the fishery of the archipelago. Surprisingly, quantitative fishery models, for example, of the type of surplus production models or yield per recruit models as needed for estimation of maximum sustainable yield, were not located, nor was work on indices or indicators for fishery management. The construction of a modern scientific base for fishery management in the archipelago appears to be a necessity.

A theme that is rather poorly represented in the database is Temporal Change (Fig. 2) with 7.3% of mentions. Four articles spanned decades in relation to Coral changes which is interesting. Articles related to other biotic groups referred to seasonal changes. Physical Sciences was present, with work that also spanned seasonal to multiannual and decadal observations. One point of concern, apart from the scarce number of mentions, is that just four articles may be related to global change, and of them, only one directly refers to global warming in relation to sea level rise. Global change and global warming should be approached with higher priority because these topics represent the biggest challenge to the future of the archipelago in the medium term.

Abundance occupies the seventh position in thematic mentions (Fig. 2). The bulk of mentions corresponds to Coral, the spiny lobster, and Fishes, with 54.5% of mentions. Other biotic groups associated with Abundance mostly recorded one mention each. Thus, not only is the inventory of species of the archipelago incomplete and biased, but the estimation of population sizes and density has concentrated on a few groups. This kind of estimation must be extended to many more species, in particular to key species in the trophic web if this feature is to be monitored as an indication of ecosystem and population health status.

All other themes received 6% of mentions, or less, in the database (Fig. 2). With some themes this might be expected, for example, Paleontology, but surprisingly, themes like Interactions and Coralline Affections are poorly represented. The scarcity of articles on biological interactions reflects a bias to taxonomy (Lists) and quantification of co-occurrence patterns (Assemblages) as themes of interest for the biota to date. Understanding the dynamical relations that underlie the biology and ecology of the archipelago is a necessary step both in advancing the marine sciences and in guaranteeing correct long-term management. No possibility of future scenario development is possible without an understanding of the interactions among species and populations.

Coral affections that include diseases and bleaching are also poorly represented in the articles the same is true of genetics. In the first case, the scarcity of mentions on this theme is surprising as most mentions in biotic group themes relate to Coral (Fig. 1). Research on this important topic has been sporadic and deserves more attention as affections are fundamental drivers of coral reef dynamics. The use of genetic tools should be promoted.

4 Patterns in Books and Book Chapters

Forty-one documents were located that qualified as books and book chapters. The distribution of mentions by biotic groups and thematic groups can be seen in Figs. 3 and 4, respectively. Most of these books are produced by public institutions, notably CORALINA (*Corporación para el Desarrollo Sostenible del Archipiélago de San Andrés, Providencia y Santa Catalina*, https://coralina.gov.co/) and INVEMAR (*Instituto de Investigaciones Marinas y Costeras*, http://www.invemar.org.co/) sometimes alone, sometimes in tandem.

4.1 Biotic Groups in Books and Book Chapters

The most frequently mentioned biotic groups in books and book chapters (from now on, books) were Fishes, Corals, Mangroves, Mollusks, and Birds, in that order, with a cumulated 59.6% of mentions (Fig. 3). All other biotic groups received less than 6% of mentions each. In the case of Fishes, most mentions referred to fishes in general.

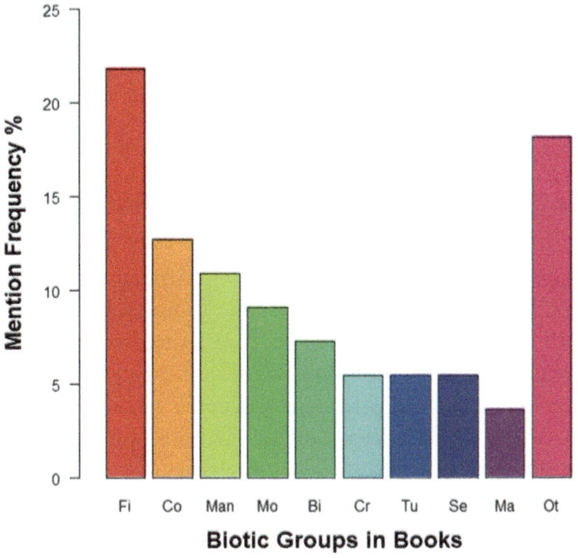

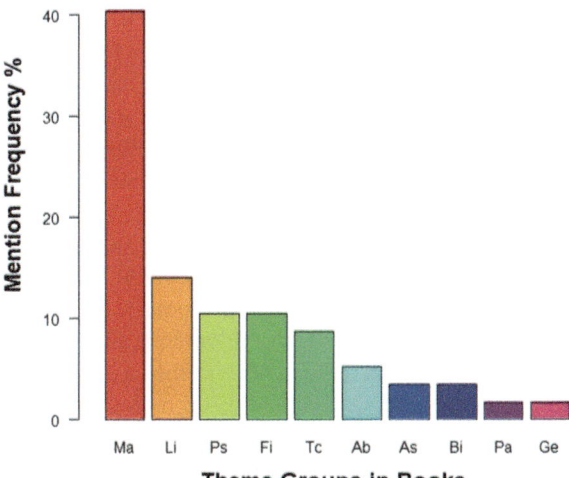

Fig. 3 Percentage distribution of biotic group mentions in books and book chapters referring to Marine Science in the Archipelago of San Andrés, Providencia, and Santa Catalina. Fi = Fishes, Co = Coral, Man = Mangroves, Mo = Mollusks, Bi = Birds, Cr = Crustaceans, Tu = Turtles, Se = Seagrasses, Ma = Macroalgae, and Ot = Others. The biotic group Others includes Echinoderms, Sponges, Polychaeta, Octocorals, Plankton, Hexacorals, and Mammals, each with less than 4% of mentions in the database

Fig. 4 Percentage distribution of theme group mentions in books and book chapters referring to Marine Science in the Archipelago of San Andrés, Providencia, and Santa Catalina. Ma = Management and Environment, Li = Lists, Ps = Physical Sciences, Fi = Fisheries, Tc = Temporal Change, Ab = Abundance, As = Assemblages, Bi = Biology, Pa = Paleontology and Ge = Genetics

Three mentions concern cartilaginous fishes and only one mention was recorded for the invader lionfish. The same pattern was found for the other biotic groups, i.e., mentions for the general biotic group and just one or two mentions of species, in this case, all of them mollusk (the queen conch, the magpie shell *Cittarium pica*—Linnaeus 1758—, and the black crab). No specific mention was recorded for the spiny lobster.

The pattern of documents concerning a limited group of taxa is also shown here, as was the case for articles. It is worrisome that there are no monographic studies of important species, be it for ecology or for fishery, which is also consistent with the findings for articles, regarding the scarcity of studies on the population biology of species of the archipelago.

4.2 Thematic Groups in Books and Book Chapters

Given that most books come from official institutions, it is not surprising that the bulk of mentions by the thematic group is for Management and Environment, with 40.3% of mentions (Fig. 4). Topics dealt with include plans aimed at conservation and management, and global descriptions of the reef landscape. An open question is whether these plans are evaluated in terms of achievement of their goals, and whether they are kept current. No book was found related to a critical evaluation of management, that is, contrasting objectives and achievements.

The second thematic group in terms of recorded mentions was Lists (14.0% of mentions, Fig. 4). Species lists are part of diagnostics and baselines. Various books covered several biotic groups, including groups not mentioned in articles like Birds, as they were aimed at inventorying the species richness of the archipelago. Interestingly, in this case of fishes, two books referred to cartilaginous fishes.

One work stands out (Vides et al. 2016) as the most inclusive biotic list covering all marine groups. The lack of taxonomic work on complete major groups identified for scientific articles is confirmed here: for phytoplankton and zooplankton, most records are labeled as "morphotypes"; peracarid crustaceans have not been identified; there is no record on marine fungi; Scyphozoa and Ctenophora are poorly known; Platyhelminthes, Nemertea, Nematoda, and Priapulida have no records; Polychaetes are poorly known; and finally, Tunicates and Sipunculida have no records.

Physical Sciences and Fisheries account for 10.5% of mentions each. In this case, within Physical Sciences, most mentions referred to geomorphological descriptions. In the case of Fisheries, the mentions refer to descriptions of the artisanal fleet and general considerations. As was the case with articles, no book was located that focused on fishery assessment based on formal models.

Temporal change received less than 9% of mentions (Fig. 4). It is worrisome that the focus has been on retrospective analysis of the monitoring of certain habitats (reefs, seaweeds, mangroves). No prospective studies in relation to climate change were found. The formulation of future scenarios is a necessary condition for rational planning that considers climate change and global warming if preservation of the biological values that sustain the creation of the Seaflower Biosphere Reserve is to be pursued.

All other themes (Abundance, Assemblages, Population Biology, Genetics) received less than 5% of mentions each. This reflects the need to broaden the palette of approaches in the marine sciences in relation to the archipelago.

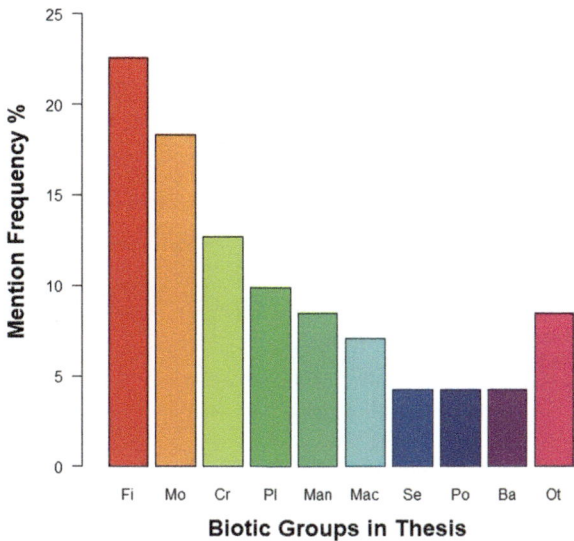

Fig. 5 Percentage distribution of biotic group mentions in theses referring to Marine Science in the Archipelago of San Andrés, Providencia, and Santa Catalina. Fi = Fishes, Mo = Mollusks, Cr = Crustaceans, Pl = Plankton, Man = Mangroves, Mac = Macroalgae, Se = Seagrasses, Po = Polychaeta, Ba = Bacteria and Ot = Others. The biotic group Others includes Sponges, Turtles, Coral, Echinoderms, and Octocorals

5 Patterns in Theses

5.1 Biotic Groups in Theses

The frequency of mentions is shown in Fig. 5. Between them, Fishes, Mollusks, and Crustaceans received 53.5% of mentions.

The first thing to notice is that Corals received less than 2% of mentions in contrast to their high percentage of mentions in articles and books. In general, the mentions associated with the keyword Fishes referred to fishes globally. One specific mention was made of the barracuda *Sphyraena barracuda* (Edwards 1771), the invader lionfish, and the yellowtail snapper *Ocyurus chrysurus* (Bloch 1791). The genera of parrotfishes *Scarus* and *Sparisoma* were mentioned one time each. In the case of Crustaceans, the spiny lobster was the only species mentioned (twice) with the other mentions referring to broad groups. For Mollusks, mentions of the queen conch amounted to 46.2% of all mentions in that group. For the other biotic groups, just one mention of a species was located for the octocoral *Pseudopterogorgia elisabethae* (Bayer 1961). Thus, as was the case with articles and books, work on the biology of specific species is scarce and biased toward certain species.

5.2 Thematic Groups in Theses

Distribution of mentions by thematic groups is shown in Fig. 6.

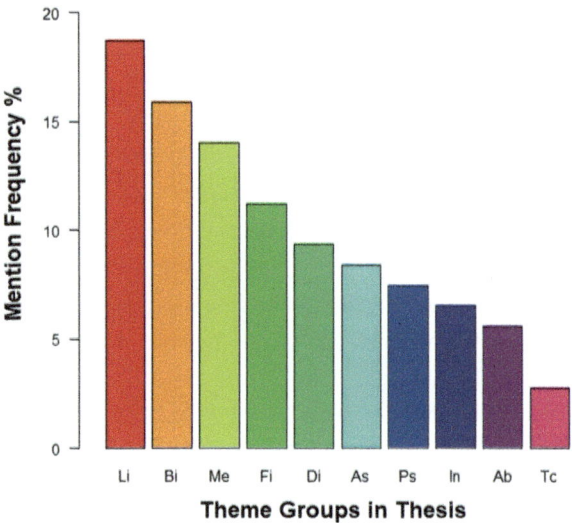

Fig. 6 Percentage distribution of theme group mentions in theses referring to Marine Science in the Archipelago of San Andrés, Providencia, and Santa Catalina. Li = Lists, Bi = Biology, Me = Management and Environment, Fi = Fisheries, Di = Distribution, As = Assemblages, Ps = Physical Sciences, In = Interactions, Ab Abundance, and Tc = Temporal Change

Lists represent 18.7% of mentions (Fig. 6). So, in this case also, much work has been dedicated to taxonomic identification of species. Interestingly, Plankton stands out with the same number of mentions as Mollusks, and one more than Crustaceans (Fig. 6). Population biology is the next theme with the most mentions (Fig. 6). Mentions at the species level in this theme refer to the queen conch. Most other mentions in this theme refer to reproductive aspects and ontogenic changes for groups of species. Management and Environment come next (Fig. 6) with three main topics, economic valuations, contamination, and water quality. As for Fishing, two theses were found that alluded to fishery models, that is, proper fishery assessments, with most other work referring to descriptions of capture and effort. All mentions of Temporal Change referred to seasonal change in assemblages, that is, no attempt to explore global change was found in this type of document.

6 Conclusions and Final Reflections

The detection and analysis of patterns and tendencies undertaken in this chapter lead to several conclusions and, accordingly, to the formulation of a number of suggestions as to where to focus future research in the marine sciences in the archipelago.

Although much effort has been made in completing the species inventory, it is clear that many taxonomic groups are poorly known or not known at all. This situation follows, in part, the available taxonomic expertise in the country and, in part, the small physical size and cryptic habitat of the poorly known groups which make them difficult to visualize as the components of reefs and ecosystems in the archipelago that they are. Training of biologists in taxonomy, on the one hand, and dedicated surveys, on the other, should alleviate this situation. If the Seaflower Biosphere Reserve is to maintain its status as a hotspot of marine diversity, complete species inventories are a necessity.

Little is known about the biology of the species present in the archipelago, even for those included in taxonomic lists. Not one complete account of the life history of a particular species could be located. Apart from some partial work on the queen conch and the spiny lobster, for most other species their biology is not known, or only limited features have been studied. Thus, it is not surprising that so few documents on interactions among species were located. The understanding of the biological and ecological role of individual species and their interactions is paramount to conservation efforts and planning. For instance, the identification of key species, be it simply because of their abundance and extended distribution or because they exert an influence disproportionate to their abundance, may allow focused management plans in the Seaflower Biosphere Reserve.

Unfortunately, beyond descriptions of species co-occurrences (assemblages), to our knowledge, no models of functioning at the level of communities exist. The trophodynamics of ecological communities, that is, the characterization and quantification of fluxes of matter and energy via the food web, is an endeavor to be undertaken. Such models could be used to assess the relative direct and indirect impact of one species on others and thus help in identifying key species. Monitoring initiatives have been concentrated on structural aspects of certain habitats, but they should be extended to the monitoring of key species identified by ecosystem models.

Fishing in the archipelago is an activity of great economic and social importance. However, apart from some descriptions of capture and effort, a description of the fishing fleet, and a couple of studies on the biology of the queen conch, the spiny lobster, and some fish that may be useful for their management, very few documents were found with comprehensive diagnoses on the fishery in the archipelago. A thesis by Castro (2005) describes the fishing regime and goes into the perception of fishermen, but it is clearly outdated. As far as we can perceive from the analysis of scientific articles, books, and thesis works, the management of fishing and fisheries in the archipelago is lacking a scientific base. Scientific fishery assessments in the archipelago are urgently needed to estimate maximum sustainable yields compatible with conservation plans. Monitoring of fisheries in terms of indices and indicators should be put in place beyond traditional estimations of capture and effort.

It is surprising that threats and perturbations to the functioning of the marine ecosystems in the archipelago, for example, the invasion of the lionfish, coral diseases, and bleaching events, have received so little attention. The ecological role of the lionfish and hence its effect on the ecology of the archipelago is unknown, apart from expectations from the relevant literature. No inclusion–exclusion experiment with the lionfish has been performed and the lack of trophodynamic models leaves us with only speculation as to the role and impact of this invader. The health status of coral species should be given a more prominent part of regular monitoring plans.

It is worrisome that no dedicated work on climate change and global warming was located. Physical aspects of global change, for example, sea level rise, have been touched upon a few times but the impact and consequences of global change on the species, biology, and ecology of the archipelago have not been assessed. As noted above, there are currently no models on community dynamics nor on species distributions that may be used to formulate simulations and projections for the future regarding possible scenarios under global change. Aspects like abundance and distribution have been treated like fixed features while they are dynamic and responsive to niche shifts due to global change. Modeling efforts within rigorous mathematical frameworks are of the utmost importance.

We hope that the diagnosis presented here, and the suggestions that have emerged from said diagnosis, are helpful and will contribute to the correct management of the Archipelago of San Andrés, Providencia and Santa Catalina, and of the Seaflower Biosphere Reserve. The advance of the marine sciences is a requisite to reach this end, and the research lines suggested should be conducive to helping secure the current and future ecological integrity of the archipelago and the Seaflower Biosphere Reserve.

References

Bayer FM (1961) The shallow-water Octocorallia of the West Indian region. A manual for marine biologists. In: Studies on the Fauna of Curacao and Other Caribbean Islands, vol 12, pp 1–373

Bloch ME (1791) Naturgeschichte der ausländischen Fische, Berlin, vol 5, pp 1–152

Børgesen F (1930) Marine algae from the Canary Islands especially from Tenerife and Gran Canaria III. Rhodophyceae. Part III. Ceramiales. Kongelige Danske Videnskabernes Selskab, Biologiske Meddelelser 9(1):1–159

Castro E (2005) Caracterización del régimen de pesca en la Isla de San Andrés, Caribe Colombiano: Inferencias sobre la estructura de la comunidad íctica. Master's Thesis, Facultad de Ciencias, Departamento de Biología, Universidad Nacional de Colombia

Edwards A (1771) The natural history of Carolina, Florida and the Bahama Islands; containing the figures of birds, beasts, fishes, serpents insects and plants with their descriptions in English and French, etc. In: Catesby M (ed) 3rd edn, 2 vols, London. [A rejected work on the Official Index, ICZN Opinions 89 and 259 (but not appendix by Edwards). [Also 1st edn, 1743; 2nd edn, 1754.]]

Ehrenberg CG (1834) Beiträge zur physiologischen Kenntniss der Corallenthiere im allgemeinen, und besonders des rothen Meeres, nebst einem Versuche zur physiologischen Systematik derselben, vol 1. Abhandlungen der Königlichen Akademie der Wissenschaften, Berlin, pp 225–380

Ellis J (1786) The Natural History of many curious and uncommon Zoophytes, collected from various parts of the Globe. Systematically arranged and described by the late Daniel Solander, vol 4. Benjamin White & Son, London, pp 1–206

Gavio B, Reyes-Gómez VP, Wynne MJ (2013) *Crouania pumila* sp. nov. (Callithamniaceae: Rhodophyta), una nueva especie de alga roja marina de la Reserva Internacional de la Biosfera Seaflower, Caribe colombiano. Revista de Biología Tropical 61(3):1015–1023

Latreille PA (1804) Histoire Naturelle Generale et Particulière, des Crustacés et des Insectes, vol 7, pp 1–413

Linnaeus C (1758) Systema Naturae, 10th edn, vol 1. Laurentii Salvii: Holmiae (Stockholm, Sweden), 824 p

Vides M, Alonso D, Castro E, N. Bolaños N (eds) (2016) Biodiversidad del mar de los siete colores. Instituto de Investigaciones Marinas y Costeras – INVEMAR & Corporación para el Desarrollo Sostenible del Archipiélago de San Andrés, Providencia y Santa Catalina - CORALINA. Serie de Publicaciones Generales del INVEMAR No. 84: Santa Marta, Colombia

Open Access This chapter is licensed under the terms of the Creative Commons Attribution 4.0 International License (http://creativecommons.org/licenses/by/4.0/), which permits use, sharing, adaptation, distribution and reproduction in any medium or format, as long as you give appropriate credit to the original author(s) and the source, provide a link to the Creative Commons license and indicate if changes were made.

The images or other third party material in this chapter are included in the chapter's Creative Commons license, unless indicated otherwise in a credit line to the material. If material is not included in the chapter's Creative Commons license and your intended use is not permitted by statutory regulation or exceeds the permitted use, you will need to obtain permission directly from the copyright holder.

Appendix

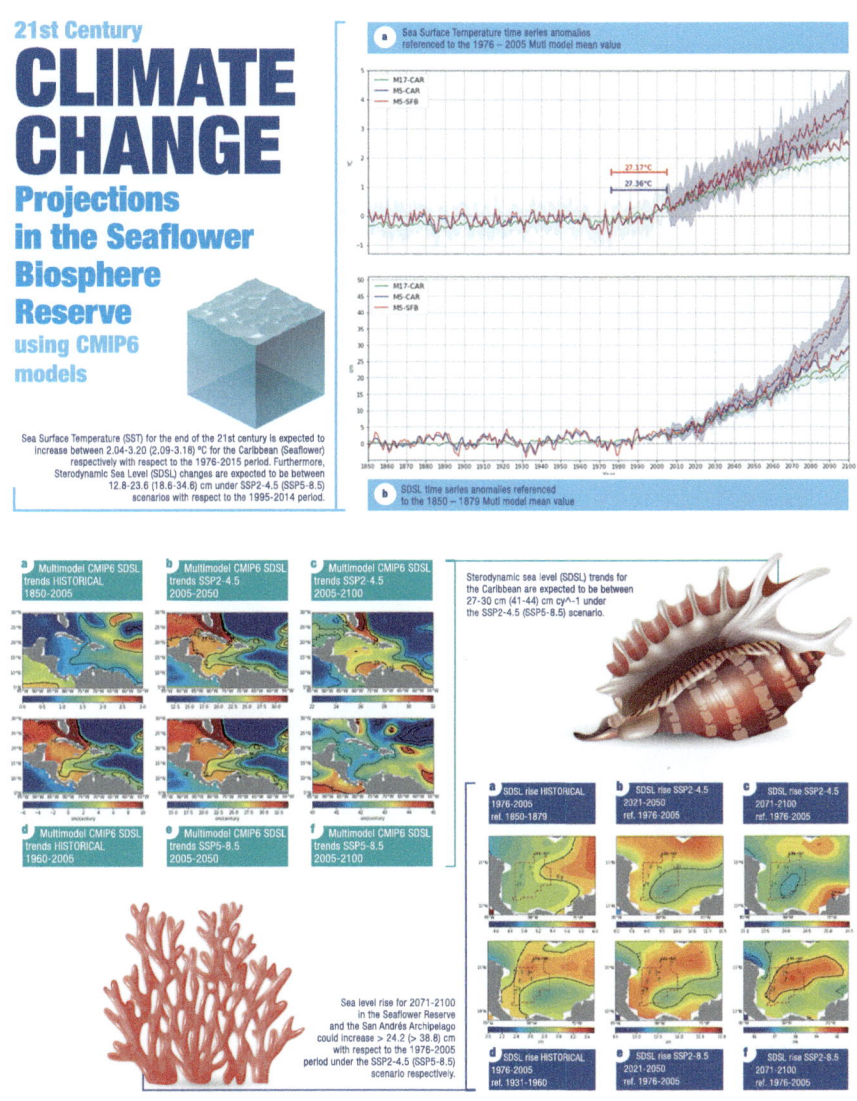

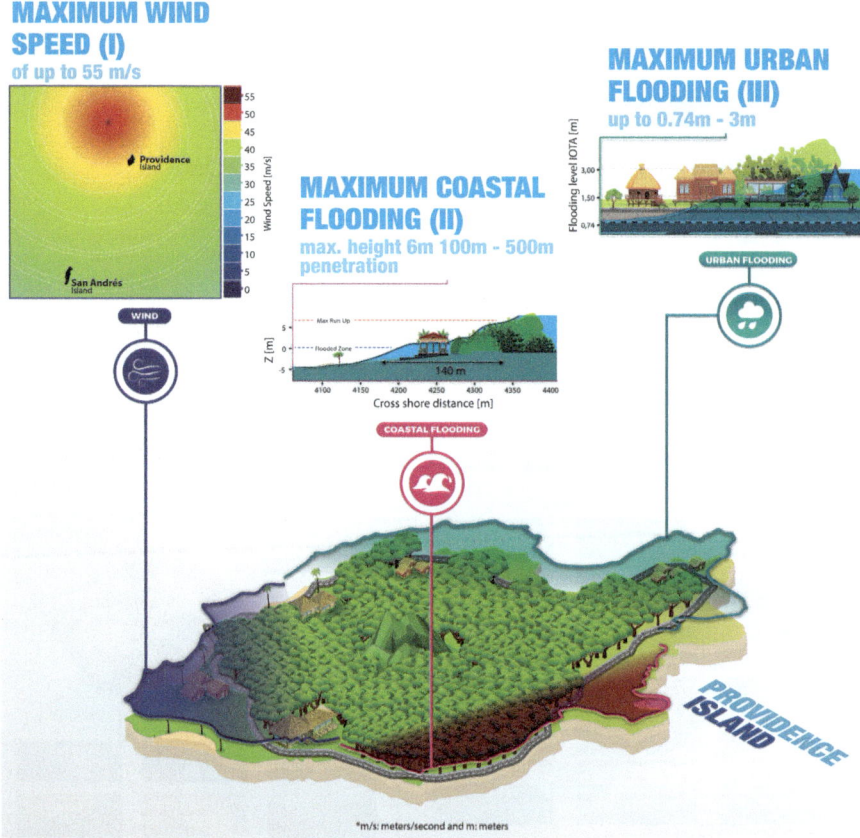

Appendix

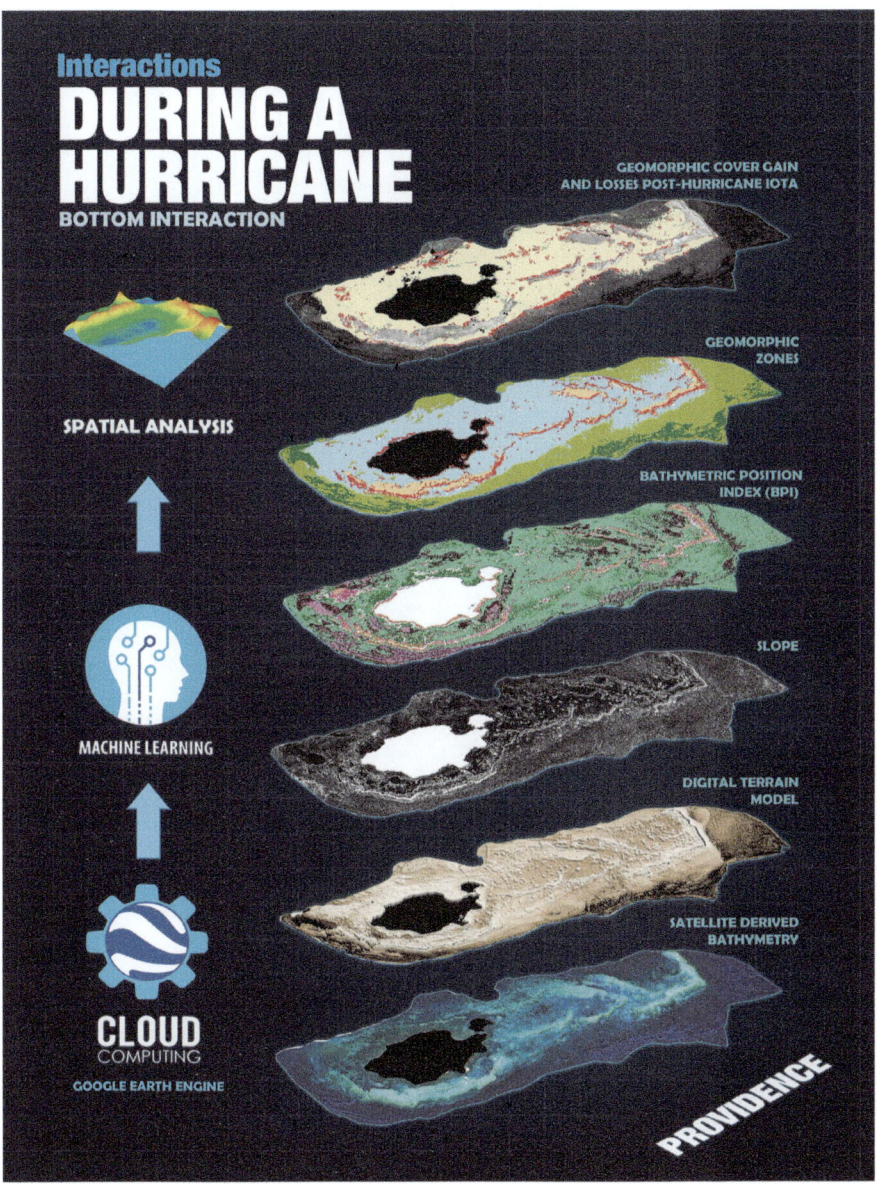

Appendix

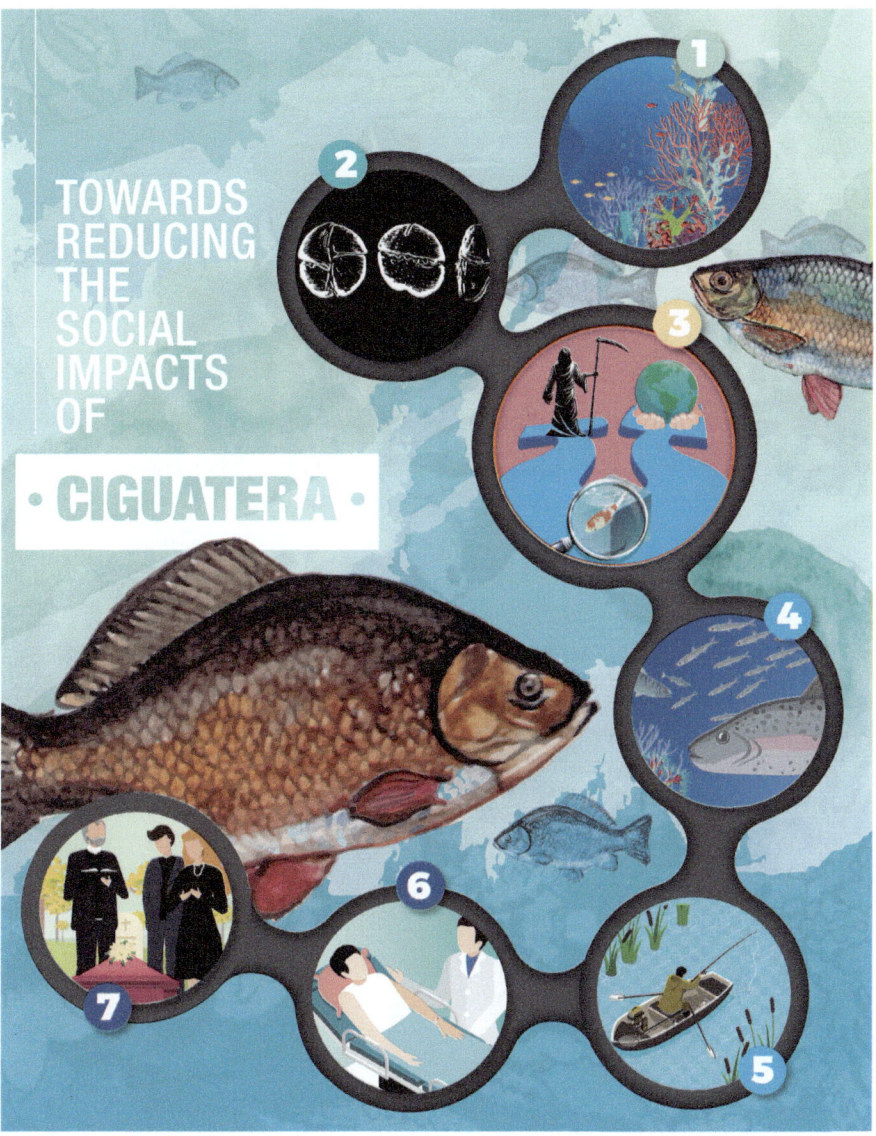

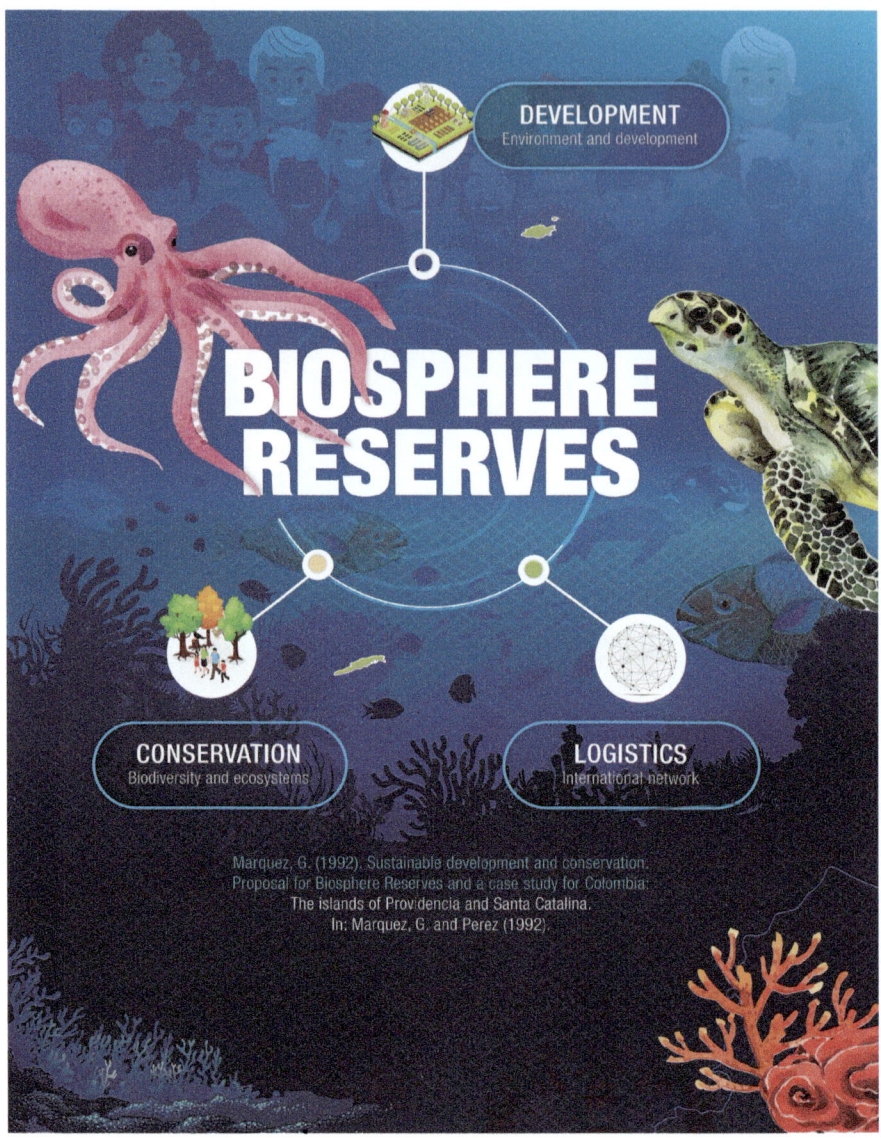

Appendix

Prato et al., 2021. MARINE ECOSYSTEM SERVICES FOR CLIMATE CHANGE ADAPTATION AND MITIGATION STRATEGIES IN THE SEAFLOWER BIOSPHERE RESERVE: Coastal protection and fish biodiversity refuge at Caribbean insular territories

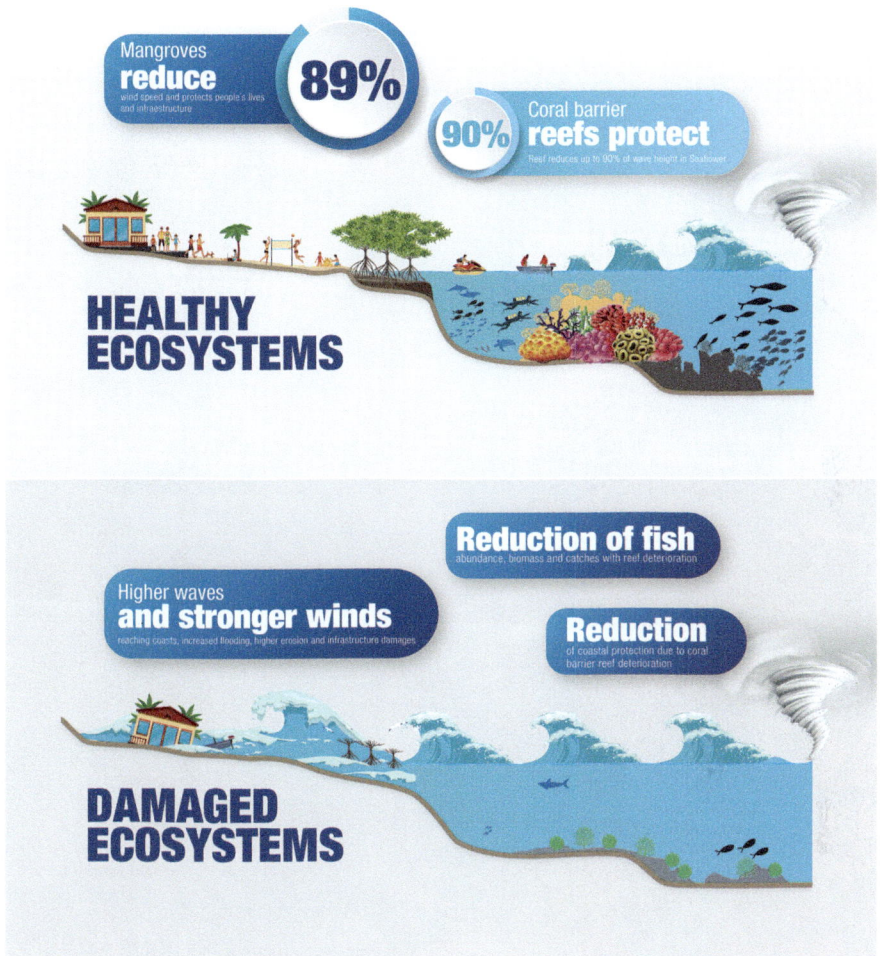

Appendix

Appendix

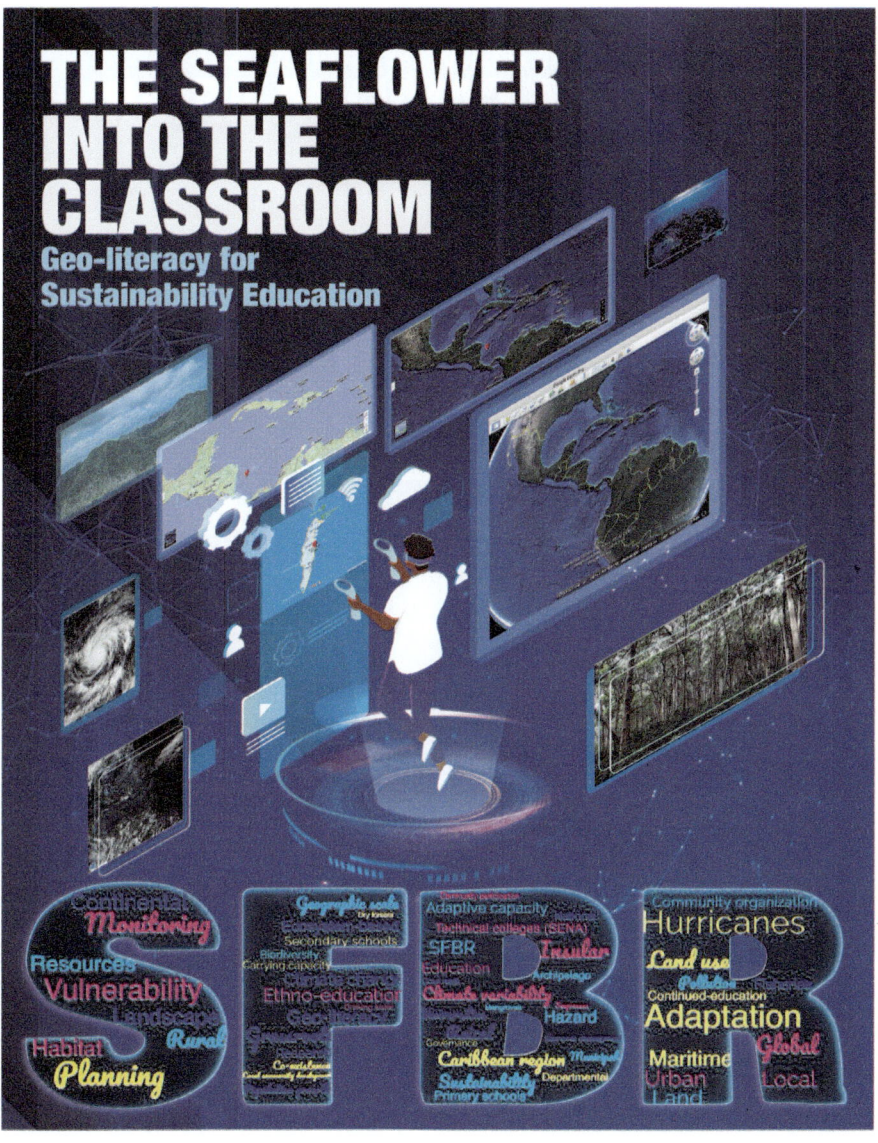

Appendix

Printed by Printforce, the Netherlands